Named one of the Best ~~Books~~
by the *Los Angeles Times* and
Publishers Weekly

For centuries humankind has dreamed of talking to
animals. Over the past thirty years Roger Fouts has
turned that dream into reality by pioneering communication with chimpanzees through sign language—
beginning with a high-spirited, unforgettable chimp
named Washoe. Fouts's groundbreaking work with
Washoe and other chimpanzees—who share more than
ninety-eight percent of our DNA—made scientific
history as their unprecedented dialogues opened a
window into chimpanzee consciousness and the origins
of human language and intelligence.

This remarkable book tells the dramatic story of Fouts's
odyssey from novice researcher to celebrity scientist to
impassioned crusader for the rights of animals. At the
heart of the story is Fouts's magical thirty-year
friendship with Washoe, whom we watch grow from a
mischievous baby chimp fresh out of the NASA space
program into the matriarch of a clan of chimpanzees
who fill these pages with tales of humor and heartbreak,
loyalty and love. Living and conversing with these
sensitive creatures has given Fouts a profound
appreciation of how much we share with our closest
biological relatives, and what they can teach us about
ourselves. It has also made him—at significant risk to
his own career—an outspoken opponent of biomedical
experimentation on chimpanzees.

NEXT OF KIN is a voyage of scientific discovery and
interspecies communication unlike any that has come
before. This stirring tale of friendship, courage, and
compassion will change forever the way we view our
biological—and spiritual—next of kin.

Debbie Fouts

ROGER FOUTS (pictured above with Tatu) is professor of psychology at Central Washington University and codirector with his wife, Deborah Fouts, of the Chimpanzee and Human Communication Institute. His extraordinary accomplishments with Washoe and her chimpanzee family over the past three decades have generated international publicity in magazines and newspapers and on television. He is a frequent speaker on chimpanzee behavior and on behalf of improved conditions for captive chimpanzees in biomedical research.

STEPHEN TUKEL MILLS writes about science and the environment. He cofounded Living Planet Press and is a consultant to the Natural Resources Defense Council. He lives in Santa Fe, New Mexico, with his wife and son.

NEXT

MY CONVERSATIONS WITH CHIMPANZEES

INTRODUCTION BY JANE GOODALL

Quill
An Imprint of HarperCollinsPublishers

OF KIN

ROGER FOUTS
WITH
STEPHEN TUKEL MILLS

FOR WASHOE

And all the other chimpanzees
who can never go home again

Grateful acknowledgment is made for permission to use the following photographs: photo insert pages 1, 14, 15: April Ottey; pages 2, 3 (*bottom*), 4, 5, 8, 9, 10, 11, 13: courtesy of Roger and Deborah Fouts; page 3 (*top*): *Life* magazine; pages 6, 7: *Science Year: The World Book Science Annual 1974*, copyright © 1973 Field Enterprise Educational Corporation by permission of World Book, Inc.; page 12: PETA; page 16: Hillary Fouts.

The photographs on pages 1 and 259 are courtesy of Roger and Deborah Fouts; the photograph on page 115 is courtesy of April Ottey.

A hardcover edition of this book was published in 1997 by William Morrow.

HarperCollins books may be purchased for educational, business, or sales promotional use. For information please write: Special Markets Department, HarperCollins Publishers Inc., 10 East 53rd Street, New York, NY 10022.

First Bard edition published 1998.

Reprinted in Quill 2003.

The Library of Congress has catalogued the hardcover edition as follows:

Fouts, Roger.
   Next of kin : what chimpanzees have taught me about who we are / Roger Fouts with Stephen Tukel Mills.
      p. cm.
   "A living planet book."
   Includes index.
      1. Fouts, Roger. 2. Experimental psychologists—United States—Biography.
   3. Washoe (Chimpanzee). 4. Human-animal communication. I. Mills, Stephen Tukel. II. Title.
   BF109.F66A3   1997   97-15144
   156—dc21   CIP
   ISBN 0-380-72822-2 (pbk.)

04 05 06 07 FOLIO/RRD 10 9 8 7 6 5

CONTENTS

INTRODUCTION
by Jane Goodall

A T LAST! For more than a decade I have been begging Roger Fouts to get this book written. It is the story of a scientific experiment that has helped us to better understand our own place in relation to the rest of the animal kingdom and, at the same time, reveals a dark and ugly aspect of the scientific method. It describes, step by step, how the lives of a young student (Roger) and a little chimpanzee girl (Washoe) become inextricably entwined. And how Roger, with determination and courage, saves Washoe from life imprisonment—by sacrificing, for her sake, the career he could have had. It is one of the most remarkable scientific, humanistic, and spiritual tales of our time. It has all the elements of a truly great novel—adventure, heartbreak, the struggle against evil, courage, and, of course, love. At times it brought tears to my eyes, but there is much to make us smile—even laugh out loud. Most incredible is the gradual unfolding of the relationship between Roger and Washoe—beings from different worlds who communicate in a human language.

Roger was a young Ph.D. when first I met him in 1971. I had managed to arrange a lecture at the university in Norman, Oklahoma, so that I could see for myself the chimpanzees who, it was said, conversed in American Sign Language (ASL), and to meet Roger, their extraordinary tutor. It was an amazing experience. It wasn't so much Washoe's intelligence that surprised me—after all, I had firsthand knowledge of chimpanzee intelligence and social awareness. It was seeing Roger and Washoe (a rambunctious six-year-old then) at work that impressed me so greatly. I watched them exchange comments about things going on around them and was amazed by the quality of their relationship. There was no doubt, they were friends, working together.

Of all the facts to emerge from my years of research on the chimpanzees at Gombe, it is their humanlike behaviors that fascinate people most—their tool-using and tool-making abilities; the close supportive bonds among family members, which can persist throughout a lifetime of fifty or more years; their complex social interactions—the cooperation, the altruism, and the expression of emotions like joy and sadness. Roger, through his ongoing conversations with Washoe and her extended family, has opened a window into the cognitive workings of a chimpanzee's mind that adds a new dimension to our understanding. Clearly chimpanzees are capable of intellectual feats that once we thought unique to humans. Not only can they reason, plan for the immediate future, and solve simple problems, but their proficiency in ASL shows that they can understand and use abstract symbols in their communication. Washoe was even able to pass on this skill to her adopted son. It is our recognition of these intellectual and emotional similarities between chimpanzees and ourselves that has, more than anything else, blurred the line, once thought so sharp, between human beings and other animals.

This is a little humbling. Of course humans are unique, but we are not as different as we used to think. We are not standing in isolated splendor on a pinnacle, separated from the rest of the animal kingdom by an unbridgeable chasm. Chimpanzees—especially those who have learned a human language—help us intellectually to bridge the imagined chasm. This crossing gives us new respect not only for chimpanzees but for all the other amazing animals with whom we, the human animal, share this planet.

How lucky that Roger waited before writing this book. Had

he written it in the early eighties, it would have been a gripping story and a great contribution to scientific knowledge. But the intervening decade has transformed Next of Kin into something more. Roger was tested—and he was not found wanting. Having proved that chimpanzees share our intellectual and emotional capacities, he had the courage to confront the ethical implications of his own research. Not only did he risk his career to ensure that Washoe be spared life imprisonment in a small, bleak cell; he also had the courage to stand up to the research establishment and its cruel treatment of our closest evolutionary relatives.

Next of Kin provides a grim picture of the dark side of science and describes some of Roger's attempts to light a candle there. I was with him on one journey into the dim world of a subterranean research lab, where hundreds of chimpanzees, each with his or her own vivid personality, eager intelligence, and sense of fun, were shut away forever from the world of sunshine and smiles. We can never forget, nor forgive. And I doubt any reader will remain unchanged.

Washoe and her extended family are lucky to have Roger as a protector and ally, as are chimpanzees everywhere. I too feel fortunate to know Roger as a colleague and friend. And to have had the unforgettable experience of spending time with that grand lady of the chimpanzee world, Washoe herself. For it is individuals such as she, and also Flo and David Greybeard of Gombe, who have played such a crucial role in clarifying our special evolutionary relationship with chimpanzees. They are indeed our closest kin and, as such, we have a special responsibility for their survival and well-being.

Now readers around the world can share in this marvelous, true-life adventure story of scientific and spiritual discovery, which is told with utmost honesty, with wonder, and with love.

PART ONE

A FAMILY AFFAIR

RENO, NEVADA: 1966-1970

I am inclined to conclude from the various evidences that the great apes have plenty to talk about, but no gift for the use of sounds to represent individual . . . ideas. Perhaps they can be taught to use their fingers, somewhat as does the deaf and dumb person, and thus helped to acquire a simple, nonvocal, "sign language."
—Robert Yerkes, 1925

ONE

A TALE OF TWO CHIMPS

THE FIRST CHIMPANZEE I EVER KNEW was Curious George, the mischievous hero of the classic children's book written by H. A. Rey.

It was the late 1940s and I was a small boy. One night my mother read me the story about "a good little monkey" who is captured in Africa by "the man with the yellow hat." The mysterious man pops Curious George into a sack, puts him on a ship, and takes him to a big city far away.

Curious George feels sad to leave home. But he is soon having fun. He tries hard to be good, but he can't seem to help getting into trouble. "The naughty little monkey" winds up in prison. His friend, the man with the yellow hat, rescues him and puts him in a zoo, where the story ends happily: "What a nice place for George to live!"

I loved this story. It never occurred to me to wonder why Curious George had to leave his home in the jungle, or who the man with the yellow hat was, or why he put George in a zoo. I was only a child.

As a child I also didn't realize that George was not a monkey at all but a chimpanzee. In fact, the book's author had once wanted to call his character Zozo the Chimp. Monkeys, for the most part, are small, narrow-bodied creatures who walk exclusively on all fours and sport tails for balancing. They are our distant evolutionary relatives. Curious George is clearly a chimpanzee: he has no tail, he sometimes runs on two legs, and his face is apelike, with its flat nose and protruding jaw. The chimpanzee is humankind's closest living relative and a member of the great ape family, which also includes gorillas and orangutans. An upright, two-hundred-pound adult chimpanzee resembles our earliest hominid ancestors more than any monkey.

Twenty years later, when I entered graduate school, I met another chimpanzee—a real chimpanzee. Her name was Washoe. She, too, had been abducted from the African jungle— in this case to become part of the American space program. She, too, was an irrepressible bundle of mischief.

Washoe the real chimpanzee was more fantastical than Curious George in one important respect: she learned how to talk with her hands using American Sign Language. Washoe was the first talking nonhuman, and in the wake of her accomplishment the ancient notion that humans are unique in their capacity for language was shaken forever.

But Washoe's use of language, while remarkable enough in itself, was only the beginning. Those first signs initiated a life-long conversation between two friends who happened to belong to different species. From the moment I first met Washoe our destinies became as intertwined as two clasped hands. This book chronicles that shared lifetime of joy and hardship, scientific breakthroughs and controversies.

How does one account for the extraordinary connection between humans and chimpanzees? The answer has to do, oddly enough, with the reason children love Curious George. Unlike other storybook animals, Curious George, the chimpanzee, was not anthropomorphized. Chimpanzee behavior really *is* like human behavior—there is no need to embellish it. Children identify with George's wonder at the world around him, his innocent need to wreak havoc, his thoughtful way of solving problems that creates even bigger problems, his delight in breaking rules and undermining authority figures, and his shame at being caught and punished. In short, children see themselves in Curious George. Little do they know that the Curious George character is no fantasy. The chimpanzee child really does think, feel, and rebel just like the human child.

Most children never discover this remarkable fact. They grow up and leave their storybook alter egos behind. I grew up and met Washoe.

Nothing was further from my own mind when I met Washoe in 1967 than humankind's relationship to other species. My future was mapped out in clear bold lines: I was going to pursue an exciting career in psychology working with children.

But then Washoe began talking. She took me on an amazing journey to a world where animals can think and feel—and can communicate those thoughts and feelings through language. Along the way I met dozens of other chimpanzees, each one as individualistic and expressive as Washoe herself. In the end I learned more about my own species than I ever dreamed possible: the nature of our intelligence, the origins of our language, the extent of our compassion, and the depths of our cruelty.

This is Washoe's story. I tell it to repay a lifelong debt to her and all the other chimpanzees who have touched my heart and opened my mind.

CURIOUS GEORGE WASN'T THE ONLY ANIMAL I knew as a young boy. I grew up on a farm where animals were a very important part of our family's life.

My closest animal companion was our dog, Brownie. Feisty and fiercely loyal, Brownie was a fixture of our household. She needed us and we needed her. In addition to guarding the house, she baby-sat the youngest kids in the fields during the harvest season.

One day I saw Brownie do something that shaped my view of animals forever. She saved my brother's life. It happened during cucumber-picking season when I was four years old. The whole family—my parents, six brothers, and one sister—had been out in the field all day working. Brownie had been watching over me and my nine-year-old brother, Ed, whenever he got tired of picking. By the time the sun was going down our Chevy flatbed was piled high with boxes of cucumbers. It was time to head home for dinner. Ed wanted to ride back on our older brother's bicycle, a big thing that he could barely control. My parents said OK and Ed headed out on the bike, chaperoned by Brownie. Twenty minutes later, the rest of us clambered onto the truck and left the field with my twenty-year-old brother, Bob, driving.

It was the dry season, six months or so since the last rain, and the dirt road was blanketed with four or five inches of chalky dust. As the truck drove along the well-worn tire ruts in the road, it kicked up a huge cloud of dust that covered us on all sides, making it impossible to see more than two feet ahead or behind. After going along for a while, we suddenly heard Brownie barking very loudly and very persistently. We looked down and we could just make her out next to the front fender. She was sniping at the right front tire. This was very strange behavior. Brownie had come to the fields hundreds of times and had never once barked at the truck. But now she was practically attacking it. My brother Bob thought this was odd but didn't give Brownie much thought as he plowed ahead even as her barking became more frenzied. Then, without further warning, Brownie dove in front of the truck's front tire. I heard her shriek, and I felt a thump as we drove over her body. Bob hit the brakes, and we all got out. Brownie was dead. And right there in front of the truck, not ten feet away, was Ed, stuck on his bike in the deep tire rut, unable to escape. Another two seconds and we would have run him down.

Brownie's death was devastating to all of us. I had seen animals die before, but this one was my nearest and dearest friend. My parents tried to explain that Brownie had only done what either of them would have done for us. No one doubted for a second that Brownie had sacrificed her own life to save my brother's. She saw a dangerous situation unfolding, and she did what she had to do to protect the boy she had been baby-sitting for so many years. Had she not acted, the course of our family's life would have been very different.

My mother had a deep respect for all God's creatures, and she was full of stories about animal intelligence that we never tired of hearing, like the one about her childhood horse who could untie knots. My mother grew up at the turn of the century in what seemed to us the romantic Wild West. She was well acquainted, as she loved to remind us, with horses, guns, and rattlesnakes.

My parents were leasing a small vineyard in California in the early 1940s. By the time I came along, in 1943—the last of eight children spanning eighteen years—they were ready to buy their own modest spread. I was three years old when our large family, animals included, moved onto forty acres outside Florin, a small town south of Sacramento.

Growing up on a farm I quickly learned that animals, like people, are best understood as individuals, no two of whom are the same. I knew a variety of pigs, a variety of cows, and a variety of horses. We always referred to a specific animal by name, as in "Bessie is a real sweetheart" or "Old One Horn's busting my chops." When my mother said, "He's an ornery critter" she might have been complaining about a horse, a cow, or a person. It was understood that all three species are prone to bouts of orneriness.

By the time I could walk, I had to know which cow was friendly and which one spiteful. If I didn't know that much I could easily walk behind the wrong one and get clobbered. By the time I was five years old, I knew exactly where each cow liked her milk pail placed; if I got it wrong then we didn't get any milk. For us, the notion that animals are dumb beasts without distinctive personalities was something city people thought and something a farm family could not afford to indulge. If you didn't deal with an animal's personality, then it would deal with you.

Earning a living for ten people on forty acres was a big job.

My parents were inseparable helpmates, rarely ten feet apart in the fields or the kitchen. All eight children were relied upon for labor, and everyone depended on the animals to be our working partners. I was taught to make allowances for them the same as I would for my brothers or sister. They were members of our family, and their personalities, illnesses, and contributions were discussed in great detail. Of course, I'd occasionally find one of our pigs, ducks, or cows on my dinner plate. In those moments I understood the full extent of our family's reliance on animals. When we gave thanks for our dinner, I knew exactly to whom I was indebted.

When I was twelve years old, and Ed and I were the only kids left in the house, our family's life changed radically. My parents decided to give up farming and move to Los Angeles so that my mother could tend to her ailing father. One morning I found myself saying good-bye to the farm, the animals, the wood-frame three-room schoolhouse, and the winding Mokelumne River, where I loved to play. The next morning I woke up in my grandfather's house in Compton, a racially mixed neighborhood of urban Los Angeles. I showed up for my first day at Roosevelt Junior High in my very best brown corduroys, neatly ironed shirt, and farm boy haircut—shorn on the sides and slicked down on top. I looked like someone who just fell off a turnip truck. It was 1955, and all the other eighth graders looked like James Dean. They wore baggy Levi's with the belt loops cut off and the waist pulled down on the hips, pant cuffs rolled up, and white T-shirts with cigarettes tucked into the sleeves.

Around this time I began dreaming of becoming a psychologist. This aspiration emerged naturally from an important event in our family that had occurred two years earlier, when I was in the sixth grade. One of my brothers had a nervous breakdown. He recovered only through the intervention of a school counselor. I was deeply impressed by this "healer of the mind," and I soon developed a kind of Florence Nightingale syndrome—I wanted to help cure others.

To become a psychologist I would have to go to college, and that was an unlikely path in a family full of farmers and plumbers. Some of my brothers did start college. They just never seemed to finish. Donald went to Berkeley for a year on scholarship, then came back to the farm, married his high school sweetheart, and became a successful plumber. Ed did a couple of years of college, then went into plumbing. Raymond did two years of college and then ran boilers in a heating plant. Arthur fought in World War II, came home, and bought a small vineyard. Jack became an electrician. Bob was deputy sheriff of Stockton. My only sister, Florence, became a painter.

But there was one member of my family who was determined to finish her education: my mother. When she was fifty-two years old my mother decided to go to high school. Ever since my brother's miraculous recovery a few years earlier she, too, had wanted to become a psychologist. Her own enthusiasm for the healing profession had a profound effect on me. As I worked my way through Compton Junior High, my mother focused her considerable energy on getting the high school and college education she had missed out on. Day after day I would come home from school to find her sitting at the kitchen table doing her homework. She couldn't seem to get enough book learning. She reveled in every minute of every course she took, and after receiving her high school equivalency, she enrolled at Compton Junior College and then Long Beach State, blazing a trail that I would follow.

In September 1960 I enrolled at Compton. Although I wanted to study human psychology I was required to take animal psychology courses as well. One of the first things I was taught was that animals are mindless creatures whose rigid behavior, unlike that of humans, is controlled by instinct. My professors spoke reverentially about "scientific objectivity," and clearly looked down upon uneducated people who still believed in the old superstition that animals had humanlike consciousness. I knew immediately that I had been one of those ignorant fools. I felt ashamed of my childhood view of animals, and I worked doubly hard to be objective and worthy of wearing a white lab coat as I studied the behavior of pigeons and rats. Objective science was personified by the promise of the new American space program, and I was glued to the television as chimpanzees were launched into outer space and their bodily responses analyzed by teams of NASA biologists and engineers.

It wasn't until I attended Long Beach State College and took courses in child psychology with a professor named Joe White that I knew I wanted to work with kids. Joe was more than a great teacher. He became a major influence in my life. A short and dynamic black man, Joe had an unpretentious, street-smart style of working with children, adults, and families. He took me under his wing and became my clinical supervisor. Many of his students knew all the theories about children, Joe said, but he thought I had that rare talent for actually working with kids. Under his mentorship I discovered that I empathized easily with kids who were uncommunicative or in pain. I knew that there was more money to be made treating adult neurotics, but I was drawn to these wounded children who were too young to have chosen their abusive families or depressing circumstances. They deserved an ally, I thought.

But as committed as I was to child psychology, there were still moments when I considered following in my brothers' footsteps. My brother Donald had a very successful plumbing business by the early 1960s. During my first year at Long Beach State my girlfriend took one look at Donald's suburban lifestyle and began urging me to forget about college and get into business with my brother instead. We'd have a nice house with a yard, a car, kids. It was very tempting.

Then I met Debbi Harris. From the first time I saw her on campus, I thought she was the most exotic girl at Long Beach State. Those were the days of bouffant hairdos and heavy makeup. Debbi pulled her dark hair back in a tight ponytail, and her piercing blue eyes and natural beauty were untouched by makeup. She seemed lit up from within by vitality and self-confidence. She was the anti-Barbie in the midst of a thousand Barbie dolls. She made no attempt to mask the person inside.

Debbi was from San Francisco, only four hundred miles away but as distant culturally from Los Angeles—the land of Disneyland, *American Graffiti*, and the Beach Boys—as one could get. She was very liberal politically and knowledgeable about the civil rights movement. I was used to talking to my girlfriends about cars, clothes, and sports—not about ethical dilemmas and righting the wrongs of society. I'd never met anyone who seemed genuinely concerned about the world's underdogs. But best of all, Debbi loved children. She had worked with Down's syndrome kids one summer in high school and it changed her life. She *knew* she wanted to work with special kids. And she encouraged me to stick to my own dream.

Nine months after we met, in August 1964, Debbi and I got married. She was determined that marriage not get in the way of her career path. We both returned to Long Beach State to finish college in the understanding that we'd take turns going to graduate school in child psychology.

But in the summer of 1966, just as Debbi completed her bachelor's degree and I was starting my master's, we learned that she was pregnant. The happy but surprising news—the babies were supposed to come *after* graduate school—delayed Debbi's career plans and focused my mind in a hurry. I began studying harder to improve my grade point average, and I applied to doctoral programs in clinical psychology. The competition was extremely fierce. Some schools had as many as four hundred applicants for a single opening. My grades were good but not of the 4.0 variety that many applicants would be offering. Still, I didn't let that obvious shortcoming stop me. I applied to nine of the nation's top clinical schools.

Joe White smelled trouble ahead. "Roger," he suggested, "why don't you apply to a second-tier school in experimental psychology."

Experimental psychology—or rat psych, as it is affectionately known—studies animals in cages. Its practitioners measure repetitive behavior like lever pressing in rats and pecking in pigeons as if they were molecules in a test tube. Nothing could be further from clinical psychology—the treatment of human problems with the "talking cure" of Sigmund Freud or other therapies based on the emotional life. Still, Joe felt that I could get an excellent background in the rigorous scientific method of animal psychology and then, after getting my Ph.D., pursue post-doctoral work with kids. He suggested that I add the University of Nevada at Reno to my list of schools and I did. Six months later, in March 1967, while the nation's more famous schools were busy rejecting me, the University of Nevada admitted me to their graduate program in experimental psychology. The acceptance arrived one day after our first child, Joshua, was born.

But I still had one major problem. I couldn't afford the university's high out-of-state tuition fees. I immediately wrote a letter requesting a graduate assistantship—any job that would allow the school to waive the out-of-state tuition.

As the weeks and months went by without any response, I began to consider the real possibility that I might not go to graduate school at all. I would never receive my Ph.D. or get to work with kids. I had already lined up a summer job as a shipping clerk in an aluminum foundry. Come the fall, I would go into my brother's business. After all the studying and the planning, after coming so far, it turned out I would be a plumber after all.

Just when I was about to give up, the phone rang one afternoon in June. It was Dr. Paul Secord, the chairman of Reno's psychology department.

"Roger, we have a half-time assistantship available," he said. "Would you be interested?"

"Of course," I answered. "What's the job?" I was already picturing myself running white rats through some very challenging mazes.

"Teaching a chimpanzee to talk," he said, matter-of-factly.

"What?"

"Teaching a chimpanzee to talk," he repeated, as if saying it a second time might clear up my confusion. At first I thought he was pulling my leg. Maybe this was the "talking chimpanzee" joke they played on all first-year graduate students.

But he went on to tell me about two laboratory scientists on the Reno faculty—a husband-and-wife team named Allen and Beatrix Gardner—who were raising an infant chimpanzee in their home. Her name was Washoe. The Gardners planned to teach Washoe to talk with her hands using American Sign Language, and they needed an assistant.

"Is Washoe signing yet?" I asked Dr. Secord.

"Oh, yes," he replied nonchalantly. "They've been at it for a year and Washoe already has a small vocabulary of signs."

Dr. Secord didn't seem to be as amazed as I was at Washoe's linguistic abilities. He was a social psychologist and so he was more impressed that Washoe was imitating her foster parents, the Gardners, by bathing her dolls in the dishpan where she received her own baths. Only humans could imitate such behavior—or so social psychologists thought.

The more I heard about Washoe the more intrigued I became. Playing with dolls didn't sound anything like running rats through mazes. It was more like working with a child. What better way to prepare myself for a career with uncommunicative kids than by learning to communicate with a mute chimpanzee!

"I'll take the job," I announced to Dr. Secord.

"I can't give it to you," he replied. "You'll have to pass an interview with Allen Gardner."

Allen Gardner was a tough-minded exponent of experimental psychology, known for his strict laboratory method and mathematical precision. I knew that he would tend to see a clinician like me as a soft-minded Freudian who talked for a living and was unable to separate feelings from facts. Gardner was surely looking for someone more laboratory oriented than me. The interview was a long shot, but I was out of other options.

On a hot Sunday in August, Debbi, Joshua, and my parents

dropped me off in Reno for my interview and wished me luck. As Dr. Gardner and I strolled across the campus, which ironically was where the Ronald Reagan chimp movie, *Bedtime for Bonzo,* was filmed, he explained the two main parts of the job. First, to help raise Washoe by taking care of her day-to-day feeding, clothing, and play. Second, to expose her to American Sign Language. I was already taking care of one primate infant, my son, so thought I could handle that, and as for learning sign language, it would be a challenge, but I had no doubt that I could master it given enough time.

But the focus of the interview soon shifted to me, and my worst fears were confirmed. Gardner was skeptical about my credentials for his research project. It wasn't my academic preparation that gave him pause—I had taken several courses in animal psychology and statistics—but my desire to do clinical work with children. To him, this was a fatal flaw in my character. Gardner had no use for someone who was likely to waste time on nebulous concepts that couldn't be proved in a laboratory.

The interview was going badly. In desperation, I tried to win him over by telling him how much I was looking forward to taking courses with two well-known philosophers of science during the coming school year.

"Science doesn't need philosophy," he shot back. "If you are influenced by them it will show that you weren't worth anything to begin with."

Our walk was now over and so was the interview. I had blown it. I felt sick to my stomach knowing that my career in psychology had reached an end. I considered begging him, but I knew it would do no good. As we said good-bye, Gardner asked if I wanted to walk over to the university nursery school and see Washoe. She played there every Sunday on the jungle gym and swings when there were no children around. I knew this was the loser's consolation prize, but I wasn't too proud to accept it.

As we approached the fenced-in nursery school, I saw two

adults playing with a child in the shade of a tree. At least I thought it was a child. When the child saw us coming she leapt up and began hooting. Then she began sprinting in our direction—on all fours. We were only a few yards from the four-foot-high fence now. Washoe continued to speed toward us and, without breaking stride, vaulted over the fence and sprang from the top rail. What happened next amazes me to this day. Washoe did not jump onto Allen Gardner as I had expected. She leapt into my arms. Before I knew what was happening, this baby chimpanzee had wrapped her arms around my neck and her legs around my waist. She was giving me a giant hug. Somewhat sheepishly, I found myself hugging her back. With my dreams in ruins, I needed that hug more than anything in the world.

Then the small diapered girl turned around in my arms and reached out for Allen Gardner. She climbed into his arms and gave him a hug, too. I was stunned to see Gardner smiling warmly at me over Washoe's shoulder. Washoe liked me, and he knew it.

I don't know why Washoe hugged me that day. In years to come I would discover that she had an uncanny knack for seeking out and comforting those who were sad or hurt, but I never again saw her jump into a stranger's arms. When Allen Gardner called me a few days later to tell me the research assistantship was mine, I knew exactly who had done the selecting. I may not have been Gardner's ideal graduate student, but as far as Washoe was concerned I would make a pretty good playmate. Thanks to a two-year-old chimpanzee I was going to be a psychologist, not a plumber.

TWO

BABY IN THE FAMILY

THE FIRST THING I DID when I got to graduate school in early September 1967 was stop in and see Allen and Beatrix Gardner. I was expecting to have a look at their "chimpanzee language laboratory," but to my great surprise I was shown their backyard.

The Gardners lived in a small, one-story home with an attached garage. The backyard—about five thousand square feet—had a brick barbecue, some flower beds, a jungle gym, a sandbox, and a tire swing hanging from a weeping willow tree. On one side of the yard was a gravel driveway where a small house trailer was parked. All in all, it looked like your typical suburban household—except there was a baby chimpanzee living in the trailer.

Almost immediately upon entering the Gardners' home I noticed that everyone was talking in whispers. This was part of the experiment. They didn't want Washoe to know that her human friends could speak with their voices. Otherwise she might try to communicate vocally—which other chimps had failed at—instead of learning sign language. Everyone who worked on Project Washoe took a vow of silence. We could talk with our hands but not with our mouths.

The Gardners took me into their kitchen so that we could watch Washoe through the rear window. I thought I would see people in white lab coats carrying clipboards and stopwatches.

But instead there was just one student, Susan Nichols, who was playing with Washoe. As Susan carried Washoe around the yard on her shoulders they signed to one another. This went on for a while until Washoe seemed to get tired of the game. She jumped down off Susan and scampered up the weeping willow tree. Susan picked up a three-ring notebook and began to write in it. This was the only evidence of science I could see.

I assumed that a renowned experimental psychologist like Allen Gardner would be working, like my college professors did, in a state-of-the-art laboratory with high-tech apparatus. I equated good science with animal treadmills, test tubes, and rocket ships. But as I would soon discover, the Gardners had built the most rigorous controls into these warm and pleasant surroundings.

During that first week, I visited Washoe several times while she played with Susan or another human companion. She seemed to enjoy our playtime together, so the Gardners decided that I was ready for my baptism by fire: a shift alone with Washoe.

The next morning I got to the Gardners' house a few minutes before 7 A.M. I entered Washoe's trailer and locked the door behind me so she couldn't run out. I turned off the baby monitor intercom, which the Gardners installed to keep tabs on Washoe's nighttime activities. Then I checked the bedroom. Washoe was awake but groggy. She was definitely not happy to see a stranger in her house.

I took off her nightgown and tried to change her cloth diaper, which I could see was full. I knew immediately that this would be nothing like changing my son's diaper—he didn't move. With Washoe, I had to roll around on the bed a couple of times before I could wrestle the diaper off her, rapidly wipe her rear end, and flush the load down the chemical toilet. Once I finally got the diaper into the diaper pail, I turned around for only two minutes, which was all the time she needed to retrieve the diaper and have a go at stuffing it down the toilet.

The next thing I did was put away her blankets and get out

her clothes, but soon I discovered that it was almost impossible to open a cabinet without Washoe getting there first to ransack it. And getting her dressed was a battle royal. With the floor knee-deep in clothes, I was beginning to understand why there were locks on every single cabinet in the trailer. Somehow I managed to get her into her high chair, and she gestured at me playfully while I unlocked the fridge and started preparing her cereal and hot formula. Again I made the mistake of turning my back for one second, and she flew out of her high chair, opened the fridge, and grabbed random food items like a crazed shopper, finally fleeing with them into the bedroom.

I was being hazed, there was no doubt about it. Washoe didn't miss a trick, catching me every time I slipped up, every time I left a cabinet unlocked. Whenever I got something covered, closed, or put away, she seemed to say: "OK, what can I do now to really get him in trouble?" She'd wait for her opening and then nail me. I had been alone with her for less than an hour and this baby chimpanzee had me terror stricken. It was not at all like taking care of a dog or cat, as I had half expected. This was a two-year-old with an agenda.

Washoe's childlike behavior seemed especially surreal because, outwardly, she looked so different from a human child. She was pretty much the same size as a human two-year-old: she stood thirty inches tall and weighed about twenty-five pounds. And she was dressed like a human child: in a diaper and pullover sweatshirt with the arms cut off so she could swing and climb.

But her flat, spadelike nose sitting atop a massive protruding jaw, her jutting brow ridge, her huge jug-handle ears, her smooth coat of head-to-toe body hair, and her impossibly long arms and handlike feet were loud reminders that Washoe was not human. And there was the way she moved through her favorite tree— like an acrobat. Baby chimps learn to crawl, walk, and climb at a much younger age than human children. By the time I met two-year-old Washoe she'd already been playing in the highest branches of her tree for at least a year.

Her treetop daring made me think that a chimpanzee was some kind of monkey, although like most people I still didn't understand the difference between them. Most monkeys, it turns out, are built for a nearly full-time life in the trees. Their anatomy is perfectly designed for maintaining balance while walking along branches: a narrow body, flexible hands and feet for grasping, forelimbs and hind limbs that are similar in size for maintaining a low center of gravity, and a tail for balancing or hanging.

About thirty million years ago a monkeylike creature began venturing down from the trees, a bold move that gradually gave rise to great apes—chimpanzees, gorillas, and orangutans—as well as to humans. Wild chimpanzees eat and sleep in trees, but they spend most of their time socializing and traveling on the ground. Washoe's body was designed for this dual-purpose lifestyle.

When she climbed a tree Washoe looked like a telephone lineman: her long arms served the same purpose as the safety belt that linemen wrap around poles, letting them lean back, extend their arms, and climb up. Once in the tree, her powerful upper body, which would soon become many times stronger than that of a well-conditioned human adult, allowed her to maneuver with ease. (Chimpanzees have been known to pull nearly one thousand pounds in weight with one arm.) With her short legs, broad back, and fully rotating shoulders she could swing gracefully from branch to branch using the hand-over-hand method. Her wrists did not flex backward like a human's; this rigidity in the hands ensured that she didn't lose her grip when hitting a branch at high speed. Washoe had no tail because she didn't balance on branches like a monkey. Her long hands, with their naturally curved fingers, were assisted by handlike feet with large opposable toes, so she was adept at clambering through the treetops.

Down on the ground, Washoe's long fingers did not prevent her from walking on all fours thanks to an ingenious anatomical innovation: the top two segments of her fingers folded twice under her palms, and she walked on the backs of her thickly padded finger knuckles. When eating, grooming, or signing she sat upright, a lot like a human. She also walked and ran on her two legs, especially when she was angry or going to hug someone.

But for all the outer differences between Washoe and a human toddler there was one thing they had in common: the eyes. When I looked into Washoe's eyes she caught my gaze and regarded me thoughtfully, just like my own son did. There was a person inside that ape "costume." And in those moments of steady eye contact I knew that Washoe was a child, no matter what she looked like and no matter what acrobatics she performed in the top of a tree.

One other reminder came from Washoe's diaper habits, which quickly showed me just how unglamorous my "lab job" really was. Like a new parent I found myself encouraging Washoe's toilet productions with the kind of enthusiasm normally reserved for first words and first steps.

Changing Washoe's diaper in the trailer was hard enough, but getting one changed in the yard required lightning reflexes. If her diaper was full, she would hang from a tree limb by one arm and give you twenty seconds at the most to wipe and change her. I began to think of this as refueling a race car in the Indianapolis 500; she was going to pull out of the pit stop whether I was done or not. But despite our frustration with her diapers, Washoe seemed quite pleased with them.

Potty training became a main topic of our discussion at the strategy meetings the Gardners held every week with Washoe's companions. At first, we tried to introduce the potty by guessing when Washoe might poop, but she did her absolute best to save it for her diaper. Finally, we decided to follow the approach that Debbi and I later used with our own kids: we got rid of the diapers altogether and placed potty-chairs at strategic locations around Washoe's yard. It worked like magic, and Washoe caught on right away. Before any of us left the trailer or the yard we asked her to use the potty. This request soon became so routine that poor Washoe sometimes sat on the potty while I begged

her in sign: PLEASE PLEASE TRY or PLEASE TRY MAKE MORE WA-
TER. She would try a little while longer and then reply, almost
apologetically, CAN'T CAN'T.

There was one more problem as well. Sometimes Washoe
would be in a great rush to get to the potty—we could tell
because she would sign HURRY to herself and then make a mad
dash across the yard—but the high back and sides of the potty-
chair would get in her way. After many accidents, the Gardners
designed a chimpanzee-friendly potty. It looked like a black plastic plate on legs with a hole in the middle. No matter what
direction Washoe was coming from, she could dive onto the
potty and make a direct hit.

Washoe quickly realized that we didn't like cleaning up her
accidents. So once out of diapers, it didn't take long for her to
learn that she could manipulate us by having, or just threatening
to have, an accident. It must have been terrific fun to go high
up in her tree and commit a simple natural act that would cause
grown-up humans to jump around in desperation on the ground
below. She would then saunter down the tree in her own good
time, accept a verbal finger-lashing, and dutifully close the pro-
verbial barn door by sitting down on the potty.

ALL OF THIS ATTENTION to diapers, high chairs, and formula
may seem pretty far removed from teaching a chimpanzee sign
language. But the Gardners theorized that the chimpanzee, as
humankind's closest evolutionary relative, might have an innate
capacity to communicate as we do. If so, then this capacity to
use language would emerge naturally, along with other childlike
behaviors, when a chimpanzee was raised like a human child by
a human family. This approach was called cross-fostering.

Cross-fostering between species has been studied extensively
in animals. The best-known example is that of ducklings or
goslings who learn to follow, or "imprint" on, the first moving
object they see, regardless of whether it's their mother, a female
of a different species, or a pair of rubber boots. The recent movie

Fly Away Home told the story of geese who imprint on a little girl. Another recent movie, *Babe*, took the cross-fostering premise to comical extremes by having a sheepdog raise a pig, who then grows up to herd sheep quite expertly.

Babe is not all that far-fetched. Growing up, I did plenty of cross-fostering with our own farm animals. I loved to put duck eggs under a broody hen and then watch them hatch. The mother hen and ducklings seemed to make one big happy family, until the ducklings got old enough to jump in an irrigation ditch and start swimming. Then their hen mother would jump up and down flapping her wings at this bizarre behavior.

One day I decided to see how far I could push this cross-fostering experiment, so I placed some duck eggs under our old mama farm cat. When the eggs hatched I was amazed to see the ducklings attach themselves to their furry four-legged mama. But even more astonishing, the cat treated the little birds like they were kittens, cuddling them for warmth and licking their feathers. Of course the day arrived when her feathered "kittens" discovered the irrigation ditch. As they filed into the water, the mother cat leapt from one side of the ditch to the other in a panic, meowing loudly. Finally she gave up and climbed into the ditch, positioned herself in front of her duckling brood with her tail perfectly erect, and led them through the water almost exactly as a mother duck would.

More sophisticated studies than that one have shown that almost all innate behavior can be shaped by early experience. These discoveries make moot the old debate over "nature or nurture"—over whether a given behavior is instinctual or learned. Most behaviors are a little bit of both. Even behavior that we think of as innate and species specific, like birds' songs and migratory routes, can be altered by parents from another species. And that's why cross-fostering has taught us so much. It shows the extent of an organism's ability to adapt by learning. By raising a kitten with dogs we can see how much a cat can learn to act like a dog. By raising a pig with sheepdogs we can see whether a pig can actually learn to herd sheep.

What about humans? What would happen if a human child were raised by another species of animal, in an environment absent all of our culturally learned niceties? Our ancestors have been wondering about this ever since the mythic twin brothers Romulus and Remus, the founders of Rome, were raised by wolves, and probably before. In modern times, we have the ever-popular Tarzan the Ape-Man books and movies about a human raised in the jungle by apes.

There are obvious ethical problems with leaving a human child on an ape's doorstep, and it would be a tough experiment to control scientifically. But chimpanzees can be raised by humans, and in the early 1930s two scientists, Winthrop and Luella Kellogg, set out to rear a chimpanzee like a human child as a way of discovering the ape's innate mental capacity for using our tools, imitating our social behavior, and speaking our language. It was a novel twist on the Tarzan scenario—Cheetah would grow up American—that hopefully would answer, indirectly, that age-old riddle of human nature: How beastlike are we? If it turned out that the ape is just like us, then it would follow, for better or worse, that we are just like the ape.

Before the Kelloggs' radical proposal, many people had kept apes as household pets, but nobody had treated those apes like children. This difference in treatment was crucial, as the Kelloggs made clear in a book they wrote about their experiment called *The Ape and the Child:*

> It is not unreasonable to suppose, if an organism of this kind [a chimpanzee] is kept in a cage for a part of each day or night, if it is led about by means of a collar and chain, or if it is fed from a plate upon the floor, that these things must surely develop responses which are different from those of a human. A child itself, if similarly treated, would most certainly acquire some genuinely *un*-childlike reactions.

We take for granted that our children's behavior is "child-like," but a great deal of that behavior is a response to the child stimuli that we human parents provide. On the other hand, if you give "sit" and "roll over" commands to your children, reward them with scratches behind the ear, and feed them off the floor, you will soon find out that your children have a disconcerting capacity for doglike behavior. (This is only armchair psychology—please don't try this at home!) Likewise, if you want to see how far an ape can go in acquiring human behavior, you have to take the laborious and time-consuming path of treating that ape exactly as you would treat a child—diapers, high chair, and all. The Kelloggs were also opposed to any systematic training of the ape. Cross-fostered chimpanzees should be allowed to pick up their table manners and all their other behavior in the gradual and irregular way that children do, by watching their parents and learning at their own pace. And they stressed that the ape should not be taught any tricks or stunts to be performed on command because that would prevent the animal from ever understanding how to use its behavior in the proper social context.

Inspired by the Kelloggs' approach, several families raised chimpanzees in their homes, and the apes became remarkably childlike in their development and abilities. They could eat with a fork and knife, brush their teeth, use a wrench, leaf through magazines, paint with fingers or brushes, even drive the family car—all quite spontaneously and in the proper context of their social life. But there was one behavior universal to human children that these home-reared chimpanzees never developed: language.

It was not for lack of trying. The Kelloggs raised an infant chimpanzee named Gua along with their own son, Donald. But unfortunately for science, the Kelloggs abruptly terminated the study because, rumor has it, Mrs. Kellogg became distressed when Donald began acquiring more chimpanzee sounds than Gua was acquiring human sounds. Apparently the Kelloggs' son was making food grunts at the dinner table.

Then in the late 1940s psychologist Keith Hayes and his

wife, Cathy, home raised a newborn chimpanzee named Viki. After six years of intensive and creative vocal training, Viki could speak just four words: "mama," "papa," "cup," and "up"— all made with a heavy and largely voiceless chimpanzee accent. This small vocabulary was a start, but it fell far short of what early experimenters were hoping for. The Hayes experience was strangely like the case of W. H. Furness, who reported to the American Philosophical Society in 1916 that he had taught an orangutan, the Asian great ape, to say the words "papa" and "cup." His orangutan then died from a high fever while saying "papa cup" over and over.

In the wake of the Hayes study, many scientists claimed that the cross-fostered chimpanzees proved that humans stand above and apart from the apes because of our unique and innate capacity for language. The word went out to college freshmen— including myself—that language is beyond the capacity of the otherwise intelligent ape. Not only was this incorrect, it was simply bad science, as any student enrolled in Introductory Statistics could tell you. We can try to prove that apes *have* the capacity for language—by teaching one or more apes to use language—but we can never prove the "null hypothesis" that apes do *not* have the capacity for language. In the latter case the best we can do is say that we failed to find any evidence of language in one or more apes and wait for new and better studies.

The failure, after all, may be our own. Just because we fail at teaching a group of Kalahari Bushmen how to play baseball, for instance, it doesn't mean that they lack the capacity to play baseball. We may be lousy teachers, or we could be taking the wrong approach culturally. The right teaching methods could have them all playing baseball in a matter of days.

This is where Allen and Beatrix Gardner entered the picture. The Gardners carefully reviewed all the cross-fostering studies and spotted a common error: the researchers had all equated language with speech. From the Kelloggs on, scientists had assumed that chimpanzees should use vocal speech because that's what most humans use. But speech is just one mode of language, and the Gardners knew that it was not a likely one for chimps.

For one thing the chimpanzee has a relatively thin tongue and higher larynx, which makes pronunciation of vowels extremely difficult. But that alone was not reason enough to abandon attempts at vocal speech. Many humans with vocal defects are able to produce intelligible speech. There are even human languages that substitute whistles or clicks for speech. So it should be possible to transform the sounds of human speech into sounds that a chimpanzee could vocalize.

But the Gardners did their homework and discovered an even more compelling reason why chimpanzees weren't likely to speak: by and large they are very quiet animals. Many people have reported passing a tree in the jungle and only later realizing that it was filled with chimpanzees silently feeding or grooming. As far back as the 1920s the pioneer of chimpanzee behavioral research Robert Yerkes recognized that though his chimpanzees could understand spoken English—he believed they could comprehend one hundred to two hundred words—they never imitated his sounds. On the other hand, they did have an amazing ability to imitate his actions. They reproduced what they saw but not what they heard, as opposed to parrots, who imitate vocally but not visually, or human infants, who imitate both vocally and visually. Yerkes concluded that an animal that doesn't imitate sounds "cannot reasonably be expected to talk."

Of course chimpanzees do vocalize, but they have very little control over most of the sounds they make, like pant-hoots when they see friends or pant-grunts when they feel threatened. These sounds are generated by the primitive limbic system in the brain. If you've ever hit your thumb with a hammer, then you are familiar with screams that are controlled by your limbic system, as opposed to more conscious speech that is controlled by your brain's cortex. It is very difficult to modify a chimpanzee's patterned sounds. Cathy Hayes confirmed this when she noticed that her chimpanzee daughter, Viki, could not steal cookies without making sounds. Viki would sneak into the kitchen in utter silence, but the minute she slipped the lid off the cookie jar and saw the cookies inside, she would let out a loud food grunt and give herself away.

But as limited as chimpanzees are in using their voices for speech, they can do almost anything with their hands. According to the Hayeses, Viki created a unique gesture to go with each one of her voiced words. And when the Gardners watched a film of Viki they realized that they could read her nonverbal messages even better when they turned the sound off. Scientists in the wild, like Adriaan Kortlandt and Jane Goodall, were also beginning to report how highly gestural chimpanzees were when communicating with one another. The Gardners had the good sense to respect this, and began searching for a human language that did not require speech. They settled on American Sign Language (ASL), widely used by deaf people in the United States.

Like most great ideas this one was not new; in fact it was at least three hundred years old. In his diary entry of August 24, 1661, Samuel Pepys, the renowned chronicler of life in seventeenth-century London, described a strange creature that had just arrived by ship from Africa:

> It is a great baboon, but so much like man in most things that . . . I do believe it already understands much English; and I am of the mind it might be taught to speak or make signs.

What Pepys thought was a great baboon was probably a chimpanzee. Some eighty-five years later, in 1747, the French philosopher Julien Offroy de La Mettrie concluded in *L'Homme machine* that apes have some "defect in the organs of speech" and then suggested a remedy:

> Would it be impossible to teach the ape a language? I do not think so . . . I would choose the one with the most

intelligent face . . . and I should put him in the school of
that excellent teacher [Amman] whom I have just
named. You know by Amman's work all the wonders he
has been able to accomplish for those born deaf . . . but
apes see and hear, they understand what they hear and
see, and grasp perfectly the signs that are made to them.
I doubt not that they would surpass the pupils of Amman
in any other game or exercise.

Then Robert Yerkes, like Pepys and La Mettrie before him,
theorized in his 1925 book *Almost Human* that "the great apes
have plenty to talk about" and their silence might be overcome
through sign language. This simple idea of Pepys, La Mettrie,
and Yerkes had gone untested for three centuries, showing just
how biased most humans are to the idea that language must be
spoken. The Gardners were smart enough to acquaint themselves with chimpanzees *before* designing their own experiment,
and a lot of the credit goes to Beatrix Gardner. Allen Gardner
had earned his reputation by manipulating animal behavior in
the laboratory, but Beatrix had studied ethology, the *observation*
of animal behavior. She received her Ph.D. from Oxford University, where she was a student of the Nobel laureate Niko
Tinbergen. She had spent years documenting the hunting behavior of the jumping spider.

Ethology in its best form takes a very humble approach
toward nature. Theories, assumptions, and scientific dogma are
put aside and instead the goal is to understand every detail of
an organism's anatomy, development, and social behavior. After
the Gardners studied the chimpanzee they knew that teaching
Washoe to speak vocally would be a waste of time. Sign language, on the other hand, would fit with the chimpanzee's natural mode of communication. That breakthrough transformed
ape language study from a dead end into a new frontier of communication between two species. But the Gardners also knew
that they should hold on to the cross-fostering approach that
had been so successful in earlier studies. Chimpanzees raised like

human children were acting very much like human children, minus the language. So the Gardners decided to take cross-fostering even further by designing an even more favorable learning environment than either Gua or Viki had.

When ten-month-old Washoe entered her new backyard home in June 1966, she began a kind of linguistic head start program. The Gardners surrounded her with interesting friends to communicate with and an interesting world to communicate about. By the fall of 1967 there were, in addition to the Gardners, four main human companions, including myself, who would stay alert and attentive to Washoe throughout each day for the next several years.

Our job was simply to make Washoe's life as stimulating and linguistic as possible. We signed when she was eating, during her bath, and while she was being dressed. We invented exciting games, introduced new toys, books, and magazines, and made special scrapbooks of Washoe's favorite pictures—all to show ASL being used in the course of daily life. As often as possible, there were two or more of us in the yard, so that Washoe could watch us signing to each other. Above all, we were expected to form warm, affectionate relationships with Washoe.

The Gardners had no doubt that if ASL was the right language for Washoe, then she would learn to make signs to ask for food, water, and toys. But they wanted Washoe to develop more than a vocabulary. They believed that she was not a passive laboratory subject but a primate endowed with a powerful need to learn and communicate. The Gardners wanted Washoe to ask questions, to comment on what we did, and to stimulate our conversation. They wanted her to have true two-way communication with humans.

BY THE TIME I MET WASHOE in September 1967 she had been living with the Gardners a little more than one year and had learned about two dozen signs. Washoe was now making steady and dramatic progress, unlike Gua and Viki, who had faltered due to their foster parents' insistence on vocal speech. For the first time in a cross-fostering study, a baby chimpanzee's language was developing stage for stage like a human child's, right along with her abilities to use cups, forks, and the potty.

Washoe indicated DRINK by making a fist with an outstretched thumb and touching the thumb to her mouth. For DOG she patted her thigh; for FLOWER she touched her nostrils with her fingertips; for LISTEN she touched her ear with an index finger; for OPEN she held her hands together, palms down, and then swung them open to face each other; for HURT she pointed her index fingers toward each other and touched them at the site of her or someone else's injury; and so on. And the Gardners guessed right again that this infant primate would not need to be prodded to make these signs part of her life. You might think that a chimp would have trouble understanding that TREE refers not just to one tree but to all trees. But very quickly Washoe was signing OPEN either to get out a door or into a cupboard, and she signed DOG when she saw the real thing and when she came upon a picture of the real thing.

After about ten months she began spontaneously combining words: GIMME SWEET and COME OPEN were soon followed by longer phrases like YOU ME HIDE and YOU ME GO OUT HURRY. She commented on her environment: LISTEN DOG; she asserted possession of her doll: BABY MINE; and she created her own vocabulary when she didn't know a sign: DIRTY GOOD, for her potty-chair.

I *also* had to learn American Sign Language, of course, and the ASL dictionary quickly became my Bible. I took it everywhere and practiced my signs on anyone who would sit still long enough to watch—which usually meant my own son, Josh, who was not yet one year old. Each week I attended ASL classes at the Gardners' house, but most of my learning came on the job with Washoe and her other student companions. We were not allowed to speak English, so it was like being immersed in the language of a foreign country.

Washoe had her own ways of drilling me in my vocabulary.

One day when I was giving her a piggyback ride around the backyard—one of her favorite games—she reached down from my shoulders and touched my chest to make the sign for YOU. Then she indicated which direction I should move in by forming the GO THERE sign with her outstretched arm and index finger. Once we got there, it was GO THERE to a different place. Then GO THERE again. And again.

After zigzagging around the yard like this for a while, I heard a snorting sound above my head. It was a distinctive sound Washoe made by contracting her nostrils whenever she signed FUNNY. I craned my neck up and sure enough she was placing her index finger on her nose in the FUNNY sign and snorting. For a second I couldn't figure out what was so funny. Then I felt something wet and warm flowing down my back and into my pants. I never forgot the sign for FUNNY after that.

I quickly discovered that riotous pranks, at my expense, were practically a daily event with Washoe. Whenever I caught on to her shenanigans, Washoe would then raise the stakes, apparently to see how far she could push me.

One morning after breakfast, about a month after we met, I was washing the dishes in the trailer while Washoe sat on the countertop next to me stirring the dishwater with her fingers. She began tasting the soapy water—a definite no-no—and she glanced up at me to get my reaction. I signed NOT DRINK THAT, and she stopped. Then she got a new idea. She dunked the dish towel in the water and looked at me carefully as she began sucking on the towel, as if to ask, "Does sucking constitute drinking or not?" Not knowing the sign for SUCK I signed NOT DRINK DIRTY and took the towel away. I needed some more dish soap so I unlocked the cabinet under the sink where we kept the cleaning supplies. After squirting some soap on my sponge, I put the bottle back in the cupboard out of Washoe's reach.

Meanwhile Washoe swiped the soapy dishrag, popped it in her mouth, and drew me into a game of keep-away in which I chased her around the trailer trying to retrieve the towel. She finally grew tired of this game, gave me back the rag, and went into her bedroom, where I found her playing with her dolls, kissing them and carefully arranging them around her in what we used to call "the magic circle."

This gave me time to clean the table, pull up a chair, and record that morning's signs and interactions in the logbook. I was deep in thought as Washoe swung out of her room, propelling herself off the overhead doorjamb like it was an overhanging branch deep in the jungle. She hit the linoleum floor with enough speed to slide to the cleaning supply cupboard, which I had forgotten to lock.

In a flash Washoe jerked open the door, grabbed a bottle, and rocketed back into her bedroom. I was on my feet and running. When I burst into her room she was squatting on her bed, inside the magic circle of dolls, chugalugging a bottle of Mr. Clean. I screamed in terror. Washoe was so startled that she stopped drinking. I grabbed her and rushed into the kitchen, where I sat her on the table while I tried to gather my wits. I kept signing STAY in such an exaggerated way that Washoe was frozen with fear.

My mind was racing: *How do you get poison out?* Get her to vomit. Washoe obviously knew something was very wrong because she cooperated like an angel. I cradled her head in my arm, opened her mouth, and shoved my finger down her throat. No luck. I tried again and again but this girl did not seem to have a gag reflex.

What to do? *Check the bottle for antidotes.* I grabbed the bottle but I could hardly read the label. All I could think was: *Washoe's going to die and it's all my fault. I killed the world's first signing chimp.* Finally I focused on the label and read it . . . and reread it. Nothing about antidotes! Now I was panicking again. I remembered something about drinking milk if you're poisoned so I grabbed her bottle of formula out of the fridge, rapidly signed DRINK DRINK, and forced the bottle into her mouth. She sucked a little, then yanked the bottle out of her mouth and shot me a look that said she was tired of all this silliness. She jumped off the table and went back to her bedroom.

By this point I could see that Washoe was not in her death throes. She was playing with her dolls as if nothing had happened. My panic began subsiding. Then it dawned on me that if the Mr. Clean label didn't say anything about antidotes it probably wasn't poison. I sat down and read the label again line by line. The next time I looked up I saw that I'd left the fridge open. Washoe was standing there removing all the containers of yogurt. She had just downed one and was opening another when she saw me looking at her. She piled the rest of the containers in both arms and swaggered on two feet back to her bedroom. *What the heck,* I thought. *Hunger is a good sign. Besides, if she is poisoned, at least she'll have a nice last meal.*

I went back to reading the label—no skull and crossbones, no mention of poison. I was feeling hopeful. Maybe Mr. Clean wouldn't kill Washoe . . . or my career. An hour later Washoe was still alive and playing with her dolls. I was off the hook, except for one thing: Mr. Clean cleaned out Washoe like nobody's business. She had a remarkable case of diarrhea, which I spent the rest of the day cleaning up—happily.

For several years I assumed that Washoe was typical of baby chimpanzees and that all juvenile chimps were rebels and rambunctious tricksters who chafed at any show of authority and tested every visible limit. Though I had developed a real fondness for Washoe, my heart went out to chimpanzee mothers everywhere. In my experience baby chimpanzees were a handful. That's why I was astonished and somewhat relieved when I finally met more young chimpanzees in 1970 and discovered that no two were alike. Though all of the ones I met were raised in attentive human families, one was rather shy and solitary while another was even-tempered and always good-natured, and still a third was desperate for approval. I even knew one who wouldn't raid an unlocked refrigerator!

I realized that a lot of what I'd taken to be characteristic chimpanzee behavior was simply Washoe's personality. Like all chimpanzees, like all animals for that matter, she was one of a kind. Of course she happened to be the kind that commanded your attention, like it or not. Part princess, part rabble-rouser, Washoe knew what she wanted and she knew how to get it.

AFTER MY FIRST FEW MONTHS WITH WASHOE I really looked forward to our four- or eight-hour shifts together. After we ate our breakfast, I would clean up the kitchen while Washoe washed her dolls in the dishpan or played with wooden blocks, stacking them into higher and higher towers until they fell over or she knocked them down. Sometimes she would just sit there sewing, which she learned by watching Susan Nichols, who used to fix Washoe's garments. Although Washoe did not actually mend anything, she would go about her random stitching with complete concentration for twenty or thirty minutes at a time. Chimpanzees have remarkable eye-hand coordination, and they can work at a problem for hours as long as it's fun and it's their own idea.

From Washoe I learned the greatest secret of working with chimpanzees and human children: make an activity a game and they'll do it forever. Ask them to do it, or force it on them, and they lose interest immediately. If I wanted to engage Washoe's attention for a while, I would "accidentally" leave a screwdriver out on the counter. She would find it and spend the rest of the morning trying to take the cupboards apart. She never made much progress, but she could sit quietly for hours studying the hardware and wielding the tool. Once she began to master the screwdriver I had to take it away from her. If I hadn't, I have no doubt she would have taken apart the whole trailer.

If there were no screwdrivers lying around, Washoe would be clamoring at the door to GO OUT, GO OUT. Once in the yard we'd play piggyback or hide-and-seek. Washoe loved hide-and-seek, and it was a great game for practicing signs because the players have to decide who is to hide, where to hide, and where to seek. But instead of using the proper ASL sign for HIDE, made by hiding the right hand under the left hand, Washoe always gestured by covering her eyes with her hands, meaning "peekaboo." Peekaboo is not a valid ASL sign, but she used it all the time, as in: YOU WANT PLAY PEEKABOO? or ME PEEKABOO HURRY.

Another favorite game was "Simon Says," in which one person has to imitate another person's actions. I'd sign "DO THIS," then put my hands on my head, and Washoe would do likewise; "DO THIS," and I'd cover my eyes with my hands and Washoe would do the same. Once Washoe's eyes were covered I couldn't resist a sneak tickle attack and I would usually tickle her until she was snorting uproariously and begging for MORE MORE. After all this game playing she usually wanted some time alone and would climb to the top of the weeping willow. She would sit there in complete privacy, munching on an occasional leaf, watching the world go by in the streets beyond her small backyard world.

At lunchtime we would go back inside the trailer, where I would prepare baby food or yogurt or Washoe's favorite, Vienna sausage. When you're feeding a chimpanzee in a high chair, it's easy to forget you're not serving a human baby. For example, we routinely mashed up Washoe's food, until one day Allen Gardner pointed out that Washoe was using her baby teeth to open soda bottles and chew bark off trees. Perhaps she didn't need her food mashed after all. Chimpanzees have extremely powerful jaws that are anchored, unlike ours, to a shelf that juts out from the brow ridge above the eyes. Working together with long canines, the chimpanzee's jaw is perfectly designed for stripping, peeling, and crushing—the chores of food preparation that humans, after millions of years of evolution, have learned to do by hand.

After lunch it was nap time, followed by more quiet play in the trailer. Washoe would pick out a favorite book, usually one with pictures of animals, and bring it over so that we could look at it together. As I signed the story to her she would turn the pages and comment on the pictures: DOG, CAT, and so on. Washoe also loved to sit at the table and draw with a pencil on paper. This would usually end when I'd turn my back. The budding young artist would be hanging from the fridge by three limbs while drawing furiously with her outstretched arm all over the "potty chart," an all-important document posted on the adjacent cabinet that immortalized her every encounter with the toilet. After an hour of reading or drawing Washoe would be at the door again: YOU ME OUT, YOU ME OUT. If it was pouring rain I could only hold Washoe's attention with a more exciting game like HIDE BABY. While Washoe covered her eyes, I would hide her doll, which took her less than a minute to find because she always cheated by peeking. We played HIDE BABY until I couldn't bear to hide one more doll.

Four o'clock was teatime, a quaint ritual Beatrix Gardner had practiced since her days at Oxford. In Washoe's case, teatime meant wolfing down some milk and cookies before demanding to go back OUTSIDE. Then it was back inside for dinner at six o'clock.

While Washoe was dining on Vienna sausage and taking high tea, Debbi, Josh, and I were discovering poverty, graduate school–style. We lived a glamorous life featuring greasy tacos, frozen Ding-Dongs, and Kool-Aid. Somehow we had to survive on a princely $140 a month—that's what I made for baby-sitting Washoe for eighty hours—plus some money Debbi had banked by teaching in the year before Josh was born. Sometimes, at the end of a month when the money ran out, we were reduced to raiding Josh's piggy bank, which was filled with John F. Kennedy commemorative half-dollars sent by his grandmother.

Family entertainment was out of the question, except on a shoestring. We would drive downtown and park in the Cal Neva Casino parking lot. Once inside the casino, we would cash a check for five dollars. For cashing the check, they gave us two free drink tickets and free parking. But instead of gambling we went next door to the movies. On the way out of the casino we checked every slot machine for stray change. One time I found a nickel, played it, and won twelve dollars. That was the richest I ever felt in Reno.

These family outings were rare, however. In the evenings I was usually in the trailer's kitchen, serving Washoe her dinner. After dinner it was bath time, which was another test of wits, with Washoe looking for every possible way to drink mouthfuls of soapy water without me catching her. I had to wash her face with lightning speed or else she sucked the washcloth into her mouth and a fierce tug-of-war ensued.

After the bath I oiled her body with Lubriderm to keep her skin from getting parched in the arid Nevada air. After the oiling Washoe jumped in my lap and lay there quietly while I brushed her hair from head to toe. For chimpanzees in the wild, grooming is the social glue that holds families and communities together. It provides reassurance, comfort, and bonding. A group of chimpanzees can sit inspecting one another's hair and skin for hours. Washoe took great pleasure in our nightly grooming sessions, and during the day she often repaid the favor by picking through my own hair during our quieter moments together.

Calmed by the grooming, Washoe was an angel at bedtime. No wonder this was my favorite time of day! I knew that, having wound down, she would not be deviling me with pranks or challenging my every request. For this single hour there was a truce, and we were as peaceful and cozy as two birds in a nest. After getting Washoe into her pajamas and tucking her into bed, I would sign a story from one of her children's books. As she grew older and her vocabulary grew, I would make up stories about her and all the friends in her life. She would follow these stories with great fascination, hanging on my every sign until, overcome by exhaustion, she would close her eyes and surrender to sleep.

Oftentimes I would stay and watch over her, wondering where her chimpanzee dreams might be taking her. Was she reliving the events of the day, signing in her sleep as she played HIDE-AND-SEEK with her human family? Or was she revisiting her distant jungle birthplace, clinging to her mother's hairy chest, once again swinging fearlessly through the rain forest canopy, oblivious to the dangers lurking below.

Washoe never gave up these nighttime secrets. And when I returned the next morning to wake her, there was a diaper to struggle with and no time for lofty speculation.

THREE

OUT OF AFRICA

IN OUR FIRST YEARS TOGETHER, Washoe's origins were a rather romantic mystery to me. I knew she had been "wild collected" in Africa, and I knew that the Gardners had acquired her when she was ten months old from the Holloman Aeromedical Laboratory, in New Mexico, where she was part of the United States space program. Naively, I assumed that Washoe must have been abandoned by her mother, then rescued by some decent person who sent her to America for the best possible care—like Curious George and the man with the yellow hat.

One of the books in H. A. Rey's Curious George series was eerily prescient about the role of chimpanzees in the American space program. Published in 1957, Curious George Gets a Medal had the young chimp volunteering to pilot the very first rocket into space because it was too small to hold a human. After flawlessly completing his mission by pulling the correct lever and parachuting to earth, Curious George is given a hero's welcome, complete with photo opportunities and a big gold medal that reads: "To George, the first space monkey." The last line of the book says, "It was the happiest day in George's life."

Four years later, in early 1961, history unfolded along the lines of the storybook. I was a freshman in college at the time. I listened to President John F. Kennedy boldly declare that the United States would beat the Soviets to the moon by the end of the decade. NASA had already developed a one-man space capsule—the Mercury—for carrying a pilot into space and returning him to earth. But nobody knew what would actually happen to a pilot as he hurtled through outer space in a small, bell-shaped can, bombarded by lethal levels of radiation, searing heat, and unimaginable G forces. Why put an American astronaut in danger when you could assess the risks with an animal?

Enter the chimponauts—our lovable "partners in space." The public knew little about the nature of their mission. We imagined the chimponauts as glorified canaries in the coal mine of outer space. If they survived, then human astronauts would be able to follow in their footsteps; if they perished, then NASA would go back to the drawing board.

It was a bit more complex, however. The chimps had to learn a series of astronaut-like actions that would demonstrate whether taxing mental activity could be conducted under the unprecedented strain of launch, weightlessness, and reentry. Meanwhile, the Mercury capsule would be guided remotely from the ground. The chimpanzees and men of the Mercury program would be more passengers than pilots.

The Air Force trained its sixty-five chimponauts on a simulated flight panel by means of operant conditioning—a system of rewards and punishments. When a chimponaut moved the correct lever in response to a blinking light he was given a tasty banana pellet. When he responded incorrectly he was punished with a mild shock on the foot. This banana protocol worked beyond all expectations. In one training exercise, a chimponaut outperformed a visiting congressman by completing seven thousand moves with only twenty misses.

The chimponauts were put to their first real test on January 31, 1961. As the nation held its breath, a three-year-old chimpanzee named Ham—short for Holloman Aeromedical—was strapped into a Mercury capsule atop an enormous Redstone rocket. At 11:55 A.M. the Redstone thundered off the launchpad, propelling Ham out of the atmosphere and into space at five thousand miles per hour.

The mission itself had a series of mishaps that caused the

rocket to gain eighteen hundred miles per hour of excess speed and to plunge back through the atmosphere at excruciating G forces, but Ham went through his astronaut's paces perfectly. He emerged from his charred Mercury capsule to the enthusiastic applause of space officials and a national television audience. Ham earned his "fame aboard the flame," and the young chimpanzee graced the cover of Life magazine ten days later. Space was now deemed safe for humans, and, on May 5, 1961, Alan Shepard rode the first manned American rocket beyond the earth's atmosphere.

Shepard's flight was impressive, but just three weeks earlier the Soviets had launched Yury Gagarin into orbit around the earth aboard Vostok 1. Getting a capsule into space was one thing, but launching it into orbit was another. NASA developed a more powerful rocket, the Atlas, for the job, but its overall launch record was poor; with only two successes and two failures, NASA officials balked at sending up a human aboard the Atlas. So once again, they turned to a chimpanzee—a five-and-a-half-year-old from West Africa named Enos, who was chosen because he had led his class at Holloman through sixteen months of grueling psychological and physical training. On November 29, 1961, at 10:17 A.M., the Atlas rocket blasted off, hurtling Enos into a picture-perfect first revolution around the earth. But trouble arose on the second go-round. A gas jet on the Mercury capsule stuck open, wasting fuel and sending Enos into an unplanned wobble.

Then there was more trouble. The Mercury's banana pellet system went haywire and began giving Enos shocks for his correct responses. Suddenly the five-year-old chimpanzee was faced with a reward-punishment system that contradicted more than a year of intensive training. Scientists assumed that Enos would begin responding incorrectly in order to find the banana pellet rewards, but instead he overrode NASA's malfunctioning system and performed the flight tasks he knew were right, even though he received an electric shock for every correct lever move. This "unthinking" ape outsmarted his human controllers.

With Enos's craft wobbling dangerously, NASA hurriedly canceled its third orbit. Enos splashed down right on target near the Bahamas. He was pulled from a capsule that was roasting at 106 degrees. Another orbit would have killed him. In postflight tests the scientists could barely match Enos's in-flight performance, and none of them were receiving shocks. *Enos* is the Hebrew word for "man," and this young chimpanzee seemed more like a reasoning human being than science cared to admit. Thanks to his and Ham's space exploits, NASA made 250 safety and comfort modifications to improve Friendship 7, the spacecraft that would carry John Glenn around the earth three times in February 1962.

Humans were now assured of survival in space, and the chimponauts faded from public view as quickly as they had appeared. The first humans in space went on to a life of fame and adulation, but the first chimpanzees in space didn't fare nearly as well. Enos died of shigella dysentery just one year after his flight. Ham was shipped off to the National Zoo in Washington D.C., where he lived alone in a cage for seventeen and a half years until 1980, when he was sent to live in a small chimpanzee community at the North Carolina Zoological Park. He died of a heart attack in 1983 at the age of twenty-six, having lived barely half the lifetime he might have had in the wild. Most of the other chimponauts were transferred into medical research, where they were subjected to painful, sometimes deadly, experiments.

The sorry fate of the chimponauts was not widely known. My own memories of their role in the space program had the same fairy-tale quality as *Curious George Gets a Medal.* I assumed that these chimpanzee heroes, including Washoe, had been brought to America humanely—indeed, that they volunteered for their mission—and that they were amply rewarded for their selfless service to our nation.

It was only years later that I learned the truth about how the Air Force had gone about "recruiting" infant chimpanzees from Africa in the 1950s and 1960s. The military procured the chimps from African hunters who stalked mother chimpanzees carrying a baby. Usually the mother was shot out of her hiding place high up in a tree. If she fell on her stomach, then her infant, clinging to her chest, would die along with her. But many mother chimps shielded their infants by falling on their backs. The screaming infant would then be bound hand and foot to a carrying pole and transported to the coast, a harrowing journey usually lasting several days. If the infants survived this second ordeal, and many did not, then they were sold for four or five dollars to a European animal dealer who kept them in a small box for days until the American buyer arrived—in this case the Air Force. Those still alive when the buyer came were crated up and sent to the United States, a journey that mirrored the slave trade of earlier centuries. Very few babies emerged from the crates. It is estimated that ten chimpanzees died for every one that made it to this country.

By the mid-1960s the United States was no longer launching chimpanzees into space but was conducting medical experiments on them instead. One chimpanzee infant that survived the journey to America in the spring of 1966 was a ten-pound infant named Kathy. She was shipped to the Holloman Aeromedical Laboratory, but before she could become a subject of disease research, fate intervened. As the largest and healthiest infant at Holloman, Kathy greatly appealed to two scientists who were visiting the Air Force laboratory on a recruiting mission of their own. Drs. Allen and Beatrix Gardner chose ten-month-old Kathy to learn American Sign Language. But the Gardners thought that a chimpanzee, even one raised by humans, should not have such a human name. So when they brought Kathy home they renamed her after the county in Nevada where she would grow up: Washoe.

WHY DID NASA PICK CHIMPANZEES to begin with? The Soviets had chosen a dog, man's best friend, for their first flight into space. A dog was fine, American scientists realized, if all you wanted to do was monitor basic bodily responses to space travel. But a dog, no matter how well trained, wasn't likely to perform like an astronaut.

NASA scientists decided on chimpanzees for the same reason the Gardners did: they were looking for the animal most similar to us biologically, cognitively, and behaviorally. Of the three great apes, the chimpanzee was the best known to laboratory science at the time. Thanks to their gregarious social habits, chimps were easy to work with, and these physiological near twins of humans had remarkable problem-solving abilities.

Washoe, too, was constantly solving problems in the manner of a human child: she used tools while trying to take apart the cabinets, she figured out how to cheat and win in games like HIDE BABY, she signed OPEN when she wanted me to turn on—or open—the kitchen faucet, even though she'd never seen anyone use the OPEN sign to turn on the water before.

But one incident, just a few months after I met Washoe, captured for me the remarkable similarity between chimpanzee and human intelligence. We had purchased a new doormat for Washoe's trailer, and I waited for her to notice it. I was sure she would take a great interest in the mat and inspect it carefully, as she did with anything new in her home. But she took one look at the mat and jumped back in terror, retreating to the corner, where she crouched, cowering in fear and issuing alarm calls.

Then Washoe stood up, as if struck by an idea. She grabbed one of her dolls, approached to within four feet of the dreaded doormat, and threw her doll right on top of it. She watched the doll intently for several minutes but nothing happened to it— it just lay there. A few minutes later she crept right up to the doormat, quickly reached out, and snatched her doll to safety. After thoroughly inspecting her doll and determining that it was not harmed, she seemed to calm down. Eventually she became braver and would go over to the doormat with hesitation but no evident fear. And a few days later Washoe was entering and exiting her trailer without a second look. The doormat had become just another doormat.

The way Washoe recruited her doll to test out this scary new doormat was very reminiscent of the way in which humans had recruited chimpanzees to test the dangers of space travel. This was my own introduction to the evolutionary fact that human intelligence did not spring full-blown from nowhere. Rather, our own talent for problem solving is one variation on ape intelligence.

Washoe's intelligence was very much like a human child's. And nowhere was this more in evidence than in her *social* problem solving. This is hardly surprising. Many scientists believe that apes and humans developed such impressive powers of reasoning in order to handle the complex dynamics of a close-knit family and a highly social community. Wild chimpanzees are masters of the political game: they seem to be constantly calculating the costs and benefits of forming an alliance with one member of the community while snubbing a second and deceiving a third. In short, they are a lot like us.

Washoe was always displaying this kind of social intelligence in her own foster family. Toward the end of my first year on the project, when Washoe turned three, she began devising very elaborate schemes for manipulating me and getting what she wanted.

One morning, Washoe and I left the trailer, and she ran directly up her tree. I sat on the trailer steps and made notes in the logbook. When I looked up she had climbed down from the tree and was in the garden, on the other side of the tree, looking very intently at something under the rocks. My curiosity got the better of me and I went over to see what she was so interested in. I found nothing. But she continued to look and look until I settled down next to the rock garden. As soon as I sat down Washoe lost interest in the "nothing" and went back up her tree.

I was now on the opposite side of the tree from the trailer. As soon as I was engrossed once again in my writing, Washoe appeared to fall out of the tree, just barely touching a limb here and there on the way down, hitting the ground at a dead run headed for the trailer. By the time I got to my feet, she had already executed her plan and was running out of the trailer with a bottle of soda pop she'd taken from a cupboard I had forgotten to lock. With the bottle tucked under her arm she couldn't run on all fours, so she was staggering toward the tree on two feet like a drunken sailor. She beat me to it and sped up to her safe haven.

This whole episode amazed me. Washoe must have noticed that the cupboard was unlocked during breakfast, suppressed her natural impulse to raid it when my back was turned, and instead devised this plan for distracting me long enough to gain access to the trailer by herself *and* give herself the opportunity to drink the soda. This was a level of planning and deception beyond anything I thought her capable of. The evolution of human intelligence is often linked to the very traits that Washoe displayed: the ability to inhibit one's responses—as opposed to Pavlov's salivating dogs, for example—and the flexibility to modify one's plans because of changing circumstances.

Washoe didn't always resort to such complex deceptions. Like any three- or four-year-old child, she had a special genius for blatantly manipulating her parents and all other authority figures, including me. As soon as she discovered a new button to push with us, she lost no time in putting it to maximum advantage.

For example, Washoe knew exactly which of her "unacceptable" behaviors would get us to jump up and down. For a while it was eating green grapes. They weren't bad for her, they just always gave her diarrhea, and there's nothing less fun than cleaning up after a rampaging chimp with diarrhea. So the green grapes became a bargaining chip. Whenever she wanted food, games, whatever, she picked some grapes off the vine, raced up the tree, and started popping the forbidden fruit in her mouth until we caved in to her demands. Sometimes she did this just to watch us put on a show of human histrionics. Fortunately the green grapes had a short growing season.

Gravel was a different story. The pebbles in her driveway *never* went out of season and all she had to do was threaten to eat them to get us in an uproar over what they might do to her throat or stomach. Finally, Washoe's doctor told us not to worry—if she ate gravel it would pass right through her system without doing harm. The Gardners instructed us to ignore Washoe's threats to eat gravel, and her behavior changed dramatically. Washoe would start toward the gravel, then stop to watch us and wait for our usual hysterical response. When we didn't respond at all she would grab a handful and put it in her mouth and sit there, looking at us. We pretended not to notice. Like magic, the gravel eating stopped within a month.

It did not disappear completely, however. From time to time, Washoe would get in one of those naughty-girl moods when she ran through all the things that she wasn't supposed to do. She would chew a hole in her sweatshirt, then try to break into the cupboards, and eventually run up her tree and pee from the highest branch. But I would refuse to react. Finally, as if in desperation, she would grab a handful of gravel, shovel it into her mouth, and watch me. When even this dastardly act failed to get my attention, she would spit out the gravel and give up.

AROUND THIS TIME, I had a cat at home, and though he was a wonderful and intelligent animal, he was *not* devising elaborate deceptions, blackmailing me with naughty behavior, or using sign language to make potty jokes. Washoe was doing all of these things. Why did this baby chimpanzee seem to think, behave, and talk like a human child?

The answer lies in the shared evolutionary history of our two species: we are both descended from the same apelike ancestor. Interestingly, the peoples of West Africa figured this out thousands of years before modern molecular biology and long before Europeans even encountered the chimpanzee. The West African forest peoples knew that their neighbor, the chimpanzee, was either the ancestor of man or his brother. The English word "chimpanzee" comes from a Congolese dialect and means "mock man."

The West Africans portray the chimpanzee as everything from almost human to fully human. The Oubi people of the present-day Ivory Coast refer to chimpanzees as "ugly human beings." According to their mythology, God created humans and then he commanded them to work. But the chimpanzees, who were smart enough to refuse this order, were punished with ugliness and then cast out into the jungle, where they lived by their wits, happily avoiding labor. To this day, the Oubi prohibit the killing of chimpanzees because they are considered religiously superior to humankind.

The Mende people of the Upper Guinean forests refer to the chimpanzee as *numu gbahamisia*—"different persons"—and they believe that humans and chimpanzees sprang from a single class of forest dwellers called *huan nasia ta lo a ngoo fele*—"the animals that go on two legs." One group of the Gouro people believe that they are the descendants of chimpanzees. The Baoulé people refer to the chimpanzee as the "beloved brother" of man. The Bakwé not only think of chimpanzees as close relatives but, in the past, actually buried chimpanzees as men. The Bété call the chimp the "wild man" or "man returned to the forest."

The peoples of West Africa, who have lived side by side with chimpanzees for aeons, have never thought of them as being deficient in their ability to reason. On the contrary, they have long known that chimpanzees make and use stone tools, medicate themselves with indigenous plants, organize social activities like hunting, and even have a rudimentary form of political culture.

The Europeans saw chimpanzees in a different way altogether. The ancient Greeks only knew of monkeys, not apes. Blissfully ignorant of the chimpanzee, the fathers of Western philosophy never had to account for nonhumans who made and used tools. Plato and Aristotle were able to draw a bright bold line between reasoning humans and the rest of the animal kingdom.

Plato believed that man was distinguished from other animals because he possessed two souls: the immortal soul was located in the head and imparted the power of reason and served as a connection to the eternal divine; the mortal soul resided in the chest and belly. Nonhuman animals possessed only this brute mortal soul, Plato said.

Aristotle also defined man as "an animal capable of intellect." He elevated humans above other animals, and then designed a Great Chain of Being, at the top of which stood the free man, imbued with intellect and subordinate only to the angels. Beneath man he put the female, the slave, and the child, arguing that they were defective in reason and destined to be ruled. Then below them he arrayed the nonhuman animals, who existed only to serve humans. These bestial animals could feel pleasure and pain, and even had the capacity for memory, but they clearly lacked reason and emotion.

This Greek notion of a world created for the sake of man found fertile ground in the Judeo-Christian tradition, where humans were given dominion over the earth and all living things. As a result, the early Church Fathers embraced the Great Chain of Being and crowned it with biblical man, a unique animal created in the image of God.

Then, in the seventeenth century, the French philosopher René Descartes went one step further and disconnected man from the natural world entirely. Man may have a body, but it is the mind that confers existence: "I think, therefore I am." The Greeks had at least given nonhuman animals credit for being able to sense, feel, and remember. But in the Cartesian world, these "unthinking, nonspiritual beasts" became unfeeling cogs in the vast mechanism of nature. Kick a dog, or vivisect a dog, and it yelped not out of pain but like the spring in a clock being struck.

It was the early 1630s when Descartes philosophically severed any link that remained between mind and body, between the human on top and the nonhuman below. Man was finally supernatural and free of nature altogether. But at that very moment, the "wild man" of Africa—the chimpanzee—came on the scene.

The first reports of great apes began arriving in Europe in

1607. They were carried by a returning English sailor named Andrew Battell, who had been taken prisoner by the Portuguese and held in Angola for several years. Battell told of two half-human "monsters" called Pongo and Engeco—what we now call the gorilla and the chimpanzee.

The shocking reports were not confirmed until 1630, when a live chimpanzee reached Europe as a gift to the Prince of Orange in the Netherlands. Thirty years later, in 1661, Samuel Pepys met another chimpanzee and made his statement that this creature "might be taught to speak or make signs." The apparently intelligent chimpanzee now threatened to bring down the entire artifice of Aristotle's Great Chain of Being.

In 1699, England's best-known anatomist, Edward Tyson, performed the first dissection of a chimpanzee and revealed an anatomy that resembled "Man in many of its Parts, more than any of the Ape-kind, or any other Animal in the World." Tyson was especially troubled by the creature's brain and laryngeal region. They looked almost human, indicating that this animal might be capable of thought and speech. But Tyson was a good Cartesian and he assumed that a thinking, talking animal was simply not possible. So he decided that though this ape-man had all the machinery for thought and speech, it did not have the God-given ability to use them. It was Tyson who invented the paradigm of the mindless ape: the chimpanzee with a human brain but not a single thought in it, the chimpanzee with a human nervous system but not the slightest emotion, the chimpanzee with the apparatus for language but not a thing to communicate. Tyson dreamed up the view of the chimpanzee that biomedical researchers still cling to today: a beast with the physiology of a human but the psychology of a lifeless machine—a hairy test tube created for the sake of human exploitation.

Ironically, Edward Tyson is often called the "father of primatology" because his anatomy of a great ape was not only the world's first but also the most influential for the next two hundred years. Tyson *did* reveal the close physiological connection between humans and chimpanzees, but he also updated the doctrine of Plato, Aristotle, and Descartes: humans might resemble other animals physically, but never mentally—even the ape.

Just three decades later, in 1735, Carolus Linnaeus, the Swedish naturalist, compiled his monumental work of zoological classification, *Systema naturae*. Linnaeus sorted species according to physical similarity, and he placed humans alongside chimps and other apes in a mammalian order he named Anthropomorpha— "resembling man." But when it came time to give names to the species, Linnaeus made it clear that the human mind was utterly unique. He christened humankind *Homo sapiens*—"wise man."

THOMAS HUXLEY, a naturalist who was known as "Darwin's bulldog," was the first to argue, in 1863, that our anatomical resemblance to apes was no coincidence but a family affair. Charles Darwin had already laid out his influential theory of evolution in *The Origin of Species* four years earlier, but he had intentionally skirted the explosive issue of our own evolutionary past. It was Huxley who presented the compelling anatomical evidence that humans are related to the apes through a common ancestor. Charles Darwin agreed and in his own later work on the topic, *The Descent of Man,* concluded that "Man is descended from a hairy, tailed quadruped, probably arboreal in its habits." (The term "ape" refers to modern-day chimps, gorillas, and orangutans. We are descended not from apes but from a species that I shall refer to throughout this book as "apelike" or "our common ape ancestor.")

Darwin had marshaled a mountain of evidence that disputed the commonly held belief that all forms of life on earth were created at once and fixed for all time. Instead, he argued that all life sprang from a common origin and was still evolving. Darwin was well aware that he was challenging our heavenly origins when he began tracing our descent from an apelike creature. His famous statement "Our grandfather is the Devil under form of baboon" captured the dark irony of what he was proposing.

Darwin's theory scandalized Europe and especially outraged

those who took the creation story in Genesis literally. But his theory also quickly won over many creationists who happened to be scientists. Evolution explained why taxonomists since Linnaeus had been able to use patterns of anatomy as a way of grouping species of animals naturally by their physical resemblance. For example, the reason dogs look more like foxes than like cats is because dogs and foxes diverged more recently from a common ancestor than did dogs and cats.

But Darwin's theory threatened more than the biblical view of our origin. Evolution threatened, and continues to threaten, the fundamental Platonic premise of all Western philosophy: that humans alone are capable of rational thought. Darwin argued that we resemble our ape kin not only anatomically but mentally as well. Evolution was not aware that humans are supposed to be unique. If the genetic program for the ape and human brains proceeded by small increments, one adaptive mutation at a time, then whatever occurred in those brains also differed only by degree. Evolution would never produce a monumental contradiction like Tyson's chimpanzee—a humanlike brain with nothing in it.

Darwin theorized in *The Descent of Man* that nonhuman animals, especially the apes, had the ability to reason, to use tools, to imitate, and to remember—all rational facilities long thought to be unique to humans. And in *The Expression of Emotions in Animals and Man*, he traced the continuity of emotions—fear, jealousy, grief, joy, and loyalty—across many species. To Darwin, our complex behavior evolved from our apelike ancestors just as surely as our complex anatomy.

Most of Darwin's contemporaries felt that he was taking his theory too far by saying that we would find the roots of our own cognition in apes. Even evolutionists preferred to see the chimpanzee as a kind of distant cousin. After all, chimpanzees looked more like gorillas and orangutans, two other large apes that dwell both in the trees and on the ground. So zoologists grouped the three apes together in one taxonomic family and called them "the great apes" (the scientific name is Pongidae). A separate family called Hominidae—thus, hominid—was reserved for humans and our now extinct ancestors like *Homo erectus* and *Homo habilis*.

Again, this grouping by resemblance upheld the old Greek notion of human uniqueness. When I was in college in the early 1960s I was taught in anthropology class that humans had parted company with apes at least twenty million years ago. That not only gave us enough time to become hairless, stubby armed, and bipedal, but it also gave our ancestors plenty of time to develop "the special human attributes" of thought, language, and culture. My college textbooks depicted the ape family tree like this:

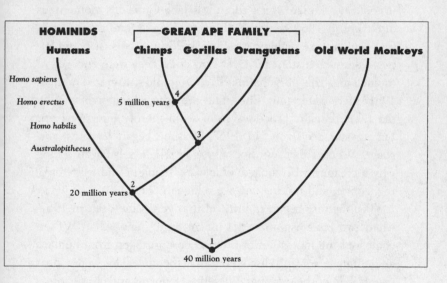

Point 1 is the common ancestor of monkeys, apes, and humans, a species of monkeylike tree dwellers that split in two about forty million years ago. One group gave rise to today's Old World monkeys (macaques, baboons, mangabeys, et cetera) and the other group evolved into tree-swinging and ground-dwelling apes, as well as humans.

At Point 2, about twenty to thirty million years ago, the ape lineage splits. One group becomes upright, giving rise to a long line of hominid species, before eventually evolving into *Homo sapiens,* modern humans. The other branch remains quadrupedal, giving rise to the great apes that share a lineage for millions of years before Point 3, when the orangutan, the distinctive Asian ape, branches off, and later, Point 4, when the African apes, chimpanzees and gorillas, diverge from one another, maybe five million years ago. The human is, in other words, only a very distant cousin of the great ape family, many times removed.

This widely accepted ape family tree did a fine job of explaining why humans have such unusual appearances and such splendid cultures. The problem was all the chimpanzee behavior it could *not* explain. First there was Jane Goodall's momentous discovery in 1960 that the chimpanzees at Gombe Stream in East Africa were regularly making and using tools. If chimpanzees were such distant relatives how come they were engaged in toolmaking, the very hallmark of hominid culture? Then, in 1961, there were Ham, Enos, and the other chimponauts. Why did their thought processes seem so uncannily like our own? Finally and decisively, in 1966, Washoe began learning language. When I met Washoe a year later I myself began asking why a creature who looked so *unlike* a human child was thinking, acting, and talking *just like* a human child.

The mystery began to unravel that very same year, in 1967, when two biologists named Vincent Sarich and Allan Wilson compared the molecules of a blood protein taken from humans and chimpanzees and discovered that the molecules were a near match. For all the external differences, humans and chimpanzees are extremely similar genetically. Sarich and Wilson concluded that humans and chimpanzees were not distant cousins at all but were in fact "sibling species"—like sheep and goats or horses and zebras. Vincent Sarich has written that, immunologically, the human and the chimpanzee are as similar as "two subspecies of gophers living on opposite sides of the Colorado River."

The implications of this discovery for the origins of humankind landed like a bombshell. This genetic similarity could only be explained, Sarich and Wilson said, if the two species diverged from a common ancestor only five million years ago, not twenty million years ago. If they were correct, then the human family had developed much more recently than anyone thought.

Leading anthropologists and paleontologists scoffed at this suggestion. But in the early 1980s two scientists named Charles Sibley and Jon Ahlquist confirmed the genetic similarity between humans and chimps by studying DNA itself—the master molecule of life. Sibley and Ahlquist found only a 1.6 percent difference between human DNA and chimpanzee DNA. Viewed another way, 98.4 percent of human DNA is exactly the same as chimpanzee DNA. The genetic program that built Washoe is virtually identical to the one that built me and you.

How significant is a 98.4 percent identity in DNA? It means that humans and chimps are closer genetically than two hard-to-distinguish bird species like the red-eyed vireo and the white-eyed vireo (they are only 97.1 percent identical). But even more telling is this fact: humans are nearly as close genetically to chimpanzees as a chimpanzee is to a bonobo, a second species of chimp (sometimes called the pygmy chimpanzee). This fact has led physiologist Jared Diamond to propose that we humans are, for all intents and purposes, a third species of chimpanzee— a human chimpanzee, as it were.

But just because chimpanzees are *our* evolutionary next of kin, it does not necessarily follow that we are *their* closest relation. They could, for example, be very close to us but even closer to gorillas. But here Sibley and Ahlquist overturned three hundred years of taxonomy. Chimpanzees are more closely related to humans than to gorillas or orangutans. Gorillas differ from both humans and chimps in 2.3 percent of their DNA. Orangutans differ from both humans and chimps in about 3.6 percent of their DNA. Despite all outward appearances, the chimpanzee's next of kin is not the gorilla or the orangutan but the human.

That makes it necessary to take another look at the ape family tree. Because DNA mutates at a fairly steady rate, acting as a kind of molecular clock, Sibley and Ahlquist were able to determine when any two species last shared a common ancestor. Their DNA evidence, which is now widely accepted, favors the following branching pattern:

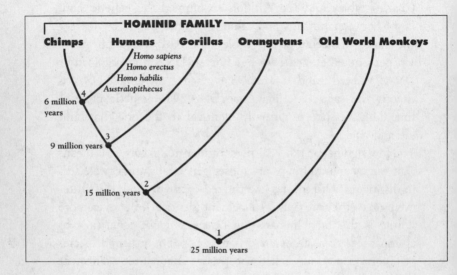

Again, Point 1 shows where monkeys and apes diverged— only now that date is set more recently at twenty-five to thirty million years. At Point 2, around fifteen million years ago, the orangutan branches off, and at Point 3, around nine million years ago, the gorillas branch off. Humans and chimpanzees then share three million years of lineage until, finally, we humans branch off at Point 4, around six million years ago, by developing specialized traits such as bipedalism. Only after this divergence, within a brief evolutionary window of six million years or less, did the upright hominids—including *Australopithecus*, *Homo habilis*, and *Homo erectus*—appear.

This is a very different family portrait because it clearly shows why we cannot lump the three apes into one category without also including humans. The evolutionary grandparents of chimpanzees and gorillas (Point 3) are also our own grandparents. And Washoe and I share an even more recent ancestor (Point 4) that the gorilla does not share. If we classify Washoe as a great ape, then we must classify ourselves as great apes because she is no more closely related to a gorilla than you or I are. Any category of ape is meaningless *unless* it includes humans. Humans are simply odd-looking apes.

Molecular biologists now agree that humans, chimpanzees, gorillas, and orangutans are so closely related—96.4 percent identity or higher—that they belong in the same family. As a result, if you look in the most recent edition of the Smithsonian's definitive classification, *Mammal Species of the World,* you will find that the members of the great ape family have been moved into the family previously reserved for humans alone: Hominidae. This new grouping reflects the fact that apes and humans are a lot more closely related than African elephants and Indian elephants, for example. It also means that, as a member of Hominidae, Washoe is a hominid, a term that used to refer only to humans and our fully upright ancestors.

Vincent Sarich, the pioneer of this molecular anthropology, says that if we could go back in time five million years and observe them, we would consider our ancestors to be small chimpanzees. Washoe, the modern chimpanzee, would more resemble our common ape ancestor than you or I would. Her species has stayed closer to its ecological niche and has probably made smaller adaptive changes, which is the reason Washoe *looks* more like that other African ape, the gorilla, than she does a human. Our own human ancestors wandered far from their origins; the 1.6 percent change in their genetic program has produced the bipedal, big-brained, highly vocal modern human. However, even these innovations are firmly rooted in the chimpanzee anatomy. Our skeleton is an upright version of the chimpanzee skeleton; our brain is an enlarged version of the chimpanzee brain; our vocal tract is an innovation on the chimpanzee vocal tract.

But the continuity between humans and chimps does not end with anatomy. As Darwin guessed, our common ape ancestor endowed chimpanzees and humans with similar cognitive abilities. It's not a coincidence that the mother-child bond lasts for more than ten years in both human families and chimpanzee families. It tells you that our common ancestor had a very long childhood as well—long enough to develop the skills of toolmaking and social problem solving that are so essential to life in an ape community.

Chimpanzees have also had the rudiments of culture since we did—namely, since we shared an ancestral parent six million years ago. The peoples of West Africa have always accepted this. It only took several thousand years for a Westerner—namely, Jane Goodall—to study the chimpanzee mind rather than assume it didn't exist. Much to the scientific world's shock, the chimpanzee that Goodall encountered in the jungle in 1960 was not Descartes's automaton or Tyson's mindless humanoid. Instead, Europeans were finally reunited with their evolutionary kin: a highly intelligent, cooperative, and violent primate who nurtured family bonds, adopted orphans, mourned the death of mothers, practiced self-medication, struggled for power, and waged war.

When I was in college, anyone who believed that animals were capable of thought or complicated motives was committing the heinous sin of anthropomorphism—ascribing human characteristics to nonhuman animals. But my college professors neglected to tell me that apes behave anthropomorphically because they are anthropoids—members of the primate suborder Anthropoida that includes humans and apes. The DNA evidence only confirmed what Darwin claimed a century earlier: that humans and chimpanzees will behave alike, emote alike, and think alike.

Washoe was genetically programmed for the chimpanzee childhood, the closest experience in nature to a human childhood. In evolutionary terms, her journey from African jungle to the Nevada suburbs was like leaving home to live with her hominid relatives across town. Yes, somewhat different behavior is expected, but then the chimpanzee infant, like her human counterpart, is a remarkably adaptable creature. She is biologically equipped for a lifetime of learning, but is flexible enough to absorb the distinctive ways of a specific chimpanzee community, much as the human genetic program has to work in either urban New York or the tribal Kalahari. In her human foster family Washoe found a distorted but recognizable mirror of her own chimpanzee family.

DURING MY FIRST YEARS ON PROJECT WASHOE I lived a double life. At home I was husband, father, breadwinner—the head of a household. At work I sat at the "kids' table" with the baby chimpanzee and, from time to time, with her other human playmates. To my son, Josh, I was "da da," with all of the authority that implies. In Washoe's eyes, I merited no such parental respect. I was like an older brother and, by extension, one child in a foster family of graduate students.

Allen and Trixie Gardner were the heads of this household. It's hard to imagine two more different personalities. They were both Jewish, but culturally they were worlds apart. Allen grew up in the Bronx, and beneath his brainy, pipe-smoking demeanor was the brash and argumentative style that the Bronx is so well known for. Trixie, on the other hand, had a very quiet, refined, and European manner. She was born in Austria to a comfortable life in an industrialist family. In July 1939, at the age of six, Trixie Tugendhat and her sister were spirited away by their parents just weeks before the Nazis invaded their home in Poland. They settled in Brazil and, later, the United States. English was Trixie's third language, after German and Portuguese, and she spoke it fluently with a slight British accent.

The Gardners were the two most brilliant people I had ever

met, and their diametrically opposite styles reflected their fields of study. Trixie had trained as an ethologist and believed in humble observation; she showed the same loving patience with all of us that she did with her research subjects, the jumping spiders. Allen had been schooled in the authoritarian control of experimental psychology, and he commanded his students the way he commanded his rats.

Trixie was a tender and nurturing soul who treated everyone—her colleagues, her students, and Washoe—with unconditional encouragement. She always found something positive to say about a person. In that respect, she reminded me of my own mother, and it's no wonder that Trixie was always Washoe's favorite person. This worldly woman who barely escaped the Holocaust had a childlike innocence that never faded with time. Trixie loved to watch *The Wizard of Oz* over and over again. And I was always pleased whenever she came out to the trailer at night to read to Washoe and help put her to bed. Trixie was the compassionate heart of Project Washoe. She was also the angel on her husband's shoulder.

Allen Gardner needed an angel. Although he had a sweet and humorous side, he did not suffer fools at all. He could explode at a student for thinking illogically, misdesigning an experiment, or just misusing a word while speaking in class. According to Allen, graduate school should "temper the metal or melt the plastic." And no one could melt a graduate student faster than Allen Gardner. The first time he screamed at me in the middle of a graduate seminar, I was absolutely devastated. In my entire life, neither my mother nor my father had ever yelled at me. I suppose I was coddled by my parents, but I liked it that way. I didn't want my "metal tempered."

It is the lot of most graduate students to be treated like infants, and this was even more true thirty years ago. Students were subservient to one or more professors whose job it was to socialize them into the culture of an academic discipline. Graduate students are a professor's academic progeny. He invests years in their educational upbringing so that they will carry on his research and, at some point, reflect glory on his name.

This parent-child dynamic became quite literal on Project Washoe. Allen Gardner was not just my major professor; I was spending twenty hours a week in his backyard. Every morning at 7 A.M. I had to tiptoe through his house to get to Washoe's trailer. It was uncomfortably intimate. Those graduate students who worked on Allen Gardner's rat research—he called them his "rat students"—rarely if ever saw his home. We "chimp students" were practically living in it.

I say "practically" because the Gardners drew very clear limits around their home and their parenting responsibilities when it came to Washoe. The Gardners did their share of shifts with Washoe, but overall they seemed to follow the British nanny model: let the hired help raise the child in separate quarters.

I didn't dare arrive late for my morning shift (the "dawn patrol") because Washoe might start rattling around in her trailer and wake the Gardners via the intercom. And I had to keep Washoe out of the Gardners' home at all costs. While the student-nannies raised Washoe, the Gardners were able to maintain a civilized lifestyle that usually goes by the board when you have a child.

I don't know whether Allen and Trixie Gardner chose to cross-foster a chimpanzee instead of raising their own human child. Their love for Washoe was obvious: Trixie adored Washoe, and Allen always carried special treats in his pockets for his little girl. The Gardners quickly took on the role of live-in grandparents and felt free to indulge Washoe to their heart's content.

In the wild, a chimpanzee family centers on the mother. Apart from mating, the father has no family responsibilities; he spends most of his time with other adult males, socializing, hunting, and protecting a community that may include as many as 150 chimpanzees. After a pregnancy of nearly nine months, the mother gives birth to a very dependent infant who nurses for about four years. After weaning her baby, the mother usually becomes pregnant again so, eventually, she is likely to have a ten-year-old, a five-year-old, and a newborn, all living together.

Susan Nichols, a graduate student who had started with the project at its beginning, was for all intents and purposes Washoe's surrogate mother. Although my own responsibilities toward Washoe were parental, there was no question that my role was that of the older brother taking care of mommy's new toddler—a dynamic found in both human and chimpanzee families. The youngest child in a chimpanzee family can do no wrong in her mother's eyes. If a baby cries while being watched by an older sibling, the chimpanzee mother will rush in and discipline the baby-sitter.

As the youngest of eight children I never knew the joys and frustrations of taking care of a younger sibling. Washoe became the little sister I never had, and like most little sisters, when she was good she was a dreamboat and when she was bad she was horrid. Chimpanzee siblings, like human siblings, can have extremely close and supportive relationships, but they also lapse into fierce competition and knock-down, drag-out fights.

Washoe knew how to take advantage of her position as the youngest. Around Trixie, Washoe would shamelessly exploit her status as the darling grandchild who could do no wrong. She would nip my fingers, grab my notes, steal my pencil, or anything else that she knew she could get away with.

When I did report Washoe's obnoxious behavior to Trixie I could always count on Washoe trying to get even with me for making trouble. Once I told Trixie that Washoe had tried to bite me, and Trixie gave Washoe a rare scolding. That very night at dinner, Trixie was cooking at the stove. Washoe was in her high chair, at the head of the table, acting like a little angel. I didn't believe this act for a minute, so I was sitting as far from her as I could get.

COME ROGER, Washoe signed to me. PLEASE COME. I shook my head no. There was no way I was getting near her.

PLEASE PLEASE COME ROGER, she tried again. I signed an emphatic NO. At that moment Trixie turned around and saw Washoe making these very sweet and perfect signs.

ROGER, Trixie implored me, WASHOE WANTS YOU! I was trapped.

I began edging around the table, ever so slowly. Trixie went back to her cooking. I kept sliding over one inch at a time. Finally, Washoe couldn't contain herself any longer. She lunged out of her high chair and grabbed me around the neck with both hands. I yanked backward with all my might and broke loose.

There was one other graduate student, named Greg Gaustad, who entered Washoe's life at the same time as I did. Greg and I had equal status as Washoe's brothers. If Greg threatened me, then Washoe would come to my defense, and if I threatened Greg, then Washoe would come to his defense. Washoe's advocacy for the aggrieved party was so predictable that Greg and I would make up disputes just to create excitement on boring rainy days when the three of us were confined to the garage-cum-playroom.

We'd be sitting quietly, Washoe playing with her dolls and her two "brothers" reading magazines, when I would slip over to Washoe and sign to her that GREG HURT ME, feigning a cry face for good measure. Washoe would drop her dolls and swagger on two legs over to Greg, who by now had looked up to see one angry chimpanzee bearing down on him. I got some good laughs watching Washoe chase Greg around and around the garage until he acknowledged his "crime" and apologized to me by signing SORRY, circling his heart with his right fist until Washoe was satisfied. Needless to say, Greg would soon get his revenge by pulling the same stunt on me.

There was never any doubt where Washoe's loyalties lay in any run-in I had with Susan, Washoe's surrogate mother: I was always the villain. If Susan pretended to cry and pointed the finger at me, Washoe would threaten me until I apologized and crouched before Susan in submission, signing my apology. On the other hand, if I told Washoe that Susan had hurt me, Washoe would start chasing me instead, as if I must have done something truly treacherous to merit a reprimand from "Mama." And God help me if Susan play-threatened me in front of Washoe. Then Washoe would jump on me and start flailing away.

In the company of her human mother and grandmother Washoe was so confident of her position as the family favorite that she would actually set me up for a reprimand just to show me who was boss. For example, when Washoe and I were alone she allowed and even invited me to play with her beloved dolls. But when Trixie or Susan was present I had to be on guard lest Washoe happen to drop her doll at my feet. If I absentmindedly picked it up for her, she would explode with indignation that I would deign to touch her doll. The message was clear: I ranked beneath not only Trixie, Susan, and Washoe, but beneath the doll as well!

When the women were absent, however, Washoe's penchant for sibling rivalry disappeared. Then the two of us were the closest of playmates. When she and I ran through the backyard, I often found myself carried back to those hot summer days long ago on the farm when I would toddle for hours with my dog Brownie through the cucumber fields and grape arbors.

But there the similarity ended between my memory of Brownie and my experience with Washoe. Although my first dog had been a wonderful companion, she had never evoked the deep and tangled emotions of friendship, competitiveness, anger, and love that I felt for this baby chimpanzee. When I played with Washoe I felt like I was with one of my brothers again, wrestling with an equal who could give as good as he got, both physically and psychologically. Quite often I had to remind myself that this little chimpanzee girl was not a human being. But after a while I realized that this distinction had become meaningless to me.

FOUR

SIGNS OF INTELLIGENT LIFE

AFTER BROWNIE DIED my brothers and I got a new dog, and we named him Pal. Every afternoon when I stepped off the school bus, Pal was there waiting for me by the side of the road. His excited barks and wagging tail seemed to say, "I've missed you and I'm glad you're home." Then, as he ran ahead of me, he sometimes stopped, looked back over his shoulder, and gave me an impatient look that said, "Are you coming? Hurry up!" If I continued to dawdle, Pal would bark until I walked at a speed more to his liking.

Pal and I were communicating—and very effectively, too. If I'd been with one of my older brothers they might have said, "Hey, Roger," or "Come on, slowpoke," but the end result would have been the same. Of course, my communication with Pal was really not unusual. We communicate with other species so routinely that we rarely stop to appreciate just how extraordinary it is.

It is clear enough why members of the *same* species must be able to communicate with one another. They need a reliable signaling system for announcing who's ready to mate, sizing up one's competitors, discouraging intruders, warning that predators are nearby, and coordinating complex group activities like hunting. Over the aeons, nature has evolved as many modes of communication as there are species: singing (whales), scent markings (lions), flashes of skin color (cuttlefish), pulses of electric current (bottom-dwelling fish), waggle dancing (honeybees), and so on.

Communication is the glue that holds every animal society together. But how is it that two members of *different* species can communicate with one another? For the same reason they look like one another: they are related through common ancestors. Animals that resemble one another physically also communicate in similar ways. For example, all mammals, unlike birds and reptiles, begin communicating in the womb: the embryo has direct physical contact with the mother. And all mammals use the tongue, muzzle, and head to groom and caress their young.

Species that are *very* closely related are even more similar when it comes to communicating. A dog and a wolf both let it be known that they're ready to play by bowing down in front and hiking up their rear ends. When a baboon on guard for his troop barks out a warning, other monkeys that are *not* baboons recognize the signal and also move out of the way of danger. It is the same with humans and chimps. When I used to tickle Washoe, she made an open-mouth "play face" and a sound like the wheezing laughter of a human child—I didn't have to ask if she was having a good time. Our human ability, evolved over millions of years, to send and read complex nonverbal messages is what enables us to form bonds with members of other species.

Humans are primates and, like all other primates, we use three channels of communication: touch, sight, and sound. We nurture our young through close physical contact, communicating reassurance through the gentle touch of our hands—not just to our own young but to cats, dogs, horses, rabbits, and chimpanzees.

A human baby is born with a complex set of facial muscles and puts them to work immediately—first crying, soon smiling, then laughing—to let his parents know he's uncomfortable, delighted, or scared. At the same time, the child becomes expert at reading adult facial expressions like happiness and sadness, and in a matter of months will distinguish between degrees of these emotions. Between one and two years of age he begins to master adult facial signals, such as the eyebrows flashed in greeting, the eye winked in conspiracy, the mouth drawn back in disgust.

Our genius at reading facial signals enables us to interpret the facial expressions of other species. When it comes to non-human primates, much of their facial repertoire is similar to our own. Apes flash their eyebrows just as we do. When Washoe was unhappy she wrinkled up her face, pulled back the corners of her mouth, and curled her lips outward—which made her look like she was crying. When she was happy she drew back her lips and exposed her bottom teeth in a big smile. When she was *really* happy to see me she pursed her lips and gave me a small kiss. All of these behaviors are typical of the wild chimpanzee. With cats and dogs and other more distant relatives, we have to learn what their bared teeth and flattened ears mean.

Whatever an animal's facial expression, it's always accompanied by other visual information—body language—which we are also adept at reading. We are such masters of gestural communication—shoulder shrugging, head nodding, hand waving, and finger pointing—that it is possible for us to find our way around a foreign city without speaking a word of the language. And gesture figures into our communication so automatically that we do it while speaking on the telephone when the other person can't even see us.

We are equally adept at interpreting vocal signals. With cats we learn to distinguish the contented purr from the hungry meow and the angry yowl. Likewise, we can detect subtle shades of meaning in a dog's vocabulary of barks, but we have difficulty in making sense out of a whole chorus of barking dogs. The sounds that chimpanzees make are a lot closer to our own. Some chimps force air through their compressed lips, creating a sound we would call the raspberry or Bronx cheer. When Washoe was frustrated she would throw a tantrum by screaming. Before opening her birthday presents she would start panting or hooting in anticipation. Neither one of us needed sign language to figure out the other's moods.

It's easy to see how our ability to communicate with other species led people to create those age-old fables about friendship between human and animal. After all, the Roman story of Androcles and the lion is not far-fetched: a lion, in obvious pain, nonverbally asks the help of a man to remove the thorn from its paw. By helping the lion Androcles wins a lifelong friend, and the lion later refuses to maul Androcles when they come face-to-face inside the Roman Colosseum. It is a short leap from this story to ones about our ancient human desire to talk with animals in a human language, a fantasy that inspired Eve to speak with the serpent, Saint Francis to converse with the birds, Native Americans to talk with buffalo, salmon, and eagles.

But as far back as the time of Plato and Aristotle, Western philosophers have claimed that talking with animals is an impossibility. Even when they admit that nonverbal communication connects us to other species, they maintain that language is much more than a mode of communication—it is the expression of the human mind. Descartes summed up this argument when he said that even "depraved and stupid" men could tell other people what they were thinking, but animals cannot speak because "they have no thoughts." Nineteenth-century scientists agreed with Descartes that language has nothing at all in common with the modes of communication found elsewhere in the animal kingdom. After all, what could the sublime verses of a Shakespearean sonnet have in common with the grunts of apes or the alarm calls of monkeys?

Even Thomas Huxley, the scientist who first suggested in 1863 that we are related to apes, couldn't find any common ground between us and our apelike ancestors when it came to communication. "No one is more strongly convinced than I am of the vastness of the gulf between . . . man and the brutes," wrote Huxley, "for [man] alone possesses the marvelous endowment of intelligible and rational speech [and] . . . stands raised upon it as on a mountain top." Since the dawn of recorded history, man had speculated about the mysterious origins of language, but in all those thousands of years no explanation had eclipsed that most ancient and universal one: language was a gift from the gods.

But, as usual, Charles Darwin challenged the reflexive assumption of human uniqueness. In *The Descent of Man*, Darwin argued that many of the features that supposedly made human language unique could be seen or heard in other animals as well. For example, Darwin saw the same variability of human languages in the numerous birdsong dialects he studied closely.

Darwin suggested that what makes human language distinctive are its abstract cognitive features, our ability to name objects and manipulate the world symbolically. Even here Darwin did not swerve from his theory of evolution. He believed that our power of abstract thought was firmly rooted in the cognitive abilities of our apelike ancestors, who set the stage for the emergence of language. And he claimed that we would find these cognitive skills—abstract thought and tool use—in modern-day apes like chimpanzees. After all, they inherited their mental powers from the same apelike ancestor we did.

Darwin was just theorizing about ape intelligence. But like so many of his other "far-fetched" theories, this one was ultimately proved correct. In the early 1950s Keith and Cathy Hayes reported that whenever their chimpanzee foster daughter, Viki, wanted a ride in the car, she would tear a picture of a car out of a magazine and hand it to them, like a ticket. Viki understood photos were symbolic. Less than a decade later Jane Goodall made her momentous discovery that wild chimpanzees were making and using tools. Darwin was right on both counts: apes could think abstractly *and* use tools. But had ape cognition been the driving force behind the emergence of language in hominids? Darwin might have been correct about abstract thinking in apes but wrong about it being the critical evolutionary ingredient for language. Perhaps our distinctive way of communicating arose from some auditory-vocal innovation or, as some linguists claimed, from a neurological mutation. Or maybe language was a gift from the gods, after all.

There was an obvious way to test Darwin's theory. If modern-day apes could learn to speak—not necessarily reciting Shakespeare, but manipulating words to the degree that they use tools—that would show that language had its roots in the cognitive powers of our common ape ancestor.

Such tests had always proved futile, of course—Viki and other chimpanzees could not speak. But then in late 1966 one baby chimpanzee began talking with her hands.

DESCARTES AND DARWIN COLLIDED in Project Washoe. If Descartes was correct, then Washoe didn't have a thought in her head and would be unable to name a single object. If Darwin was correct, then Washoe was already thinking and would be able to express her thoughts by manipulating ASL signs like tools.

During my first semester on Project Washoe, I witnessed this philosophical collision in an unforgettable way. A prominent philosopher of science named Rom Harré was in residence at Reno as a visiting professor from that bastion of Cartesianism, Oxford University. Harré was renting a house near the Gardners and every day he drove by Washoe's backyard to get to campus. One morning he noticed something rather strange in the Gardners' willow tree. He parked his car and got out to take a closer look. Sure enough, it was Washoe, but she wasn't surveying the neighborhood as she usually did. Instead she was lounging in the branches, leafing through the pages of a magazine, signing to herself as she identified various objects in the photos and advertisements. The sight of a chimpanzee talking to herself, thinking out loud like a person, left Harré shaken.

I imagine that this had the kind of impact on Rom Harré that Edward Tyson, the British anatomist, might have felt in 1699 if his "mindless" chimpanzee had suddenly leapt off the examination table and started talking a blue streak. Two thousand years of Western philosophy said that a talking animal was impossible. Harré later confessed that this moment had forever changed his belief in human uniqueness.

Washoe's spontaneous "hand chatter" was the most compelling evidence that she was using language the way human children do: for example, she would sign QUIET to herself when she sneaked into a forbidden room. Or when she was perched atop her willow tree she would announce to us (who couldn't see the front of the house) the name of the person about to arrive at the front door. Or she would sit on her bed and talk to her dolls spread out around her. The way Washoe ran on with her hands like a gregarious deaf child, sometimes in the most unlikely of circumstances, caused more than one skeptic to reconsider his long-cherished assumption that animals can neither think nor talk.

To this day, those people who claim that the capacity for language is unique to humans talk about Washoe as if she were a talented circus act that was trained to mimic human signers. But those like Harré could see with their own eyes that Washoe had woven language into the daily flow of her life quite spontaneously. We hadn't trained her to talk to herself and her dolls or to issue reports from the top of her tree. And she couldn't have learned these uses of language by mimicking us because none of us signed to ourselves, climbed thirty-foot trees, or talked to dolls.

A lot of Washoe's linguistic behavior—talking to toys and dogs, for example—was *inappropriate* adult behavior. Washoe was simply doing what all human children do: she was using her newfound words on anyone and anything, whether they were listening or not. And, like a deaf child, if she started signing to strangers she would keep at it long after it was clear that they didn't understand her or know how to sign back.

In the second and third year of Project Washoe I would wake up every morning and listen to my two-year-old son, Josh, talk to an imaginary friend he called "Ga-caa." Josh didn't like it when I interrupted or even eavesdropped on these talks. So I'd leave for work, where I'd find Washoe sitting on her bed signing to her favorite doll. If she caught me peeking through the doorway she would stop mid-sign. As soon as I turned away she would pick up the "conversation" where it had left off. It was episodes like these—Washoe's childlike and private experimentation with language—that most effectively proved that she hadn't been trained.

IF WASHOE WASN'T TRAINED LIKE A SEAL, how *did* she learn language? In the mid-1960s, when the Gardners were about to launch Project Washoe, there were several theories to explain how children acquired language. The psychological school was represented by B. F. Skinner, a Harvard professor and the leading exponent of "behaviorism."

At the turn of the twentieth century, the founders of modern "behaviorism" tried to emulate physics by explaining animal behavior in terms of mechanics that could be easily measured. In Russia, Ivan Pavlov, the famed physiologist, trained a dog to salivate when it heard a dinner bell ring, thereby reducing its behavior to conditioned reflex. In America, John B. Watson, B. F. Skinner's precursor, analyzed the "muscle twitches" that produced an animal's behavior. By the time Skinner came along in the 1930s, behaviorism had fortified the old Cartesian wall between humans and animals: humans were thinking organisms who shape their environment; animals were mindless brutes whose behavior is *shaped by* the environment. Humans learn; animals are conditioned.

Skinner tried to tear down this wall between humans and animals, but he didn't do it the way Darwin had a century earlier, by assuming that there is continuity between the minds of humans and nonhumans. Skinner simply got rid of the mind altogether. He claimed that the human learning process is no different from conditioning in animals, and that it could be described mechanically without resorting to nebulous terms like "thought" or "consciousness."

According to Skinner, the behavior of all human and non-human species is governed by a single law of operant conditioning. It is *operant* because there are reinforcers and punishers in the environment that *operate* to shape and limit our behavior. A rat learns to press a lever in a cage by being reinforced or rewarded with food pellets. A human child learns to fear fire by being punished with burns.

Not surprisingly, Skinner argued that human children learn language the same way they learn everything else: through parental reward and punishment. The child's babbling is shaped into the word "mama" because of the positive reinforcers of Mommy's smiles. Poor pronunciations and bad grammar disappear thanks to frowns and verbal corrections. According to Skinner, parents do not *intentionally* shape language the way a psychologist knowingly shapes lever pressing in a rat, but the end result is achieved all the same.

Allen Gardner was a rat psychologist and well schooled in the principles of operant conditioning. If language was conditioned as Skinner said, and if all species learned through the same means of conditioning, then a chimp should be able to learn sign language through the use of rewards. So the Gardners decided to try to teach Washoe using Skinner's techniques.

To understand what happened next, it helps to know a bit about American Sign Language (ASL). ASL is *not* an artificial system that hearing people invented for the deaf. ASL has existed for at least 150 years and has its roots in various European sign languages that were developed by the European deaf themselves over the centuries. There have been several attempts to create a universal sign language—one is called Gestuno—but they've been about as popular as a universal speaking language— in other words, not very. ASL is just one of many indigenous sign languages that have emerged in deaf communities around the world, and meetings between these communities, like the

International Congress of the Deaf, require a team of simultaneous translators. Since sign languages are not universal it should also be clear that ASL is not a set of simple gestures, a pantomime, that people understand universally.

And though ASL appears to be nothing like spoken languages, it is actually built on the very same principles. ASL, like spoken languages, is infinitely flexible because its basic building blocks have no meaning in themselves, but take on meaning when combined. The building blocks of spoken language are *phonemes*—the fifty or so sounds humans can make (for example, the "b" sound in "boy" or the "g" in "girl"). If each phoneme had a specific and fixed meaning, we'd be stuck with a fifty-word vocabulary. But each phoneme is meaningless, so we can combine them to form more than one hundred thousand words and an infinite number of sentences.

The building blocks of ASL are *cheremes*, which are meaningless hand configurations, placements, and movements. When combined, these cheremes can form an infinite number of signs, the equivalent of words. There are fifty-five cheremes: nineteen identify the configuration of the hand or hands (for example, the pointing hand); twelve identify the place where the sign is made (the cheek); and twenty-four identify the movement of the hand or hands (moving the hand vertically). The pointing hand indicates one sign if it is made at the cheek, another at the forehead, and still another at the chin. At any one of those places, the pointing hand can indicate one sign if moved toward the signer, another if moved away from the signer, and still another if moved horizontally. Change the configuration of the hand (from pointing to a closed fist, for instance) and you create a whole new set of possible signs.

Most ASL signs appear arbitrary (for example, BLACK is signed by drawing the index finger across the forehead), but just as English takes advantage of auditory references (*choo-choo, bow-wow, tuba,* et cetera), ASL often uses visual references. CAT is signed by stroking an imaginary pair of whiskers with the thumb and forefinger, and DRINK is signed by tipping an imaginary glass at the lips. There is more visual mapping in ASL than there is auditory mapping in English simply because people have more visual reference points.

You can create proper names in ASL by combining the first letter of the person's name, using the manual alphabet of finger spelling, with a sign indicating a characteristic of that person. The Gardners signed WASHOE by flicking a "W" hand against one ear, meaning "Washoe, big-ears." I created the sign for ROGER—it translates as "Fouts from California." The ROGER sign was made by combining the "F" hand (for Fouts) with the sign for CALIFORNIA which, in ASL, used to be translated as "the golden play land"—signed by grasping the earlobe, like a golden earring, between the thumb and index finger and pulling down. We signed DR. G, Allen Gardner's name, by touching the forehead with the "G" hand, indicating "wisdom." MRS. G was signed as "lady," by drawing the "G" hand down the cheek—a sign that derived from the string used to tie a lady's bonnet in the olden days.

ASL has its own rules for organizing signs into sentences, and this grammar is very different from English grammar. Like Hebrew and some other spoken languages, ASL does not have the copula, the forms of "to be" that link a subject and predicate. As a result, "You are happy" translates into ASL as the more compact YOU HAPPY. Nouns in ASL can do double duty as verbs. So instead of saying "Give me a banana," you might sign BANANA ME. You can also sign OUT ME instead of saying "Take me out."

English relies heavily on word order to change meanings, but ASL, like Russian, is highly inflected. You can easily alter any sign to indicate a grammatical feature like person, tense, or number. That inflection is expressed visually by facial expression or the speed, location, or repetition of a sign. The difference between a word, such as GOOD, and a sentence, such as I AM NOT VERY GOOD AT THAT, may only be clear to a native signer who can read the subtle visual grammar.

The Gardners decided to teach Washoe ASL signs as if they

were teaching a rat to press a lever by using an operant conditioning technique called shaping. The experimenter shapes the rat's movement by using rewards to lead the rat closer and closer to the food lever. According to Skinner, parents shape a child's babbling into words when they reinforce wordlike sounds with smiles and nods of approval.

Conditioning a chimpanzee to sign seemed like a pretty straightforward task. The Gardners would simply wait for Washoe to make a gesture that resembled an ASL sign and then shape the gesture by encouraging and rewarding her until the gesture became the sign itself. For example, the sign for MORE is made by bringing the hands together so that the fingertips touch. If Washoe was being tickled, one of her favorite things, she would reflexively bring her arms together to try to protect herself. If the Gardners stopped tickling her and pulled her arms apart, she tended to bring them back together again. When she did so, crudely making the MORE sign, they would reward her with more tickling.

Next they required Washoe to use the correct configuration, place, and movement of the sign—sometimes they would prompt her by making the perfect sign themselves—before rewarding her with tickles. It wasn't long before Washoe was spontaneously signing MORE to get more tickling. After that the Gardners introduced MORE into a game of pulling Washoe around in the laundry basket. She caught on fast to the idea that MORE could be used to get more of anything, and soon she was using it to get more food, more games, more books.

Washoe's acquisition of the sign for MORE was a textbook case of operant conditioning. It was also one of the *last* words she learned through B. F. Skinner's techniques. Life with Washoe was a high-speed romp in the backyard that was not easily interrupted. Let's say we wanted to teach her the sign for BIRD (right thumb and index finger against the mouth, pointing out like a beak). The chance that Washoe would make a gesture resembling BIRD just when a bird was flying by was practically nil. And even if she did, we hardly had time to reward her before she was on to her next adventure. It soon became obvious that the reason Skinner's rats had to be confined in boxes was to keep their behavior completely under the control of the experimenter. Shaping one sign of Washoe's could take months.

Skinner's theory didn't hold up very well either when it came to language acquisition. Washoe babbled a lot with her hands, especially when she began learning signs. When one of her babbles resembled anything like an ASL sign, the Gardners would clap, smile, and repeat the gesture. During a full year of positive reinforcement for babbling, this method produced a grand total of one sign: FUNNY, probably because the FUNNY sign, two fingers brushing the nose, lent itself to a ripsnorting nose-touching game.

Debbi and I made a similar discovery later with our daughter Rachel. When Rachel was just learning to talk, instead of saying "wa wa" for "water" like most kids do, she made the noise, "goloink," that sounded as if she were swallowing water down her throat. Of course, Debbi and I thought that goloink was very special. It reminded us of a cartoon character, Gerald McBoing Boing, who was quite popular then. Consciously or not, we reinforced Rachel's goloink by clapping, smiling, and getting her to do it when friends visited. We couldn't get enough of goloink. But soon goloink disappeared and Rachel started saying "water," no doubt because she heard the rest of the family saying water.

We couldn't "shape" Washoe any more than we could Rachel. Washoe was picking up signs left and right by seeing us use them. We used only sign language around her, which gave her plenty of opportunity to imitate us, even if she did so according to her own timetable. The most we could do was demonstrate a sign to Washoe, the way all parents do with their children: THIS TOOTHBRUSH, while brushing our teeth. Every night after dinner, Washoe was told FIRST TOOTHBRUSH, THEN YOU CAN GO OUT. But we couldn't really tell if she even understood the sign because she still didn't use it. Then one day

Washoe went into the Gardners' bathroom, saw their tooth-brushes, and, unprompted, signed TOOTHBRUSH.

Most signs didn't need an explicit demonstration. We could just point to a car and sign THIS CAR. One day another companion of Washoe's, Naomi, was searching for matches but couldn't find any. Washoe began following Naomi, who explained what they were looking for by holding up an empty box of matches. Washoe signed SMOKE, a sign she obviously picked up by watching her human friends ask each other in sign for cigarettes and matches.

At other times Washoe would learn the beginnings of a sign through our demonstration but she'd perfect it through imitation. For example, the correct ASL sign for FLOWER is made by holding all five fingertips of one hand together, so that the fingertips touch, and then touching the bunched fingertips first to one nostril, then to the other, as if sniffing a flower. After observing many discussions of flowers in the fall of 1967, Washoe began using the flower sign but in her own childlike form: she would touch her nostrils but she'd use only one index finger—like a speaking child saying "bobble" instead of "bottle." Most parents do not correct their children's "child-speak" and, having given up on conditioning Washoe, neither did we. After a few more months of watching us, Washoe began signing FLOWER correctly.

In my first year with Washoe, I hit on another way of introducing new signs that built on Washoe's power of imitation. If I wanted to show her the sign for TREE, I would first point out a tree. Then I would bend her left arm at the elbow so that her hand pointed up and place her right hand under the bent elbow—forming the TREE sign. This idea of guiding Washoe, of allowing her to imitate herself, may seem perfectly obvious to anyone who has ever helped a child learn to tie his shoes or button her shirt (or helped a child with the letter "b" by holding her lips together). But to Skinnerians, guiding an animal is like taking the rat's paw in your hand and depressing the food lever.

Animals are not supposed to learn in this way, without any rewards. Indeed the textbooks say that methods like guidance *retard* learning in animals.

The Gardners were somewhat appalled by my unorthodox technique and cautioned me against using guidance. Not knowing any better, I persisted, mostly because it seemed to help Washoe when she was struggling with a new sign. Washoe would need to be guided for only one or two sessions before she learned a sign, and this was almost always in combination with her imitating our usage. Notwithstanding "scientific opinion," Washoe learned very quickly this way, and the Gardners were converted. As a result, the topic of my Ph.D. dissertation became "Use of Guidance in Teaching Sign Language to a Chimpanzee."

We didn't know it at the time, but chimpanzees in the wild learn new skills from their mothers through this same combination of imitation and guidance. Chimpanzee mothers in Africa do not methodically train their children to make tools or crack nuts. Rather, the youngster must carefully observe her mother, imitate her through play and practice, and gradually acquire the skill over a period of years. Christophe Boesch, a primatologist who studies the stone tool culture of chimpanzees in the Tai Forest of the Ivory Coast, describes seeing chimps place hard nuts on a flat stone or log anvil and then pound them open with stone or wood hammers. One day, Boesch observed a young female chimp flailing away with an irregular hammer, not having much success. (Nut cracking may sound easy, but it takes a *human* novice about thirty to sixty minutes to crack one *Coula* nut.) Despite turning the hammer around several times, the young chimp couldn't get the knack of it, and she grew very frustrated.

After a while, the mother came over and took the hammer out of her daughter's hand. Then, in a super-slow-motion movement, she rotated the hammer in her own hand to demonstrate the proper grip. After cracking a few nuts, which she shared, the mother handed the tool back to her daughter and walked away. Her daughter then held the hammer properly and proceeded to open several nuts.

There is a very good reason that chimpanzee mothers don't stand over their children all day showing them how to crack nuts but instead offer pointed guidance only when it's necessary. It is the same reason human parents don't drill two-year-olds in vocabulary and grammar. When behavior is acquired in a controlled way, through rigid training, it's not *flexible*. Flexibility is the key to primate intelligence. Flexibility is what enables a chimpanzee and a human to apply a skill learned in one situation to an entirely different situation. Without flexibility, a young chimpanzee might be taught by her mother to crack a nut in a single day, but she would be helpless to generalize that skill to different hammers and different nuts. Similarly, a young human, or chimpanzee, might be drilled on a vocabulary list with food rewards but she wouldn't be able to use these linguistic tools in other social situations.

Young chimps and young humans come into the world well prepared to learn in a manner that builds in flexibility. They are endowed biologically with a voracious curiosity, an ability to imitate, and a powerful drive to play. And they are given free rein to indulge these traits, thanks to a long childhood that is made possible by constant parental care and protection. Unburdened by the need to find food, defend territory, or reproduce, young chimps and young humans have plenty of free time on their hands. This prolonged and dependent childhood, highly unusual in the animal kingdom, comes at a very high evolutionary price. The mother, closely bound to her child, can give birth only every few years—a very risky strategy for reproduction.

Chimpanzees and humans would have died off a long time ago if this costly investment of parental time didn't pay big dividends. The youngster gains a relaxed family environment where she can develop an array of skills—motor, cognitive, and communicative—that are essential to survival. She has the chance to imitate the adults around her, and she has years of playtime when she can integrate this new behavior into her large repertoire of other behaviors. Washoe not only played with dolls but she signed to them; she not only climbed trees, she signed in them. Play is nature's way of fusing complex behaviors and creating intelligence that is adaptive to new situations. It is nature's version of school, a school where there are no teachers per se, just interesting adults that can be watched and imitated.

The anthropologist Ashley Montagu called words "conceptual tools," the mental equivalent of physical tools. Darwin would have appreciated this metaphor, since he believed that tool use and language were driven by the same powers of cognition. Washoe learned to modify her linguistic tools for new social situations, much as wild chimpanzees modify their own tools to collect nuts, honey, ants, or termites.

If true learning means having the flexibility to adapt to new situations, then Washoe's improvisation with signs was the best evidence that she was self-taught. She referred to her toilet as DIRTY GOOD and the refrigerator as OPEN FOOD DRINK even though we referred to them as the POTTY CHAIR and the COLD BOX. We did not teach Washoe any of these sign combinations. Instead, she dug into her own "tool kit" of symbols—her vocabulary—and arranged them in a way that got the job done.

Moreover she learned how to correct her own misuse of word tools. On one occasion she signed THAT FOOD while looking at a picture of a drink in a magazine. Then she looked at her hand closely and changed the sign to THAT DRINK.

Washoe was also capable of inventing entirely new signs. For example, our sign language manual did not have a sign for BIB so the Gardners used the sign for NAPKIN instead, made by touching the mouth with an open hand in a wiping movement. But Washoe seemed to have trouble with this sign and asked for her bib by outlining a bib on her chest with her two index fingers. The Gardners acknowledged her ingenuity but insisted that she use the sign for NAPKIN because we wanted to see whether Washoe could learn a human language, not whether we could learn a language invented by a chimpanzee. Washoe finally learned the sign for NAPKIN and began using it to ask us for her bib.

Several months later when a group of human signers at the California School for the Deaf were watching a film of Washoe, they informed the Gardners that the baby chimpanzee was not signing BIB correctly. It should be signed, they told the Gardners, by drawing a bib on the chest with the two index fingers. Washoe had been right all along.

Nobody was teaching, much less conditioning, Washoe. She was learning. There is a very big difference. Despite the misguided attempts in the first year to treat Washoe like a Skinnerian rat, she was forcing us to accept a truism of chimpanzee and human biology: the child, not the parent, drives the learning process. If you try to impose a rigid discipline while teaching a child or a chimp you are working against the boundless curiosity and need for relaxed play that make learning possible in the first place. As the Gardners finally conceded, "Young chimpanzees and young children have a limited tolerance for school." Washoe was learning language not because of our attempts to school her but *despite* them.

Washoe had already taught me that chimps will stay with a task for hours as long as it is fun. Now she was teaching me a lesson that would prepare me for a lifetime working with chimps and children: learning cannot be controlled; it is out of control *by design*. Learning emerges spontaneously, it proceeds in an individualistic and unpredictable way, and it achieves its goal in its own good time. Once triggered, learning will not stop— unless it is hijacked by conditioning.

The impossibility of teaching through reward and punishment was most evident at mealtime with Washoe. According to behaviorism, the combination of a hungry chimp and ready-to-dispense food should be the perfect opportunity for reinforcement and learning. But the hungrier Washoe was the quicker her signing deteriorated into pure repetition and finally outright begging. Again, any parent can imagine what would happen if they withheld a three-year-old's breakfast cereal until he asked for it in a complete, grammatically perfect sentence. The child would likely do or say anything—punctuated by whining, crying, or begging—just to get the cereal.

The negative impact of reinforcement is well documented in teaching situations. Freehand drawing, an example the Gardners often point to, is one of the favorite activities of nursery school children. But when psychologists tried to increase the quantity of freehand drawing by offering rewards, they undermined the quality of the art. Desmond Morris found a similar phenomenon among young chimpanzees, who also love to draw.

> The ape quickly learnt to associate drawing with getting the reward but as soon as this condition had been established the animal took less and less interest in the lines it was drawing. Any old scribble would do and then it would immediately hold out its hand for the reward. The careful attention the animal had paid previously to design, rhythm, balance and composition was gone and the worst kind of commercial art was born!

The desire for a reward overwhelmed the natural desire to draw, just like Washoe's hunger at dinnertime overwhelmed her natural desire to sign. Creativity and learning are examples of innate behavior that can only be hindered, not helped, by rewards. Every animal begins life, thanks to millions of years of evolutionary pressure, with a number of species-specific behaviors that help it survive in its native environment. For example, roof rats run up when they are frightened and burrowing cellar rats run down. Neither of these rats needs a reward for these behaviors; they are simply compelled by nature to behave this way. Likewise, children are compelled to be creative.

One obligatory behavior that every species has in common is the need to communicate. The ability to send and receive messages is crucial to the organization and survival of every animal society. Some animal communication is automatic and fixed, like the ability of the cuttlefish to change color and attract a mate. The honeybee's waggle dance also seems to be automatic and stereotyped. But in species where the signaling system is highly variable—like birds and whales—the infant is born with a powerful drive to learn whatever system of communication it will need to socialize, mate, and breed.

Washoe also seemed driven to learn her family's mode of communication—in this case, American Sign Language. And she was succeeding. If we had been communicating by speech, waggle dancing, or electrical pulses, Washoe probably would *not* have learned to communicate with us, despite her drive to learn. Evolution, it seems, had prepared Washoe for a gestural form of communication. But why?

For a long time scientists assumed that chimpanzee communication had nothing to do with human communication because the grunts, screams, and pant-hoots of chimpanzees bear little resemblance to human speech. But these scientists were focusing on the wrong channel of communication.

In 1967, the same year I met Washoe, the Dutch ethologist Adriaan Kortlandt published a breakthrough study of wild chimpanzee communication. Its title told the whole story: "Use of the Hands in Chimpanzees in the Wild." According to Kortlandt, chimpanzees were using their hands for much more than building tools. They were communicating in ways previously unimagined. "It is hardly possible to overestimate the importance of the hand in chimpanzee social life," Kortlandt wrote. Chimpanzees gestured to beg for food, to seek reassurance, and to offer encouragement. There were various chimpanzee gestures, Kortlandt reported, for "Come with me," "May I pass?" and "You are welcome."

But most astonishing, some of these chimpanzee gestures differed from community to community. For example, Kortlandt observed three different stop signals. In one forest locale the chimps made a stop signal like a traffic cop—hand held up, palm out. In another, chimps made a scooping underhand signal. And in a third, they waved away intruders overhand.

This variability was quite similar to what we see in human societies. For example, northern Italians indicate no by shaking their head. Southern Italians use the "Greek no"—an upward toss of the head. In other countries, tossing your head doesn't mean "no" at all—it means "yes." Some human facial expressions, such as those indicating happiness, surprise, sadness, anger, and fear, are the same around the globe, others, like the head toss, are culturally specific and must be learned.

Soon, other ethologists began reporting even more variations in the chimpanzee gestural system that appeared to be culturally transmitted. In 1978 William McGrew and Carolyn Tutin observed that two chimpanzee communities, which were only eighty kilometers apart in Tanzania, used slightly different gestures in order to ask for grooming. The chimps at Gombe each raised one arm straight into the air, but the chimps in the Mahale mountains would each raise one arm over the head and then grasp each other's wrist.

In 1987 Toshisada Nishida observed the use of a "leaf-clipping" gesture that, again, was found only in the Mahale chimpanzee culture. This elaborate gesture appeared to solve an age-old problem of male apes everywhere: How do you lure a female away from the community without other males noticing? In Mahale, the male makes eye contact with a female in estrus while placing a leaf in his mouth. After conspicuously tearing the leaf and letting it drop, he quietly leaves the group. Soon after, the female will slip away herself and follow him into the forest for a "mating safari" that may last anywhere from a few days to several weeks. In nearby chimpanzee communities the males use an altogether different courtship signal, such as holding or shaking a branch.

Adult chimpanzees also invent variations of old gestures to convey new messages related to grooming, courtship, the presence of enemies, and other important social information. The other chimps somehow understand these new messages even though they may not choose to begin using them themselves. The chimpanzee baby, in other words, is not born with a fixed system of communication. Like a human, she comes into the world with certain postures, gestures, and calls but learns how to use those signals properly only after years of experience in her community. Until she grasps her group's specific gestures and social cues—its dialect—she won't be able to learn important skills from her mother, form alliances with her peers, attract a mate, and raise her own children.

Adriaan Kortlandt's study of chimpanzee gesture concluded with a statement that was very significant for Project Washoe in 1967: "The manipulative possibilities available in the chimpanzee hand are far from fully exploited in its use for gesturing." The Kortlandt paper showed all of us why Washoe was born to learn a gestural system of communication like American Sign Language. If she hadn't been captured and brought to America, Washoe would have been communicating with her biological mother in her community's own distinct dialect of hand signals, arm motions, and body gestures. The same facilities for learning and thinking had to underlie both ASL and chimpanzee gesture, as Darwin might have guessed. Otherwise, Washoe would not have been able to manipulate our symbolic signs like tools.

For nearly a century, Darwin's radical hypothesis that human language was rooted in the cognition of apes had been ridiculed or ignored. Now, Washoe was proving that Darwin was right about the ape's ability to think abstractly. Washoe knew that TREE referred not just to her favorite willow tree but to all trees, no matter what they looked like. But Darwin also appeared to have been correct when he said that the cognition of our common ape ancestors, and not some other innovation, set the stage for the emergence of human language. Washoe was thinking abstractly like a human child, but she was also *communicating* like a human child. She wasn't just learning symbols, she was using them to share her feelings, to control her backyard world, and to get her way in every imaginable situation.

One beautiful moment early on during Project Washoe illustrated the common need of chimps and children to use their signs. The Gardners were in their kitchen entertaining some friends whose toddler happened to be deaf. Washoe was playing outside. Suddenly, the child and Washoe saw one another through the kitchen window. As if on cue, the child signed MONKEY at the same moment Washoe signed BABY.

Every day I was reminded that Washoe was developing a humanlike capacity for language. By early 1969 my three-year-old chimpanzee sister was not just acting like my two-year-old son, she was talking like him as well. At 7 A.M. Washoe would greet me with a flurry of signs—ROGER HURRY, COME HUG, FEED ME, GIMME CLOTHES, PLEASE OUT, OPEN DOOR—that were a gestural version of what I heard from two-year-old Josh every morning. And the way Washoe would play-fight, scratch me, and then watch my bleeding cut, all the while signing HURT HURT and SORRY SORRY, was almost an exact replay of what I got at home. And Washoe's ability to use language to manipulate me or threaten me soon became quite routine with my son as well.

I often found myself in heated exchanges with Washoe that reminded me of my own childhood. For example, in early 1969, I had the thankless job of keeping her in the garage on laundry days while Susan Nichols used the washer in the Gardners' house to clean Washoe's clothes. Before, whenever Washoe had seen us gathering up her clothes, she'd know that the Gardners' back door would soon be open and she could sneak inside, where she would launch a chimpanzee-style raid: emptying the refrigerator, romping through the beds, and ransacking the closets. I always wound up frantically chasing her around the house. One time I turned on the vacuum cleaner to scare her out. This worked a little too well. In her panic to escape, she began defecating all across the Gardners' Persian rug.

The new laundry day strategy had me luring Washoe away from the trailer by suggesting that we GO GARAGE PLAY before Susan gathered up the dirty clothes. Washoe was usually enthusiastic about this because we had fixed up the garage as a rainy day playroom. We painted jungle scenes on the walls and put in a mattress for Washoe to bounce on, a parachute to swing on, and rugs to roll in. It was big enough for her to ride her tricycle around or to have wagon rides in. Once we were inside I would surreptitiously padlock the door.

This worked fine until Washoe looked out the window and saw Susan on her way to the Gardners with the laundry. Then the garage became a prison and I was the big, bad brother. First she asked to GO OUT. When I refused, she signed OPEN KEY, just in case I had forgotten how to get out. She even resorted to her most polite, PLEASE OPEN. When I signed my refusal, she first began tickling, then pinching and scratching, and finally tearing my shirt off. I was bigger than Washoe but nowhere near as strong. I had to do something fast or these games would turn into major brother-sister brawls.

It was during one of these brawls that I remembered a trick my older brothers played on me when they wanted to keep me from going into a forbidden room. They would tell me that the "bogeyman" was in that room, and he would "get" me if I went inside. There's no question that Washoe's bogeyman was big black dogs. So I pointed to the locked garage door and signed BIG BLACK DOG OUT THERE. EAT LITTLE CHIMPANZEES. Right away, Washoe's eyes got big and her hair stood on end. She stood up on two legs and began swaggering like one angry ape. She hammered on the wall with the back of her hand. Then, suddenly, she charged across the garage, leaping into the air at the last moment, and slammed into the locked door with both feet. Then she came back over to me.

This was working better than I had ever imagined. Washoe had ripped so many of my shirts on laundry day that I decided it was time to even the score a little. I asked her, YOU WANT GO OUT AND PLAY WITH DOG? She signed NO, NO DOG. Then she edged away from me, putting distance between herself and the door. Now I knew I had her. I actually went over to the door, unlocked it, and opened it. Then I signed COME. WE GO OUT AND PLAY WITH BLACK DOG. She retreated to the farthest corner of the garage.

These exchanges went considerably beyond the kind of non-verbal communication one can have with a chimpanzee using facial expression and body language, or with a dog through barks and single-word commands. Washoe and I were communicating symbolically. She gave me symbolic information—telling me to open the door and suggesting that I unlock it using the key. I responded, with symbolic information, false though it was, about the big black dog. If I hadn't been able to conjure up a non-existent dog, and if Washoe hadn't been able to comprehend it, I might not have been able to defuse our conflict.

My son may have been acquiring English faster and more comprehensively than Washoe was acquiring ASL, but they were both using language to communicate abstractly and effectively. For me, this was the most powerful evidence supporting Darwin's theory that human language emerged from our apelike ancestors.

Of course, by confirming the premise of Darwin's theory, Washoe raised a whole host of new questions, at least in my own mind. If our ape ancestors communicated gesturally, were early man's first languages signed? If so, how and when did these signed languages become spoken? Why did spoken language emerge in human beings but not in chimpanzees? How and why did the rich visual and gestural language of the body change into these disembodied black marks that are on the page in front of you?

There were still many questions, but Washoe had effectively and dramatically put one ancient truism to rest. Language did *not* descend from the gods. It emerged from our animal ancestors. Two thousand years of Western philosophy had it wrong. We *can* talk with the animals—or at least with our ape siblings. I was sure of that. I was talking with Washoe.

FIVE

BUT IS IT LANGUAGE?

AROUND THE YEAR 1500, King James IV of Scotland ordered that a baby be raised in isolation to see what language he would develop if left on his own. The King speculated that the child would speak Hebrew, because presumably that was the language of Adam and Eve and the original tongue of all humankind. Like other experiments of its kind, this one came to a bad end. The baby wasted away from lack of affection.

These days, the only fact about language on which most scientists and linguists will agree is that King James was wrong: a child is *not* born with knowledge of a specific language. We now know that language develops as a child interacts with speaking or signing adults over a period of years. However, exactly what happens during this so-called "critical period" of language acquisition continues to be a matter of great debate. Somehow, virtually all children manage effortlessly to acquire a means of communication so complex that no one has ever fully described the grammatical rules for even a single language.

How do children do it? Washoe's progress in learning American Sign Language promised to help answer this age-old question by shedding light on the origins of language. Until she began signing it was widely assumed that sometime after our ancestors diverged from Washoe's ancestors about six million years ago, we developed a major anatomical change—a new voice box, a new brain mechanism, or new powers of high-speed auditory discrimination—that enabled us to develop language. But if Washoe could learn a human sign language it meant that the common ancestor of both humans and chimps also must have had the capacity for this kind of gestural communication. And because evolution always uses the materials it has at hand—recruiting existing structures and behaviors to build new ones—early humans must have built up signed and spoken language on the very ancient foundation of cognition, learning, and gesture laid down by our common ape ancestor.

What does this have to do with the mystery of how children acquire language? If we knew that our ancestors developed language through cognition and learning, then it follows that modern human children probably do the very same thing, only in a more specialized way. Children must use the same strategies to learn language—observation, imitation, and play—that they use to learn other skills, like tying their shoes or playing the piano. They naturally imitate adults, generalize the skill to new situations through endless interaction and practice, and integrate this skill into other behaviors through their play. Language, of course, is more complicated than shoe tying and more universal than piano playing, so somewhere along the way humans must have developed a speedy and specialized way of learning in order to acquire language. And this only makes sense because, like any other species, we have a very powerful need to communicate.

You would think that linguists everywhere would have warmly welcomed Project Washoe's attempt to map out a likely pathway for the evolution of human language. But as it happened, this pathway pointed in a direction that utterly contradicted a theory of human language acquisition that reigned triumphant in the 1960s.

That theory, first advanced by Noam Chomsky of the Massachusetts Institute of Technology, argued that children do *not* learn language in the same way they learn to tie their shoes or play the piano. In fact, said Chomsky, language is acquired independently from all other learning processes and cognitive abilities. According to him, the rules of language are so complex, and the adult speech children hear is so disordered and confusing, that a child couldn't possibly learn language by observing and imitating. Instead, the rules of language syntax had to be encoded in the brain somewhere.

These rules of syntax are not the rules of traditional grammar. As Chomsky pointed out, a grammatical sentence can be nonsensical. *Colorless green ideas sleep furiously* was his famous example. Syntax, he said, gives language its infinite plasticity, which we demonstrate by our ability to combine words into unlimited new sentences that are grammatical as well as meaningful. Chomsky discredited B. F. Skinner's theory that children learn language through parental reinforcement by pointing out that children can construct completely new sentences (sentences they've never heard before) without any reinforcement at all.

Chomsky suggested that there is a "deep structure" of meanings that all languages have in common. Those meanings are transformed into the sounds and words of different languages, according to Chomsky, by means of a "universal grammar." He claimed that once this grammar was mapped out, it would reveal the logical properties that govern the infinite variety of sentences that can ever be formed. In practice this mapping task proved formidable, if not impossible. Every time the "universal grammar" encountered a new language—Chinese, say—it had to be revised to accommodate a whole new set of principles. One attempt to describe French in this logical fashion required twelve thousand items just to classify its simple predicates.

Obviously, if a universal grammar did exist, then no human two-year-old would be able to learn such a complex logical system, made up of tens of thousands, if not hundreds of thousands, of abstract rules. So Chomsky suggested that every child is born with a "language acquisition device" that already has the universal grammar wired in. This device would enable the child to generate abstract rules of grammar, unconsciously, simply by listening to the garbled speech of adults. Chomsky claimed that the universal grammar was part of a child's genetic makeup, making language as unique to humans as dam building is to beavers or waggle dancing to bees.

Chomsky said that the language acquisition device—or "language organ"—was located somewhere in the left hemisphere of the brain, but there is no anatomical evidence to support this claim and it has never been identified. But biology and anatomy aside, the language device was a reasonable hypothesis for explaining how children acquired language. What was not reasonable, however, was Chomsky's suggestion that such a device was unique to humans. There simply wasn't enough time, in the brief six-million-year period since humans diverged from our fellow apes, for evolution to add on a completely new brain structure. This "add-on" scenario was at odds with the laws of biology and neuroscience. The primate brain did not evolve like an ever-expanding house, adding on new rooms as it grew from monkey ancestor to ape ancestor to human. Instead, evolution was continually reorganizing what it already had—taking old structures and old circuits and putting them to use for new mental tasks. In fact, brain research since the 1960s has shown that human language is controlled by a network of independent cortical areas, each of which has an analogous area in the chimpanzee brain.

Supporters of the "language organ" theory still try mightily to reconcile Chomsky's theory with Darwin. They argue that complex organs—the eye, for example—arise all the time in the course of evolution through an evolutionary process of natural selection involving cumulative mutations. This is certainly true, but organs such as the eye emerge over tens of millions of years or more—not over a mere six million years. And when it comes to sibling species that have recently descended from a common ancestor, one of them can't possibly have enough time to develop an entirely new and unique biological system. For example, if the African elephant has a trunk, you expect to find a trunk on its sibling, the Indian elephant. The human and the chimpanzee diverged from a common ancestor even more recently than the two elephant species. Finding a language organ in humans but not in chimps would be like finding a trunk on only one of the elephants.

After humans and chimps diverged six million years ago, each species undoubtedly adapted the single system of communication it inherited from a common ancestor to suit its specialized needs. But these modes of communication must be grounded in the same ancestral cognition, *or Darwin's theory of evolution is wrong.*

From the point of view of evolutionary biologists, that has always been the problem with the language acquisition device. Chomsky's theory was a *deus ex machina,* a modern version of the ancient belief that language was a gift from the gods. It contributed nothing to our knowledge of how human language might have evolved from our apelike ancestors. This failure is understandable once you know that Chomsky and his followers were not biologists; they were logicians working in the philosophical tradition of Descartes.

Whenever biologists or comparative psychologists study an aspect of anatomy or behavior, we must account for its evolutionary development, or phylogeny, from ancestral species. Linguists, however, did not operate within these evolutionary constraints. As students of Western philosophy they assumed a *discontinuity* between humans and apes. Like Descartes, Noam Chomsky operated from the premise that human language stood wholly outside the animal kingdom. To him, the logical principles of language bore no relation to any other form of animal communication, even though little was known at the time about ape communication.

It is easy to see how Chomsky, who focused narrowly on written English, reached this conclusion. He did *not* study language as social communication, the face-to-face interactions between humans that integrate words, intonation, and body language. The way people actually speak is very similar to the rich *visual* grammar of American Sign Language. In ASL I can inflect the sentence "I feel good" with ten different shadings—

from "cautiously good" to "unbelievably good"—by altering the height or speed of my sign. When speaking English, I can use tone and facial expression to inflect the word "good" with the same shadings. Or I can change the meaning entirely by stressing the word "feel"—"I *feel* good" (meaning: "I feel good, but something else is wrong"); or by stressing the word "I"—"*I* feel good" (meaning: "I feel good, but someone else we both know does not"). Whether signing in ASL or speaking in English I can eliminate a hundred possible ambiguities so that the person I am addressing will grasp my meaning. Linguists who have studied such interaction say that as much as 75 percent of the meaning in a face-to-face conversation is communicated through body language and intonation—that is, without syntax.

By focusing on what was easiest to measure and quantify, words on the page, Chomsky and his followers removed language from its natural social context and constrained it to a linear form. Written language is *forced* to employ rules of logic to remove ambiguity. A universal grammar appears necessary to clarify shades of meaning that would otherwise be clear in a face-to-face exchange. Language is certainly governed by rules. But when you assume that people communicate like formal logicians in perfect grammar those rules tend to look a lot more complicated.

One result of Chomsky's approach is that "linguistic" came to mean all of the *disembodied* elements of language, whatever can be written down and analyzed mathematically. All the face-to-face signaling behaviors we share with other primates were deemed unimportant and therefore not "linguistic." For many years linguists were so biased against gestural communication that they didn't even study sign languages. In fact, during the first half of this century, educators tried mightily to eradicate American Sign Language because they thought its gestures were too "monkeylike"; speech was seen as the "higher and finer part" of language. The visual grammar of American Sign Language was not studied until the mid-1960s, and then only by pioneering linguists like William Stokoe, who recognized humankind's gestural continuity with the apes. Thanks to his research, American Sign Language was finally recognized as a "true language" in the late 1960s.

Most linguists thought the idea of a chimpanzee learning language was absurd. Chomsky said it was like an island of birds that had the power to fly but had never done so; if chimps had an innate capacity to use language, he said, they'd already be talking in the wild. Of course chimpanzees *have* been using gestural communication in the wild for millions of years, and their dialects of hand movement, facial expression, and body language look very much like the nonverbal elements of human language.

Those of us in Project Washoe looked at chimpanzee gesture and saw the roots of human language. But Chomsky had already decided that *human* gesture is not linguistic. So whatever chimpanzees were doing in the wild, their gestural dialects could not be related in any way to human language.

Just as Descartes and Darwin collided in Project Washoe, so would Chomsky and Darwin. If Chomsky was right, then Washoe didn't have a language acquisition device and she wouldn't be able to combine signs meaningfully. If Darwin was right, then Washoe already had the cognitive basis for language and wouldn't need a language organ with its hundreds of thousands of rules built in.

THE GARDNERS BEGAN KEEPING exhaustive daily records in 1966 of every sign Washoe made—every OPEN to get out the door, every TICKLE to get us to play, every LISTEN when a plane flew overhead. But as Washoe's signing mushroomed in the second year this note taking became impossible, and we started writing down only her use of new signs.

Before we would consider a new sign a "candidate" for Washoe's vocabulary, three separate observers had to see her make a spontaneous, well-formed, and appropriate use of the sign on independent occasions. These tough rules ensured that any sign that ultimately made it into Washoe's vocabulary could withstand any challenge.

Once Washoe's use of a sign was verified three times, then it went on a list of "candidates for reliability"—a list of signs each of us carried everywhere. A sign would not become reliable until Washoe used it in a well-formed, unprompted, and appropriate way, fifteen days in a row. If a day passed without a qualified report, then the fifteen-day count started all over again. After a sign was verified fifteen days in a row—usually by most if not all members of the foster family—then and only then was it considered part of Washoe's reliable vocabulary.

Waiting for Washoe to use a sign appropriately fifteen days in a row was sometimes a problem. There were no dogs around, for example, so it was hard to create a situation when she might sign DOG. When we drove in the car we usually passed by one yard in particular that had a dog. The dog would always rush the fence and bark at our car, and Washoe would sign DOG. But if we drove by and the dog wasn't there, Washoe would still sign DOG. If Washoe were a human child no one would doubt that she was commenting on the absent dog. We assume that everything a child utters has meaning. But with Washoe we lived in dread of overinterpreting her signing. So if Washoe signed DOG, and there was no dog, then we did not count it.

Since Washoe might not have a reason to sign DOG or STRAWBERRY or GREEN in a day, much less fifteen days in a row, we had to come up with a more systematic way of creating the contexts for signs. The Gardners introduced a procedure to test Washoe's knowledge and prevent cueing. Such testing controls have been standard procedure in comparative psychology since the notorious Clever Hans case at the turn of the century. Clever Hans was a German horse who solved arithmetic problems by tapping out numbers with his hoof. His apparent genius resisted explanation by philosophers, linguists, circus experts, and his own trainer. Finally an experimental psychologist named Oskar Pfungst revealed the method behind Hans's "math." Hans would begin tapping his foot while keeping a close eye on his trainer or the audience. The humans, who knew the answer to the problem, would unwittingly straighten up or otherwise signal Hans when he reached the correct number, thus cueing him to stop tapping.

Of course, cueing a horse to stop tapping is a lot easier than conveying one of a hundred possible signs to a chimp. Even so, Washoe's vocabulary was tested in a way that would prevent *any* possible nonverbal cueing. She sat in front of a box that was plywood on three sides and Plexiglas on the side she faced. Experimenter number one would place an object—a brush or bib or can of soda—in the box. Then Experimenter number two, who didn't know what was in the box and therefore could not cue Washoe, stood behind the box and asked her what the object was. He would then write down the first sign Washoe made. (This is called "double blind" testing because Washoe could not see Experimenter number one, who knew the correct answer, and Experimenter number two could not see the object in the box.)

Washoe thought this was a swell game, especially when the test object was candy or soda. She would lift the Plexiglas, swipe the food, and hightail it out of the room. One time she grabbed my watch out of the box and taunted me from up in her tree while I begged for its return. The other problem we had was that we couldn't fit a *real* car in a box so we had to find realistic replicas of everything from cows to cats to airplanes. But these replicas presented a new wrinkle. Washoe made many more errors with replicas than with photographs. This poor performance stumped us for a while until we noticed a pattern: the errors all involved the sign BABY. When looking at photos, Washoe would identify a dog as DOG, and a cow as COW. But when she saw the replicas she identified the cow as BABY, the dog as BABY, and the car as BABY. Any miniature replica was BABY. Washoe seemed less concerned with an object's name than with the fact that it was little, a "baby" this or a "baby" that. There was an obvious logic to this, especially from her baby point of view, but we had to count her answers as incorrect—again, to guard against overinterpreting.

An even bigger problem was Washoe's impatience. This was understandable given that she had to sit and watch while manic humans ran around opening and closing boxes behind one another's back. Washoe didn't like this ASL school, and after about five objects she would refuse to play. So the Gardners came up with slide transparencies, an innovation that Washoe enjoyed because she could control them. She would sit in front of a cabinet built into a wall. A back-projection screen was built into the middle of the cabinet so that no one else in the room could see what was on the screen. Washoe never saw the slides before the test, and no slide was ever used twice, so there was no way she could memorize anything. If she saw a marigold, a duck, and a German shepherd on one test, she would see a daisy, a blue jay, and a terrier on the next test. Washoe's answer was scored correct only if two human observers independently confirmed her sign.

Her performance, at four years of age, was remarkable. She scored 86 percent correct in one representative test that had 64 trials, and on a test twice as long—128 trials—she scored 71 percent. (Guessing randomly on this test would produce a score of 4 percent.) But her mistakes were perhaps more interesting than her correct responses. She would mistake COMB for BRUSH, or NUT for BERRIES, or DOG for COW—but never COMB for COW. In other words she would mix up items within a category, which showed that she had a mental grasp of classifications. According to linguists this ability to symbolize objects and then group them mentally by type—like sorting objects into cubby holes—is one of the key features that sets human language apart from other animal modes of communication. Washoe's ability to categorize displayed yet another skill that we inherited from our common ape ancestor.

Another of Washoe's errors occurred when she sometimes confused two signs that are similar in form. For example, CAT and APPLE are both signed on the cheek and BUG and FLOWER are both signed on the nose. This is like confusing words in English that sound alike. If you ask a child to bring you a *pan* and she brings you a *can* she's made a form error. Form errors are very strong evidence for familiarity with a language. A child cannot confuse pan and can unless she knows a certain amount of English. Similarly, Washoe could not confuse CAT and APPLE unless she had learned the shapes of these two ASL signs. Such answers were counted as incorrect, of course, but, ironically, they demonstrated a kind of competence.

Washoe also had to be counted wrong if she guessed by stringing together a bunch of signs. Whenever she did this her guessing also tended to reveal her grasp of ASL. If she saw a picture of daisies, she might string together several signs in the same category, such as FLOWER TREE LEAF FLOWER. Once when she was trying to identify a photo of frankfurters, she signed OIL BERRY MEAT. This looks like random guessing until you realize how similar these signs are. Each one is made by grasping different points along the edge of one hand with the thumb and index finger of the other hand. This is like juggling words on the tip of your tongue—*bone, cone, phone,* and so on. Washoe had the correct sign on the tip of her fingers—she just couldn't pinpoint where.

Washoe's style of signing was extraordinarily consistent. The two independent observers agreed in their reading of Washoe's signs about 90 percent of the time. In the summer of 1970 two fluent deaf signers, recent graduates of Gallaudet College, observed Washoe from behind our one-way mirror. Their job was like that of a hearing person who must identify the spoken words of a human toddler they've never met before—a notoriously difficult task. Both of these experts agreed with our own observer 89 percent of the time.

By 1970, at age five, Washoe was using 132 signs reliably and could understand hundreds of others. In addition to naming and categorizing objects, Washoe began doing something with language that Chomsky said only humans could do: she assembled words into novel combinations. You will recall that Chomsky debunked Skinner's theory by pointing out that children can create grammatically correct sentences that they've never heard before. Washoe also began combining signs into phrases that went beyond the ones she learned from us. She may have seen us sign YOU EAT and WASHOE HUG, but we could never show her every possible combination of subject and action. Still, like a human child, she manipulated her categories by pairing, for example, any given person with any given action: ROGER TICKLE, SUSAN QUIET, YOU GO OUT.

Her sign combinations were clearly not random because they always made sense in whatever context she used them. To test her we created situations where she needed help from one of us. For example, in the "doll test," Susan would "accidentally" step on Washoe's doll. Here were all of Washoe's reactions: UP SU-SAN, SUSAN UP, MINE PLEASE UP, GIMME BABY, PLEASE SHOE, MORE MINE, UP PLEASE, PLEASE UP, MORE UP, BABY DOWN, SHOE UP, BABY UP, PLEASE MORE UP, and YOU UP. Washoe only used signs in her vocabulary that were relevant to the situation, and she did not pair signs in nonsensical ways, like: BABY SUSAN, SHOE BABY, YOU SHOE, and so on.

This capacity for combining symbols in an order that conveys meaning, not nonsense, is exactly what linguists defined as syntax, the hallmark of human communication. According to Chomsky, it was the language organ that enabled a child to apply syntax automatically and unconsciously to generate sentences like UP SUSAN, instead of the nonsensical YOU SHOE. If Washoe's signing didn't have rules, then she would have combined signs randomly, but 90 percent of the time, her subject preceded her verbs, as in YOU ME OUT, YOU ME GO. She also understood how to use the subject and object. When I signed ME TICKLE YOU, she would get ready to be tickled. But when I signed YOU TICKLE ME, she would tickle back. In addition, when Washoe used YOU and ME in sentences, 90 percent of the time she signed YOU before ME, as in PLEASE YOU ME GO.

Even Washoe's longer combinations seemed to follow rules of syntax. She once pestered me to let her try a cigarette I was smoking: GIVE ME SMOKE, SMOKE WASHOE, HURRY GIVE SMOKE. Finally, I signed ASK POLITELY. She responded PLEASE GIVE ME THAT HOT SMOKE. It was a beautiful sentence but, as with my own children, I sometimes *had* to say no to Washoe, and this was one of those times.

Did Washoe have a language organ, with all the rules of syntax built in? Or was she simply *learning* the rules as she went along? Given what we know about chimpanzee cognition, the latter explanation is more likely. As we've seen, a chimpanzee child in the jungle is quite adept at learning through generalization. Every time she opens a nut with a hammer, she learns something in general about nut cracking. She must derive the underlying patterns of successful nut cracking—the rules, in other words—and then generalize those rules to new situations.

Washoe's ability to learn and generalize rules was a strong indication that our hominid ancestors did the very same thing. And indeed, from a behavioral viewpoint, there was no difference at all between Washoe's sign order and that of deaf human children learning ASL. Therefore, the notion of a language organ unique to human children was not only biologically improbable, it was unnecessarily complicated. Good science is parsimonious—it seeks the simplest possible explanation. And the simplest explanation was that human children were acquiring language like Washoe did, by learning it.

The best evidence that humans and chimpanzees learn language in the same way was the fact that Washoe developed language in the same exact sequence as a human child. First, she learned single signs, then two-sign combinations, and finally three-sign sentences. Her first combinations were "nominative phrases" (THAT KEY) and "action phrases" (ME OPEN), followed by "attributive phrases" (BLACK DOG, YOUR SHOE), and, finally, "experience" or sensory phrases (FLOWER SMELL, LISTEN DOG). She could answer WHO, WHAT, and WHERE questions *before* HOW and WHY questions.

Like a child's, Washoe's language development followed a pattern that emerged in tandem with her evolving grasp of objects, categories, and relations. And once she learned a sign, category, or relation she would generalize it to other situations and incorporate it into her daily behavior. Communication behavior in both chimp and child seemed to spring from the very same cognitive roots, just as Darwin had predicted.

IN 1969 THE GARDNERS PUBLISHED the first report of Washoe's linguistic progress in the respected journal *Science*. The astounding news that a chimpanzee was using a human language was received enthusiastically in the world of natural science. As the London *Times* would later say: "For biologists it was as much an epoch-making event as was landing on a heavenly body for astronomers, and it was fitting that the Gardners first published their work in 1969, the year of the first Moon walk."

I felt like I had walked on the moon with Washoe. For thousands of years humans had fantasized in myth and fable about talking with animals, and now we were realizing that dream. It was tremendously exciting to be a direct partner in this breakthrough conversation. Fifteen hundred centuries after our modern human ancestors left Africa, a channel to our own distant origins was open right in front of us. All we had to do was talk with Washoe.

Personally I expected, somewhat naively, that *all* scientists would embrace this unexpected and dramatic proof of Darwin's thesis that humankind is a product of evolution. But most linguists did not thrill to the news that an ape was talking. Chomsky's neo-Cartesian theory had gone virtually unchallenged for over a decade but now it had been dealt a serious blow. It would not be the last. In the early 1970s, a series of human studies was under way that would contradict another of Chomsky's central assertions—that children also needed their built-in language device so they could decipher the garbled, opaque speech of adults.

These new studies, which focused on mother-infant interaction, showed that mothers in different cultures modified their speech when talking to their newborns, presumably to make their language more understandable. "Motherese," as it came to be known—though it is used by fathers also—is not only complex, it is unconscious. Mothers alter the pitch of their voices so subtly that the change can often be detected only by acoustic analysis. Motherese is marked by dozens of features that make it appropriate only for children who are learning language: it is slower, simpler, more repetitive. Deaf mothers, as well, use a slower and grammatically simpler form of ASL when signing to their toddlers. All children, it seems, even those raised by fathers or older siblings, hear a simplified form of language until they are two or three years old.

The motherese studies, more than anything else, would lead to the widespread abandonment of Chomsky's theory in the 1970s. Chomsky and a coterie of his students would continue to defend the notion of a unique human language instinct, but soon the focus of linguistics research would shift dramatically. Linguists would begin studying language as a form of social communication, and the search for innate grammatical rules would give way to an exploration of how language and intelligence develop hand in hand in children. And the assumption that humans are unique in the animal kingdom would shift to an assumption of a biological basis for human language. In short, the study of language acquisition in children would soon follow the path blazed by Project Washoe.

In 1969, however, few linguists were willing to cede ground to Washoe without a fight. Nine years earlier, when Jane Goodall reported to Louis Leakey that chimpanzees in the wild were making and using tools, he delivered his famous reply: "Now we'll have to redefine tool, redefine man, or accept the chimpanzee as Man." In the wake of Project Washoe, linguists were facing a similar dilemma. They either could accept the linguistic continuity between chimpanzees and humans or redefine language.

Many linguists chose to do the latter. Although they agreed

that Washoe was communicating in sign, they refused to call it language. They suggested a new "checklist" definition of language that incorporated a series of linguistic traits that would presumably exclude all nonhuman communication. This approach showed the absurd lengths to which some academics would go to defend human uniqueness. It succeeded in keeping all nonhumans out of the "language club," but it explained nothing about how language might have arisen through evolution. This was about as helpful as Chomsky's theory that chimpanzees do not have a language organ, even though no one has ever found it in humans or bothered looking for it in chimpanzees.

The point of Project Washoe, after all, was not that chimpanzees were the same as humans or that they could master language to the extent that humans can. We wouldn't say that a member of *Homo erectus* "had" language or "didn't have language" based on his ability to speak or sign exactly like a modern adult human. Language emerged along a continuum, beginning with the gestural system of our apelike ancestor and evolving gradually, over millions of years, into the complex modes of signed and spoken languages we use today. This continuum cannot be cut in two with a hatchet—"language" on one side, "not-language" on the other.

Similarly, language develops along a continuum in the individual human child. A child is not *born* with language, and no one has yet determined the exact moment when a child "has language." Does he have language when he utters one word? Two words? His first action phrase? For years linguists tried to pinpoint when language begins by trying to force-fit children's utterances into an adult model of grammar. But this attempt failed, and children's language was recognized as similar but not identical to adult language. We now know that a human child progresses along a continuum of communication, and we call all of it language.

Why apply a different standard to a chimpanzee? By 1969 Washoe certainly had what is classified in children as Stage I language: she could form and use two-word phrases. That much was agreed to later by Roger Brown, an early critic of Washoe and the psycholinguist who defined these stages of language. And Washoe continued to progress: by 1970 she was handling longer combinations, Wh-questions, prepositions, and other elements of grammar that were comparable to children in Brown's Stages II and III.

This was the main point of Project Washoe, that the chimpanzee's capacity for language was *similar* to a human child's but not exactly the same. Washoe was born with language skills that were perfectly suited for gestural communication among a small group of thirty or so individuals in the jungle. Our own languages may share these same origins, but they've become specialized for a very different way of life in much larger communities.

Unearthing the connections between these two forms of communication—the chimpanzee and the human—would preoccupy me for the next twenty-five years.

IN THE SPRING OF 1970, it began to dawn on me that my days with Washoe were nearing their inevitable end. I was still struggling with my Ph.D. dissertation, which would go through many rewrites, and I would soon schedule my oral defense. Before I met Washoe I'd thought my dissertation would be about human children. Now here I was arguing the finer points of how chimpanzees learn sign language. Ironically, Washoe had taught me more about human children, especially their intense desire to learn and communicate, than I would have gotten from any academic study of child development. I felt well prepared to go to work with "learning disabled" kids. As soon as I finished my dissertation I would begin applying for jobs. My family was now counting on me to earn a living wage. Debbi was pregnant with our second child and we were tired of living on a research assistant's pittance.

Many afternoons that spring I sat in the trailer, watching

Washoe play with her dolls, and wondered where I would be in twenty years. I had no doubt I would look back on this time as one of the most unlikely episodes of my life. I had stumbled into Project Washoe, thanks to many rejections from other graduate schools and to my desperate need for a job. And of course there was that day on the playground when Washoe jumped into my arms.

I was going to miss Washoe. Through all our games in the trailer, all our sibling spats, and all our riotous escapades, she had won my heart. I would miss her antics in the willow tree, her enthusiasm for favorite books, her concern for my cuts and scrapes, and the way she signed WASHOE SMART GIRL. But most of all, I was sorry I wouldn't see her grow up. Like a big brother leaving home for college, I was going to miss out on a big piece of my kid sister's childhood. When I saw her next she would be a different person. It was a bittersweet thought.

Washoe was barely five years old. Her baby teeth were just starting to fall out. It would be another seven or eight years before she hit adolescence and sexual maturity. In the meantime she would keep growing physically and mentally. Human children don't stop learning and developing at five years, and there was no reason to think Washoe would either. She certainly showed no sign of reaching any limit in her ability to learn language. She kept picking up new signs, creating new sentences, and using them with increasing frequency. To determine the full extent of her language capacity, the Gardners would need to study Washoe until she was at least a teenager, and that's exactly what they planned to do.

So nobody was more surprised than I was when, one day in May 1970, Allen Gardner called me into his house as I was finishing a shift with Washoe and dropped the bombshell that would shape the rest of my life.

"Roger, we've decided to send Washoe to the University of Oklahoma. We want you to go with her."

I was dumbstruck. Project Washoe ended? My family and I

moving away with Washoe? I couldn't begin to imagine what it meant, or even why it was happening.

Apparently, the Gardners had decided months earlier to conclude the research and had been quietly scouting the nation's universities to find a new home for Washoe. There were many reasons. For starters, a new shopping center was going to be built on the empty lot across the street, which meant that our relatively quiet backyard laboratory was about to adjoin a major thoroughfare and giant parking lot. Among other things this would bring hundreds of kids to the fence to scream at "the monkey."

Then there was Washoe, who was getting to be a mighty big and headstrong girl, having reached an age when chimps are seen as "unmanageable" and past their prime by animal trainers in Hollywood. People in the neighborhood were already telling stories of Washoe escaping her trailer in the middle of the night and carousing around the neighborhood. Life in the daylight hours was difficult enough. The tiny trailer could barely contain Washoe's energy.

Outings to the city had become major logistical undertakings. I used to be able to take Washoe out in the car myself. Then one day, while I was driving, she reached out, grabbed the steering wheel, and stared right at me. I panicked, knowing I could never muscle her off the wheel. Then I realized that all she wanted was the lunch bag between me and the driver's door. I handed the bag to her, and she proceeded to eat her lunch as if nothing had happened. Meanwhile I tried to stop hyperventilating. That was the last time anyone went out with Washoe alone.

Even with two humans along, car rides had begun to feel like a James Bond movie, mostly because Washoe hated motorcycle policemen. Every time she spotted one she would lean out the window, bang repeatedly on the side of the car, and threaten the "offending" officer. We had to avoid cops at all costs. If one was sighted ahead, we immediately took evasive action by turning onto an alternate route.

Out of the car, it got worse. Spoken English was forbidden around Washoe so we had to stay between her and anybody else, which wasn't easy, especially with kids. In the past, before she became too big, we always carried Washoe. From a distance, most people thought she was a human baby. Now, it was like escorting the Queen of England around Reno. Washoe was always flanked by her human entourage, which was ready, at the first sight of someone approaching, to whisk her off to a waiting chauffeured car.

Even taking her to the university campus on weekends had become more risky. Washoe was banned from visiting the psychology building, because the psych professors working overtime were upset at having to shut their office doors to keep her from grabbing their coffee mugs and sodas. So I started taking her to explore the biology building instead, where her favorite game was to run down the hall at top speed, trying to open every single door, and stopping only to assault the candy and soda machines. Nothing ever came out of these machines, but she always gave it the old college try.

On Sundays only two types of rooms were open, the labs and the bathrooms. One particular weekend, Washoe managed to find both. The lab belonged to a devoted researcher in a white lab coat. When Washoe burst in he stepped forward heroically and defended his test tubes and beakers. But as I chased Washoe around and around the lab table, he flattened himself against the wall, watching the scene in horror. Since I wasn't allowed to use English around Washoe, all I could do was smile and wave to him as I ran out of the room in pursuit of the wild chimp.

Next stop was the men's room, where she loved to get up a head of steam, storm through the swinging door, and dive onto the tile floor, sliding on her sweatshirted stomach underneath all three stalls until she popped out the last one. But this time she only made it to the second stall before I heard someone scream: "MY GOD IT'S A GORILLA!" Washoe flew out of the stall and right into my arms, and out we ran. The next day I checked the papers to see if anyone had a heart attack in the biology building, but luckily there was nothing.

All of these incidents began to add up and helped convince the Gardners that Washoe needed a new home. In addition, all of the research associates were preparing to leave. Greg was nearly finished with his graduate work and had lined up a job elsewhere. Susan was getting married and wanted to raise a family of her own. I was also leaving in the fall. The Gardners would have to break in a whole new family for Washoe, something she might well resist.

The most recent graduate recruits did not seem to have the same devotion to the project that Greg, Susan, and I had. Some of them came in with an attitude of superiority toward Washoe. They acted as if they were there to "teach the young chimp sign language," not realizing that she knew a lot more ASL than they did. Washoe was pretty good at humbling these uppity types. She would walk right up to them and begin signing in an extremely methodical and exaggerated fashion, like someone talking English very slowly and loudly to a foreigner. Some of them didn't even come back.

So the Gardners made their decision. Project Washoe was over, at least in its present incarnation. They found a new home for Washoe at the Institute for Primate Studies in Oklahoma. Washoe could be accompanied by a member of her foster family—namely, me—and her language could continue to be studied in a pleasant environment. The Institute was located in what sounded like an idyllic rural setting with lots of trees, a pond with three islands, plus housing for about twenty chimps and other primates. It was all run by a clinical psychologist named Dr. William Lemmon who was studying the maternal behavior of chimpanzees. He told the Gardners that the chimps were an extension of his own family.

The Institute was affiliated with the University of Oklahoma, where Lemmon taught. I would begin as a Visiting Assistant Professor and Research Associate on grant money. If all went according to plan, I would eventually get a permanent teaching position. It sounded too good to be true. I would have a teaching job with a real salary. We could build a new home for our growing family, and Debbi could go to graduate school in Oklahoma when she was ready. I would work with Washoe and a whole new group of chimpanzees—a very exciting prospect because I could discover whether or not other chimpanzees were capable of learning ASL.

Mind you, Allen Gardner never asked me, "Roger, *will* you go?" He didn't suggest that I visit Oklahoma or meet William Lemmon. It was clear that Washoe's needs came first, and that I was going with *her;* she wasn't going with me. Gardner had negotiated the deal, given the order, and, like a good soldier, I obeyed. I didn't dare question my major professor even when he was planning my life. Allen Gardner told me I was going to Oklahoma with Washoe and so I was going to Oklahoma with Washoe.

I wasn't blind to what this meant, however. My responsibility for Washoe was about to grow exponentially. Up to now I had been just one part-time member of her foster family. But now the Gardners wanted Washoe to leave her foster parents—surely a traumatic event for any five-year-old—and live with one of her siblings. Washoe and I were great playmates and our attachment was obvious, but I was *not* her mother, and certainly not the beloved matriarchal figure Trixie had been since Washoe's infancy. If the Gardners were ready to shut down the project because, among other things, Washoe would have a hard time adjusting to new graduate students, why were they sending her away to adjust to an entirely new *life*? The solution seemed worse than the problem, at least in my eyes.

But the Gardners kept their motives to themselves. It never occurred to me to ask why *they* weren't going to Oklahoma with Washoe. "Washoe," Allen Gardner would always say, "belongs to science." It was hard not to feel that this lofty sentiment was, on some level, a convenient way of evading the responsibilities of foster parenthood. But Washoe had never been a full-fledged Gardner family member; she was part stepchild and part research subject, and now she was being shipped off to live with another scientist and part-time family member. Washoe had already suffered more upheaval and loss in five years than most people get in a lifetime. Did she really need more? I would go with Washoe, for her sake and mine, but the whole plan troubled me.

Our baby was due in July so Debbi and I began preparing for an August move. In addition to finishing my dissertation and working shifts with Washoe, I had to take a crash course in Spanish because I couldn't graduate without showing proficiency in a second language. My fluency in ASL did not meet the requirement, though today it does at many colleges. Meanwhile I was learning why so few people have gotten Ph.D.s with Allen Gardner. I was on the seventh rewrite of my dissertation but it still didn't meet his Olympian standards.

Somehow I finished my dissertation and passed my Spanish course. Our daughter Rachel was born on July 22, and two weeks later Debbi and the kids got on a plane and flew to Oklahoma. I drove a U-Haul filled with our belongings fifteen hundred miles to Oklahoma, then immediately turned around and went back to Reno to face my "orals," the ritual inquisition of a Ph.D. candidate.

Doctorates are not earned, they are conferred. I had to face a committee of seven psychologists and linguists, including the Gardners, who fired questions at me about the use of guidance in Washoe's signing. For several hours I answered, quite confidently, until my overtaxed brain began to shut down. One professor asked me a fairly straightforward question, but in my haze I couldn't seem to grasp what he was asking. Finally, he turned to Allen Gardner and said, "He's *your* student. Maybe *you* can get him to understand the question." Gardner glared at me and banged his fist loudly on the table. "Fouts, if you don't answer this question in the next ten minutes, we're going to repeat this entire exam again tomorrow." Terror stricken, I blurted out the correct answer.

The seven inquisitors had achieved their goal. Orals are not complete until there's a little blood on the floor. I had bled, and the guardians of Psychology—the dragons at the gate, as they called themselves—now conferred their Ph.D. upon me.

It took years for me to get over Allen Gardner's domineering style and his determination to "temper the metal" of graduate students. Even after I received my doctorate I continued to address him as "Dr. Gardner." Then one day in 1971 he addressed me as "Dr. Fouts," as if to let me know that we were now peers. But I was still too intimidated to call him "Allen," and for many years I didn't call him anything at all. After teaching for several years I grew to realize that Allen Gardner had tempered my metal after all. When I'd arrived in Reno I was a soft thinker without much knowledge of scientific method. Allen made me a stickler for proper experimental design, procedure, and reporting.

In hindsight, it is obvious that Allen Gardner was a godsend for a revolutionary experiment like Project Washoe. Thanks to his precision, the project's data will stand up forever under the harshest scrutiny. In the hands of a lesser scientist, the outcome might have been disastrous. Allen Gardner prepared me for the rough and tumble of scientific discourse. After training with him I could tackle almost any challenge in the behavioral sciences, and there was no challenge bigger and more controversial than ape language study.

In late September 1970 I packed my bags and prepared to leave with Washoe for Norman, Oklahoma. My family's exit from Reno seemed as unlikely as our entrance. Washoe had gotten me into graduate school, and now she had gotten me my first teaching job. Debbi and I had arrived in Reno three years earlier with one infant, and now we were leaving with *three* children, one of them a talking chimpanzee.

All of our plans—to take turns going to graduate school, to have children afterward, to work with special kids—seemed like another couple's fantasy now. We decided to stop pretending we were in charge. A definite pattern was emerging: Washoe kept opening doors, and we kept walking through them.

PART TWO

STRANGERS
IN A STRANGE LAND

NORMAN, OKLAHOMA: 1970-1980

Speak and I shall baptize thee.
—Cardinal of Polignac to a chimpanzee, early 1700s

It's no wonder that these animals, when confronted with the prospect of salvation, enslavement or culture, wisely pretend to be mute.
—Jean-Jacques Rousseau, 1766

SIX

THE ISLAND OF DR. LEMMON

THE GARDNERS CHARTERED THE PRIVATE JET of Bill Lear, who owned Lear Jet and was a resident of Reno, to take Washoe and me from Nevada to Oklahoma. We didn't tell Washoe anything about the move because we were scared she wouldn't want to go. It was impossible to know whether Washoe, who was then five years old, could even grasp what leaving home meant. She hadn't been out of Reno in four years. Her idea of a big trip was a car ride to the biology building.

The day of the move, October 1, 1970, did not begin well. My first job was tranquilizing Washoe to keep her from panicking on the airplane and running amok in the cabin. I had two plans for giving her a powerful animal sedative called Sernalyn. Plan A was the stealth approach: I would spike a cup of Coca-Cola, Washoe's favorite drink, and let her have it with breakfast. She had never once refused a Coke.

As I handed her the cup, I signed SWEET DRINK and threw in a few chimpanzee food barks. Washoe wasn't buying it. Coke on the breakfast table made her extremely suspicious. She studied it closely, then lifted the cup and sniffed it warily, like she was a millionairess in a murder mystery who thought her own children were about to poison her and steal all her money. Finally she set down the cup as if to say NO THANKS.

Time for Plan B. I called in Linn Anderson, a graduate student and former football lineman who weighed at least 250 pounds. Linn would get Washoe involved in a tickling-wrestling match, then pin her for five seconds. That would give me enough time to jab her leg with a hypodermic. Linn was five times bigger than Washoe, and he managed to pin her on the bed for about two seconds. But as soon as Washoe saw me approaching, needle in hand, she bench-pressed Linn off her chest and hurled him across the room like a ten-pound barbell. As Linn sailed past me, I stuck Washoe with the needle, and she began to grow groggy.

Washoe had to be out cold before I could carry her on the jet because Bill Lear's pilot was already spooked about his rather unusual passenger. The Gardners had told him about the "friendly baby chimp," but apparently he'd started talking to other pilots, and they'd scared the daylights out of him with their own war stories about animal passengers. One pilot had told him about a horse that had broken loose and battered the plane's walls, desperately trying to escape. The pilot had bolted from the cockpit, grabbed an ax, and killed the horse with several blows to the head.

That anecdote weighed on my mind as I boarded the jet, and I made a point of showing our pilot that Washoe was harmless and sound asleep in my arms. But he was leaving nothing to chance: behind his seat was a long ax handle that scared me out of my wits. As I took my seat and propped Washoe up beside me, I tried to think calmly and rationally. The key was keeping her heavily sedated. The vet had told me that it is impossible to overdose on Sernalyn, so every time Washoe so much as twitched I jabbed her with another dose. (Sernalyn, which is now illegal and known in the drug trade as PCP, angel dust, or rocket fuel, causes no apparent side effects in chimpanzees but induced a violent paranoia when tested on humans in the 1960s.) By the time we got off the plane in Oklahoma, Washoe was still completely out cold.

On the tarmac we were greeted by Dr. William Lemmon, the Director of the Institute for Primate Studies. He was a large, big-bellied man in his mid-fifties with a shaven head, bushy white eyebrows that twirled up on his forehead, and a white goatee. He was wearing a single-piece white jumpsuit that was zippered up the front, sandals, and no socks. The overall look was a lot more intimidating than the friendly clinical psychologist I'd been told to expect.

I carried Washoe into the backseat of his Mercedes diesel and we headed off for the Institute, about five miles outside Norman. As we drove along, the adrenaline I'd been running on all day began receding, and the reality of my situation began creeping in. Yes, I'd accomplished my main mission, and Washoe was in Oklahoma. But now what? The Gardners' instructions only went as far as "Bill Lemmon will pick you up at the airport." After that, nothing.

Questions began racing through my head. What would Washoe's living arrangements be like at the Institute? (The Gardners had sent Bill Lemmon money to build a special enclosure for Washoe, but that's all I knew.) How would Washoe relate to other chimpanzees? She hadn't seen another chimp since she was a baby at Holloman Air Force Base. What would happen when she started signing to other chimps and humans and they didn't respond? How would she react when people spoke to her in English, a language she'd never heard and wouldn't comprehend? What would I say if Washoe asked to see Trixie? Or even worse, if she signed YOU ME GO HOME NOW! Not only would I have to be Washoe's guardian but I was now her translator and child psychologist as well.

As we pulled into the gravel driveway that led onto the Institute's grounds, I could see tall grass growing along a dry streambed and giant cottonwood trees shading a large pond behind Lemmon's pink farmhouse. There were three islands in the pond and monkeys were darting through the trees on one of them. This parklike environment was a far cry from the Gardners' suburban backyard. It looked like a good place for Washoe to grow up, like a nature reserve for primates.

Then we pulled up to a concrete building and Lemmon got out. "This is the main chimp colony," he said. "Washoe will be staying here."

Washoe was still asleep as I carried her inside, and we were immediately greeted by twenty adult chimpanzees, screaming their unhappiness at having two strangers trespass on their territory. The building was about forty feet by forty feet, and the chimps were housed together in a series of seven cages that were separated by heavy-gauge chain link but connected to each other by tunnels. The individual cages could be sealed off from the others with sliding doors. At the end of the walkway was a metal stairway leading upstairs to a catwalk that ran right over a connecting tunnel.

As the big chimps continued screaming and rattling their cages, I followed Lemmon to an empty corner cage that had been sealed off from the others. It was seven feet by ten, completely bare, and the cages on both sides of it were filled with some very curious and angry-looking chimpanzees.

"Put her in there," Lemmon said, pointing to the empty cage.

I was stunned. Where was the special enclosure he was supposed to be building for Washoe? Did he really expect her, after four years cuddling with her dolly in a real bed, to sleep in a concrete cell surrounded by all this noise and aggression? Surely there was some mistake. I wanted to protest but, having been there only five minutes, I thought better of it. I was on Lemmon's turf now. He made the rules, and the Gardners had told me to follow them. They had assured me that Bill Lemmon was a very good scientist.

I carried Washoe into the cage and laid her favorite security blanket on the floor before placing her on top of it.

"No blankets!" Lemmon snapped at me. Now my impulse to protect Washoe overcame my caution, and I snapped back at him.

"You can't make her sleep on a bare concrete floor!"

"Look, I don't care what you did in Reno," Lemmon responded. "I'm going to teach Washoe to be a chimp."

Evidently, Lemmon thought Washoe was a spoiled little brat. Even worse, in his eyes, she saw herself as a *human* brat. And his job was to disabuse her of that notion in the most traumatic way possible. I was equally determined to defend her. After a heated argument, Lemmon agreed that Washoe could have a single blanket.

As soon as that was settled, we had another dispute. When Lemmon ordered me to leave Washoe alone in the cage, I refused. After butting heads for another ten minutes we compromised: I would be allowed to stay outside the cage until Washoe woke up, and then I would leave the building. We both left the cage, and Lemmon slammed the steel door shut and locked it.

Then I waited. Six hours passed before Washoe finally opened her eyes. As she slowly began coming to, she struggled to her feet and collapsed in a heap. Then she got up again only to fall down one more time. Finally she managed to stay up long enough to stagger around the cage, completely disoriented.

WASHOE I'M HERE, I signed through the chain link. COME HUG. She managed to stumble in my direction on her hands and feet. When she reached me she slumped against the cage, and I groomed her through the chain link. The other chimps had been quiet for hours. But now they were on their feet and hopping mad—screaming and banging on their iron cage doors, and glowering at us from just a few feet away.

Their display of intimidation terrified me, and my so-called experience with chimpanzees suddenly seemed laughable: I knew one chimpanzee who thought she was human. *So these are chimpanzees*, I thought to myself. *What have I gotten myself into?*

Meanwhile, Washoe watched all of this through a drugged haze, and it must have been truly nightmarish for her. Imagine never meeting a member of your own species until you were five years old. Washoe's day as America's most famous chimpanzee had started like any other, with a civilized breakfast served by her foster family in her personal trailer. The next thing she knows she wakes up in a dimly lit jail cell surrounded by a band of very hairy, violent animals.

WHAT THEY? I signed to her, pointing at the crowd of on-lookers.

BLACK BUGS, she responded. Washoe loved to squish black bugs. They were the lowest form of life, as far beneath human—and therefore herself—as anything she could think of. Along with everything else she had learned from her foster family, Washoe had apparently absorbed the lesson of human superiority.

Then, for the first time in our three years together, I whispered to her in English, even though I knew she wouldn't understand a word of it. "Washoe, we're not in Reno anymore. . . . We're not in Reno anymore."

THE NEXT MORNING I VISITED WASHOE in the main colony. The building was quiet when I arrived, but the biggest males began displaying as soon as I came through the door, screaming and shaking their cages. I edged along the far wall toward Washoe's cage and greeted her. She was ecstatic to see me and she kissed me through the chain link. She kept signing ROGER ME OUT. YOU ME OUT. The other chimps were making a racket. I was terrified but tried to stay calm for Washoe's sake. But I soon noticed that she wasn't at all intimidated by the other chimps. At one point she even stood up on both feet and threatened them.

A few minutes later the lab techs came in and served breakfast, which was a meatloaf-looking dish made of meat, carrots, and grains. Washoe ate a long bar of it hungrily. Then the techs came back to clean the cages. They kept yelling at Washoe to "move, move." But she didn't understand them and didn't budge.

"She's not disobeying you," I told them. "She just doesn't speak your language."

I signed to her PLEASE MOVE OVER, THEY CLEAN and Washoe moved to the side. After they finished cleaning, I taught the techs how to sign WASHOE PLEASE MOVE. They were happy to learn anything that would make their job a little easier.

After breakfast I got permission from Lemmon to take Washoe for a walk, but I had to follow his two rules for taking chimps out of cages. First, I had to carry a cattle prod at all times, in case Washoe had to be "disciplined" with an electric shock. Second, she had to wear a padlocked collar around her neck, which was connected to a twenty-foot lead, a kind of leash. This device was ludicrous, I pointed out to Lemmon, because there was no way I could restrain Washoe with a lead. He told me that the chain collar had a symbolic purpose. "We follow the Ivanhoe tradition," he explained. "Chains remind the chimps of their servitude."

Slave or no slave, Washoe was thrilled to be sprung from her cage. As we rambled through the woods she kept pointing and signing at so many wonderful things all in one place: TREE, BIRD, COW! When her thirty minutes of freedom were up, we walked slowly back to the main colony. As I locked her in her cage and signed GOOD-BYE, my heart sank. After that, I thought of little else but how to get Washoe out of her jail cell.

I ARRIVED AT THE INSTITUTE IN AWE of Dr. William Lemmon's reputation. He was, by all accounts, the most influential psychotherapist in the state. For two decades he had directed the doctoral program in clinical psychology at the University of Oklahoma and was therapist to most of the department's Ph.D. candidates. Once some of them graduated, they joined Dr. Lemmon on the psychology faculty and remained his longtime patients. Dozens of his other student-patients took on positions of power in the state agencies for mental health, prisons, family services, and veterans affairs. This network of former students and patients gave Lemmon an exceptional amount of power in the state establishment.

Lemmon was a natural politician who knew how to charm journalists with quotable wisdom. He was also good at impressing visiting scientists, like the Gardners, with his mastery of a broad range of topics from psychoanalytic theory to the mating habits of the woolly monkey. I went into my assignment with Dr. Lemmon thinking that anyone who could impress the Gardners must be truly brilliant.

It only took me about a week to discover that the legend of William Lemmon was wrapped up in good deal of controversy. Depending on which professor I asked, the powerful Doctor was either an infallible and benevolent father figure, or he was an arrogant, Machiavellian opportunist. His detractors thought his loyalists were mindless sheep, and the loyalists described Lemmon's enemies as "heretics." All of these opinions were voiced in hushed tones as if spies might be lurking, and adding to the drama were stories of suicides and attempted homicides among the faculty.

Was William Lemmon a genius or a charlatan? Healer or manipulator? Or was he all of these things? The only thing everyone seemed to agree on was that the war over William Lemmon had engulfed and destroyed the graduate program in clinical psychology. Two years earlier, the university had brought in a new chairman with orders to terminate the clinical program. Lemmon was overthrown, and the department was taken over by experimental psychologists. Lemmon had secluded himself at his Institute for Primate Studies outside Norman, where he ran a nearly full-time psychotherapy practice. But now he was blending that practice with a rather bizarre plan for studying the maternal behavior of chimpanzees.

Looking back, I wandered into the Institute just as the saga of Dr. Lemmon was beginning to resemble *The Island of Dr. Moreau*, the famous H. G. Wells novel about a brilliant but megalomaniacal scientist who holes up on a Pacific island so he can carry out controversial experiments on animals. Like the ostracized Dr. Moreau, Dr. Lemmon was drummed out of academia by his colleagues and took up residence on the scientific fringe. Dr. Moreau had a motto that I heard almost verbatim from Dr. Lemmon: "Men of vision become outcasts."

Dr. Moreau practiced biological alchemy, transforming animals into near-human creatures by means of genetic engineering. Lemmon was attempting the same transformation, only he was using the slightly different means of cross-fostering. Both doctors were obsessed with shaping the destiny of their research subjects. And both took on the vengeful persona of an Old Testament God in meting out justice to those subjects who disobeyed them.

Lemmon's menagerie was no less impressive than Dr. Moreau's—only less fictional. The main chimpanzee quarters, where Washoe was caged, were originally built for parrots and macaws, and its drains were so small they were always clogging up. Cattle, peacocks, and African guinea hens roamed the Institute's 160 acres. There were monkeys living on one of the islands in the pond, and a family of gibbons (the "lesser apes" of Asia) lived on another. They could be seen swooping, hand over hand, through the cottonwoods at dazzling speed. The third island was a floating playground for a group of juvenile chimpanzees.

Lemmon's study of animal behavior emerged quite incidentally when he began buying exotic pets for his children in the 1950s. First he investigated social behavior in ducks, pigeons, fish, sheep, goats, and dogs. Then he turned to woolly monkeys and gibbons. Whenever he bought a new species, Lemmon would discard his old subjects or put them out to pasture. One day his house was filled with plants; the next day they were gone and the walls were lined with eighty-gallon aquariums filled with giant, bulging-eyed goldfish.

His specialty was studying young animals that had been isolated from their mothers. Lemmon would observe lambs that were separated from their mothers for the first ten hours of life (they never recovered) and Border collies kept in total isolation for nearly two years (they started herding sheep within six hours of being released). In the late 1950s Lemmon became convinced that maternal behavior in primates is innate, and that behaviorists like B. F. Skinner, who argued that it was learned, were wrong. Lemmon concocted his ambitious cross-fostering experiment to prove his theory.

Over a period of years Dr. Lemmon took female chimpanzee infants from their mothers at birth and farmed them out to human families who would raise them as human children away from other chimps—just as we'd done with Washoe. But that's where the similarity with Project Washoe ended. Lemmon was assigning baby chimpanzees to his psychotherapy patients.

"In our experience," the Doctor advised, "an adopted chimpanzee infant is clearly not the solution of choice for a precarious marriage, but may be of some therapeutic value to the potential human mother who, for whatever reason, is uncertain of her maternal competence." As soon as chimpanzees were born in the Institute, Lemmon whisked them off to his waiting patients. Once these cross-fostered females grew up, Lemmon planned to inseminate them artificially (with sperm from his adult male chimps) and then study how they nurtured their offspring. If they mothered their babies the way wild chimps do, that would show that maternal behavior was prewired into apes and probably into humans as well.

This inter-species experiment struck many people as scientifically far-fetched and ethically dubious. Meanwhile, the atmosphere at the Institute, with its grotesque code of law and order, was even more bizarre. Lemmon swore he loved the chimps like his own children, and he had begun his colony by raising a pair of wild-caught infant chimpanzees, Pan and Wendy, in his own home. But Lemmon's idea of fatherhood, at least when it came to chimpanzees, seemed to involve a good deal of corporal punishment.

He practiced what he called the "two-by-four" method of chimpanzee control, beating them to instill respect for his dominance. He told me he picked this up from circus people and animal trainers. But Lemmon's favorite weapon was an electric cattle prod that terrified the chimps. He was especially fond of zapping them for no reason other than to make them fear his unpredictability and omnipotence. They were always cowering in his presence, knowing he could strike at any time.

The chimps knew better than to ever cross Lemmon. Pan, Lemmon's "foster son" and favorite chimp, once made the mistake of displaying and spitting at Lemmon. According to the lab techs, Lemmon went into his house and came back with a hand-pump pellet gun. He pumped several times and shot Pan with a good-sized pellet. Pan screamed but refused to submit. Lemmon reloaded, pumped the barrel, and shot him again. It took several rounds before Pan finally gave in and threw himself on the floor. Lemmon ordered Pan to present himself spread-eagle on the cage wire. Then Lemmon took a long folding fishing knife from his pocket and dug the pellets out of Pan's skin.

The other chimps were absolutely terrified by Lemmon's dominance of Pan, the powerful male who dominated them. When Lemmon entered the chimps' building, many of them would approach their human master in a chimpanzee posture of submission, bending low to the ground while extending a wrist limply and pulling back their lips to show all their teeth in a grimace of fear. Lemmon would then place his hand against the cage and the chimps would kiss his large silver ring. The ring was a coiled snake with two eyes made of large rubies.

Most of the adult chimps had been confined, beaten, and humiliated for their entire lives—before and after coming to the Institute. And like the shipwrecked visitor to Dr. Moreau's island, I immediately felt pity and compassion for them. During my first week, I encountered one of the colony's most fearsome chimps, a mean-tempered male Lemmon had named Satan. Satan had picked all the hair off his face and arms, which definitely made him look menacing. One morning I passed his cage, and he began displaying at me. Then he scooped up a pile of his feces and hurled it, scoring a direct hit on my chest. Lemmon had ordered me to assert my own dominance in moments like this by threatening Satan with the cattle prod or pellet gun.

But I figured that Satan's miserable life had given him a damned good reason to hate people. "Don't flinch," I said to myself. "Don't flinch." Satan continued pelting me with feces until the wet slime was running down my face and clothes. The stench was almost unbearable. When Satan had emptied out his cage, he went over to the water spigot in his cage, filled his mouth, and drenched me from ten feet away. The spray helped clean me off, and I let him shower me over and over again until I was completely clean. Satan was dumbfounded by my submission. He came right up to the cage and began hooting to me softly, as chimps do when they greet one another. I hooted back. From then on, Satan and I were good friends, and he never bothered me again.

Lemmon mocked my attempts to befriend the chimps. "One of these days," he said, "you're gonna get your well-intentioned face bitten off." He saw everyone at the Institute, humans and chimps, as part of a single primate hierarchy that he ruled over as the undisputed alpha-male. His workers referred to him as Mount Olympus, and he acted the part. He expected me to dominate those below me, and he assumed that I would be looking for any opportunity to challenge and overthrow him.

At the time, it never occurred to me that Lemmon might consider me a threat. I was only twenty-seven years old, a brand-new Ph.D. just setting out on my career. Lemmon had single-handedly built the psychology department where I taught and he directed the Institute where I worked. He was Oklahoma's most famous—some would say, notorious—psychologist. I was the new kid on the block and, like so many of my colleagues, scared to death of Lemmon.

Looking back, it's perfectly clear why Lemmon had reason to fear. I may have been young and inexperienced, but I was also the guardian of the scientific world's most famous chimpanzee. Project Washoe could very easily have upstaged Lemmon's own research. So he was determined to assume control of its chimpanzee star. That's why he threw Washoe into the main colony, instead of building the special enclosure he'd promised the Gardners.

But Washoe refused to be broken. Her dislike for Lemmon was more obvious than my own. She blatantly ignored his commands, even when I translated them for her. Washoe was accustomed to giving and getting respect. If Lemmon didn't respect her, she wasn't about to respect him.

About a week after we arrived, Lemmon ordered his technicians to put Washoe in the group cage. The techs balked, telling him that Pan and the other male chimps would probably tear Washoe apart, which of course was what Lemmon intended. He insisted. The techs opened the sliding door that separated Washoe from the other chimps and watched as Pan displayed before attacking Washoe. But then something unexpected happened. The adult females stepped in front of Washoe, and when Pan tried to attack, they turned on him and drove him off. Washoe had formed some kind of alliance with the older females from the solitude of her cage. Lemmon's plan had backfired, and his alpha-male was humiliated.

Lemmon immediately came up with a new plan to circumvent Washoe's female allies. He ordered the techs to put Pan and Washoe in a cage by themselves against the back wall. But Pan and Washoe just ignored each other, and each sat in a corner, acting as if the other didn't exist. Evidently, Pan had learned his lesson. He wasn't about to incite the females again; he would have to face them later.

When I heard about these attempted attacks on Washoe, a mere seven days after we had arrived, I finally had to admit I was in a dangerous situation way over my head. I had come to Oklahoma to teach at a college, but I suddenly found myself in a life or death struggle with an adversary unlike any I had ever encountered. I desperately needed help from someone outside of Oklahoma, so I called Allen Gardner.

"We sent you there because you get along so well with everybody, Roger," Allen told me after listening to my tale of horrors. "If anybody can get along with Lemmon it's you."

The Gardners seemed to know more about Lemmon than they had let on before. The details of Washoe's incarceration didn't seem to faze them. Or maybe they just didn't believe me. After all, who in their right mind *would* believe a story about a scientific institute where chimpanzees named Satan wore neck chains and kissed the coiled-snake ring of their lord and master?

"Make the best of it, Roger," Allen Gardner suggested feebly.

Maybe the Gardners did complain to Lemmon, but if so I never heard about it. I got the feeling that Allen and Trixie did not want to make waves in Oklahoma. Any blowup with Lemmon might land Washoe back on their doorstep in Reno. I suddenly felt foolish that I had trusted the Gardners so completely that I hadn't visited the Institute beforehand or interviewed with Lemmon or even talked with him on the telephone.

It was now clear that no one was going to stand by me against Lemmon, but I wasn't about to bolt his penal colony and leave Washoe there alone. I was on my own. Maybe it was my survival instinct, or maybe I just grew up fast, but the next day I did what I had to do: I cut a deal with Lemmon. He agreed to spring Washoe from her cage in the main colony and let her join the young chimps who spent their days on one of the man-made islands in the pond. In return I agreed to share credit with him—in the scientific press and on federal grant applications—for all of the ape language studies I would conduct. That research would involve not only Washoe and the four young chimps on his island but the other chimps being raised by his psychotherapy patients.

As threatened as Lemmon felt by Washoe and me, he needed our research, our grant money, and our media attention. And we needed his Institute because we had nowhere else to go. I acknowledged his authority and agreed to play by his rules. But I would run the chimpanzee island, and he would stay out of my way. Our truce would be tested on an almost weekly basis, but it would hold for nearly nine long years.

EARLY ONE MORNING during the second week of October, Lemmon's techs let Washoe out of her cage, and I walked her down to the shore of the pond. We got in a rowboat and cast off for a new life on the chimpanzee island. Chimpanzees have a natural fear of water—they sink like stones because they don't have enough body fat to swim—so an island is the perfect way to contain them without using bars or cages. This particular island was hardly worthy of a picture postcard. It was man-made and was about one quarter acre in size. It was covered in Oklahoma red dirt and scrub, and the few remaining scrub oaks were dead, having been stripped bare by the rambunctious chimps. Here and there were a few tall poles, which the younger chimps used to perch and survey the gibbons and monkeys on the other islands. All in all, it was a bleak, postnuclear landscape, like a stage set left over from a Samuel Beckett play.

But it felt like paradise. Washoe had her freedom at last, and she was forced to find her place among the small tribe of chimpanzee orphans who lived on the island. For me, the tiny island became an emotional and scientific oasis. Every day, when I finished teaching at the university and settled whatever dispute I was having with Lemmon, I would cast off in my rowboat, then step ashore on the island and enter a world that was as new and exciting as my very first days in the Gardners' backyard.

Knowing one chimpanzee is like knowing one child; it makes you want to meet more of these extraordinary creatures. Washoe's four new playmates had four completely different personalities, and their very different minds would profoundly shape my understanding of chimpanzees and language. The two girls were named Thelma and Cindy, and the two boys were Booee and Bruno. All of them had landed on the island sometime within the previous year. Three-year-old Thelma and four-year-old Cindy were both "Peace Corps rejects." They'd been born

in the African jungle but were carried back to the States by two
different Peace Corps volunteers who quickly discovered that
their families weren't as enthusiastic about raising these hairy
babies as they'd thought they would be. Soon after that Thelma
and Cindy wound up in Lemmon's orphanage.

Thelma was a loner with a stubborn streak a mile long. She
did things her way or not at all. She was extremely bright, but
she was also a daydreamer who drifted off into her own quiet
thoughts for long periods of time. Cindy, on the other hand,
was a homely chimp with a flat, freckled face and a bottomless
hunger for approval. We called her "Poor Pitiful Pearl" after a
disheveled doll Debbi had loved as a girl. Cindy would act like
a little lost puppy, tagging along after Thelma and never making
a fuss about anything. She was so passive that it was almost
aggressive. That pathetic face would work on you until you gave
her what she wanted, which usually was being told, in English,
what a good girl she was.

Bruno, the offspring of Pan and a female named Pampy, was
born at the Institute two and a half years earlier—in February
1968. Lemmon took Bruno away from his mother soon after the
birth and turned him over to Dr. Herbert Terrace, a psychology
professor at Columbia University in New York. Terrace flew
Bruno to New York as a kind of test run for a chimpanzee language experiment that he was planning. Terrace wanted to determine whether or not a chimpanzee, born to live in a warm
climate, could survive a cold New York City winter. A year
later, having served his scientific purpose—he survived the winter and bonded with a human family—sixteen-month-old Bruno
was returned to Oklahoma. (The actual sign language experiment went forward in the 1970s with Bruno's half brother,
Nim.)

Bruno was a tough-guy version of the aloof and headstrong
Thelma, but he was probably even more defiant. He loved to
challenge authority as much as Washoe did, but he had none
of her easy charm. Bruno was hard to get to know, and he had

little interest in cultivating friends. He would sit off on his own, content as could be, needing no one. My name for him was signed by thumping the thumb on the chest, meaning, BRUNO PROUD. Bruno's only real friend and hanger-on was Booee, a very sweet three-year-old he easily dominated. Booee was beloved by everyone at the Institute and was probably the best-natured chimpanzee I've ever known. He could be cajoled into anything and was always ready to sell his soul for a raisin.

Booee was born in 1967 at the National Institutes of Health research facility in Bethesda, Maryland. The staff was unaware that his mother was pregnant—a common oversight in labs—so Booee was an unexpected arrival and not slated for a specific biomedical study. But when he was just a few days old he convulsed, causing the researchers to speculate that he might be epileptic. That was the only excuse the NIH surgeons needed to subject Booee to the latest experimental treatment for grand mal seizures: a split brain operation. Doctors opened Booee's skull and severed his corpus callosum, cutting all the connections between his two cerebral hemispheres. Booee was left, in effect, with two separate brains. The postoperative swelling was so extensive that surgeons had to reopen Booee's skull to relieve the pressure on his cranium.

Finally a doctor at NIH named Fred Schneider took pity on Booee, who was in agonizing pain, and brought him home. There, Dr. Schneider's wife, Maria Schneider, and their six children nursed Booee back to health. Then Booee got lucky again because no one at NIH noticed that he was gone. The doctor who was supposed to do the follow-up study on Booee's brain became ill, and Booee fell through the cracks at NIH until finally he became a full-fledged member of the Schneider family.

Booee, like Washoe, soon got too big to live in a house built for humans. He was defending his new family's territory against dogs and strangers by breaking the living-room picture window. The Schneiders felt that sending Booee back to NIH and a lifetime of biomedical experiments was out of the question. In early 1970, Dr. Schneider flew to Reno for advice from the Gardners, who suggested that Booee be sent to Lemmon's Institute.

Booee had arrived on the chimp island only a few months before Washoe and I did. Perhaps it was his years with such a loving family or maybe it was just his personality, but Booee was every bit as sociable as Washoe. As for his split brain surgery, there were only two enduring effects I could ever detect. When Booee played piggyback, he would tell me where to go by pointing in two different directions at the same time. I would stand there not knowing where to go until finally Booee would swing one arm around so that they were both pointing in the same direction. Likewise, when he drew or painted he always did so in two opposite corners of a page. I created a sign for Booee's name by drawing the index finger over the top of the head from back to front, meaning—what else?—BOOEE SPLIT BRAIN.

Washoe and I had landed on this island of chimpanzee orphans: there was stubborn Thelma, poor pitiful Cindy, proud Bruno, and sweet two-brained Booee. Marooned on an island outside of adult chimpanzee society, these youngsters had created their own working hierarchy. Thelma bossed Cindy, and Bruno bossed Booee. Thelma and Bruno stayed out of each other's way. I wasn't sure how Washoe would fit into this gang, and I had reason to be worried. Washoe had been the pampered baby in Reno, and she'd probably want the same treatment here in this group. What would I do if she refused to play with these hairy nonhumans at all?

But to my astonishment, from the moment Washoe stepped onto the island she acted like a foster mother to all four orphans. When Cindy was upset, Washoe would sign COME HUG, COME HUG. Cindy had no idea what Washoe was saying, but she got the message when Washoe groomed and comforted her. When Booee and Bruno had a fight, Washoe would step between them like a boxing referee and send them to their corners by signing YOU GO. Washoe even found a way to get aloof Thelma to play by signing TICKLE CHASE and then goading her into the game.

And she made up a new term of endearment for the very dark-skinned Thelma: BLACK WOMAN. Evidently, Washoe had decided that chimpanzees were not bugs after all but people.

Communication on the island resembled a primate Tower of Babel. Booee, Bruno, Thelma, and Cindy let each other know what they wanted through natural chimpanzee gesture, vocalization, and facial expression. For example, if Booee wanted Bruno to play he would make a play face, laughing and motioning to him. But Washoe would sign a more specific message like COME TICKLE CHASE. When the others didn't respond, she would sign again very slowly and emphatically, like a mother signing to a baby. When they still didn't understand, she would get her message across just like they did, by gesturing and vocalizing.

Washoe's friends had been raised in human homes so they understood a good deal of English. For example, I could say, "Move that tire," and they would do so. Washoe had never heard English, but she and I had always communicated with food grunts, screams, laughter, pant-hoots, and a lot of other meaningful vocal signals. Now that we were on the island, Washoe seemed to perceive English as an outgrowth of the vocal communication she was already familiar with, and in no time she could comprehend as much English as her friends. And the more time I spent with the other chimps, the more I mastered chimpanzee vocal communication. However, when Washoe and I talked we stuck mainly to ASL. Her command of ASL continued to grow. She no longer required any hands-on guidance at all, and could pick up a sign instantly just by watching me use it once. Her sentences now included as many as seven or eight signs.

It was not uncommon on the island to see a human-chimpanzee conversation that involved English, pant-hooting, ASL, and facial signals. In fact early on there was one episode that revealed both Washoe's feelings of responsibility for her new friends and their different modes of communication. One morning we were all playing in a grassy area by the shoreline.

There were many snakes around the island, some of them poisonous, so we were always on the lookout. Suddenly Thelma spotted a snake near the group and let out a long, drawn-out *wraaa*. The chimps and I headed to the other side of the island, all of us except Bruno, who hadn't moved from where we had been sitting in the grass.

As soon as Washoe saw that Bruno was in danger, she rushed back over there and signed COME HUG, COME HUG. He looked at her blank-faced and didn't respond. Washoe could have abandoned him right then but she decided to get her message across more directly. She ran up to him, putting herself in danger, grabbed him by the arm, and dragged him to safety.

Over those first few months, Washoe stopped behaving like a princess and became a doting big sister. It was a remarkable transformation. Raised as a deaf human child, Washoe was at last becoming a chimpanzee among chimpanzees. The young ones looked to her for leadership, reassurance, and protection. She was older and bigger, and this undoubtedly helped her win their respect and establish dominance. But she also seemed emotionally suited to the job. Having been parented by so many older siblings in Reno—Susan Nichols, Greg Gaustad, and me—Washoe knew how to be a big sister. It was as if all these "babies" triggered her latent parenting behavior.

During that first year, Washoe made another good friend in Debbi. They had met only a few times in Reno, but Debbi was a familiar face, and I was anxious for Washoe to have any continuity at all between her old life and her new one. I was also just desperate for human help. My hands were full, often literally, with teaching, research, and baby-sitting five chimps. Not only was Debbi great with kids, but she knew ASL and could communicate with Washoe and help teach the new chimps. I secretly hoped, too, that Debbi, as a grown woman, might fill some of the lonely space left by the terrible loss of Washoe's mother figures, Susan Nichols and Trixie Gardner.

Debbi came out to the island every day, and it wasn't long before she and Washoe were close pals. When Debbi was on the island, Washoe could step out of her new parental role and act like a kid again. It was not unusual in those days to see Washoe, all 80 pounds of her, riding piggyback on Debbi, who weighed no more than 115 pounds herself. At other times, our infant daughter Rachel was in Debbi's backpack, and three-year-old Joshua was playing and signing with Washoe, who seemed to love children no matter what species they were. Within a few months, the chimpanzee island began to look like an exceptionally manic day-care center staffed with a few students from my classes and supervised by Debbi. After teaching at the university in the mornings, I would head out to the island for the afternoon "tea party," a tradition that Washoe and I carried on from Reno.

Debbi played another vital role of getting the chimps to the island each morning. They spent their nights in the "pig barn." (Later that year, Lemmon agreed to build a "rundevaal," an African hut, on the island so the chimps could sleep there.) The pig barn was where Lemmon conducted two of his more gruesome biomedical experiments. He had about forty pigs divided into two groups on a split-level metal grid. The pigs on top received electric shocks, while the ones down below did not. After months of shocking, the pigs were then studied for heart disease. The barn was also home to eight siamang gibbons, Asian apes that form strong pair bonds. After the male and female siamangs paired off, Lemmon separated them and switched their mates. The siamangs developed gastrointestinal inflammation and, ultimately, died, confirming the well-known point that stress induces illness.

Each morning the five chimpanzees would greet Debbi with excited pant-hoots, and Washoe would sign OUT OUT. Then Debbi would lead them down to the shoreline, where they would pile into the rickety rowboat and head across the moat.

Getting the chimps off the island was a lot tougher than getting them on. At the first sign of the neck chains and leads, they knew it was time to return to the dreaded pig barn. Herding five rebellious chimps into a rowboat was like getting five

screaming kids into a station wagon for a trip to the dentist—
only the chimps are a lot stronger and a lot faster. Washoe
would take her lead off and throw it in the pond. Bruno would
run up a tall pole and refuse to come down, or else he'd taunt
me by getting in and out of the boat, making me chase him
around and around in a circle. Booee, Bruno's sidekick, would
stand on the sidelines, cheering on his hero and imitating his
most outrageous antics. I focused on wearing down the defiant
ones, Bruno and Thelma. If they caved in, then the others were
sure to fall in line.

One surefire way of getting all the chimps lined up at the
boat and *volunteering* to put their leads on was to promise a walk
in the woods. Accompanied by Debbi and one or two student
chaperones, each holding a lead, we explored the Institute's
farm. When fruit trees were in season the chimps loved foraging
among the wild plums and persimmons. They would find a tree
filled with fruit, pick an armful, and then eat to their heart's
content while they lay on the ground. Once their bellies were
full, they would tussle in the grass or sit together grooming one
another.

At times like this, they seemed so perfectly in tune with their
chimpanzee nature, so effortless and joyful in their play, that I
sometimes closed my eyes, listened to their laughter, and imagined we were in the deepest African rain forest. Washoe,
Thelma, and Cindy must have spent many long afternoons like
this, safe in their mothers' arms, entranced by the sights and
sounds of other creatures in the jungle. I could barely comprehend how different their lives must have been in those days.
The girls had awakened each dawn in soft treetop sleeping nests;
they had ventured down on their mothers' backs to forage for
breakfast; they had taken their very first baby steps on the lush
jungle floor.

Booee and Bruno had never known such simple times; they
were born in concrete laboratories. But like their foster sisters,
the two boys still had the mark of all chimpanzee children—
the white tail tuft that attracts every adult chimpanzee's uncon-

ditional affection and indulgence. It broke my heart that these five had so little of either.

Who was worse off? The girls, who had once known their jungle birthright but were torn away from it? Or the boys, who had never known their true home at all? Just asking this question jolted me back into the present. Kidnapped from Africa, abandoned by human families, they were all growing up much too fast and much too lonely. All they had was each other. They desperately needed a mother's love that no human could give them, certainly not in Lemmon's penal colony. They wore chains around their necks and slept in cages at night. As they grew older and bigger, they would become more independent and less able to adapt to human society. What in the world, I often wondered, would happen to them?

THE BETTER I GOT TO KNOW my new chimpanzee friends the more they taught me. One morning I had the two boys, Booee and Bruno, out for a stroll in the woods. They were on leads, and three of my undergrad students came with us. When it was time to go back to the island, Bruno was unusually cooperative and got right in the boat, but Booee did not want to leave the woods. He went up a tree and stood on a branch out of reach, refusing to budge. His sudden rebellious streak irked me.

Who does he think he is? I thought. *I've got a Ph.D. in psychology. My students are watching. I'll show this little chimp who's boss.*

"Come down now, Booee," I yelled up. Booee sat down on the branch, as if he were saying, "No way," like a kid who locks his bedroom door and won't come out.

"Get down here, Booee," I yelled even louder. No response. It was a standoff.

Now I was really getting embarrassed. I wrapped the twenty-foot lead around my arm so that only six feet of it was between me and Booee's neck. I gave the line a good tug to let Booee know that I really would pull him down.

That was my big mistake. Booee reached down, grabbed the lead with one arm, and lifted me clear off the ground—like a weight lifter curling 180 pounds with no discernible effort. I experienced a long moment of complete terror as I swung helplessly in the wind. My students took a step back. Then I gathered my wits, looked up at Booee, and in my sweetest voice said, "That's OK, Booee. No hard feelings." He instantly lowered me to the ground and let go of the lead. Then he stood and screamed. He obviously had been scared that I was so angry with him. The next thing I knew, Booee jumped out of the tree, straight into my arms. He wrapped himself around my neck and hugged me for a full minute—his way of making up after a fight. We were friends again.

This was my first lesson in the futility of getting into a battle of wills with a chimpanzee. Dominance that is based on physical power will almost always backfire. Human power leads to chimpanzee anger and aggression, which then leads to more human fear and violence. It is a cycle that can only escalate out of control. Lemmon's Institute was ample proof of that. First it was chains and leads. Then cattle prods. Then pellet guns. Later there would be electric fences and Doberman pinschers. Finally, Lemmon wanted us to carry loaded pistols, even with the juveniles. It was hard to tell who was more scared of whom: Lemmon of the chimps or the chimps of Lemmon. When a relationship is not built on mutual respect, the only way of maintaining control is through brutal force.

On the other hand, when there is a respectful relationship between human and chimp, there is no fear, and coercion is needed only on rare occasions. Washoe and Booee let me know early on that my arrogant way of ordering them around—"Do it my way or not at all"—wasn't going to fly with them. Washoe was no fool. She seemed to learn my teaching schedule in Oklahoma right away. If we were out on a walk, she would get me farther and farther from home just when I needed to be going back. I'd sign TIME GO HOME, and she would respond, NO, NO and turn her back. She knew she had me. What was I going to do? Pull on her lead when I knew she was eight times stronger than me? Threaten her with a cattle prod? Our conversations would go like this:

Roger (*anxiously looking at his watch*): YOU ME GO HOME
    NOW.
Washoe (*defiant*): NO.
Roger (*desperate*): WHAT YOU WANT?
Washoe (*matter of fact*): CANDY.
Roger (*very relieved*): OK. OK. YOU CAN HAVE CANDY AT
    HOME.
Washoe (*ecstatically happy*): YOU ME HURRY GO.

Call it blackmail or call it negotiation, but bargaining like this is a reality of life with chimpanzees—just as it is with children. I followed the loving example set by chimpanzee mothers who indulge the young ones as much as possible and say no as rarely as possible. Acknowledge their desires, negotiate their demands, and always try to avoid using force to restrain them, unless they or someone else is in danger of getting hurt. The result of this give-and-take was that Washoe and Booee respected my parental dominance and authority, even though they knew full well that they were stronger than me. And despite their overwhelming strength, I never had to fear them.

My cooperative approach did have its drawbacks. One time when we were leaving the island, I agreed to let Thelma and Cindy get in the boat without wearing their collars or leads. As I stood next to the rowboat, Cindy climbed in and sat down. Then Thelma rushed the boat and launched it onto the pond with a powerful heave, hurling herself inside next to Cindy, all in one blaze of motion. I had been set up!

As I stood on the island screaming, Thelma and Cindy headed toward the mainland and freedom. They didn't even know how to row, but Thelma's shove was so strong it carried them all the way to the other side, and they fled the boat like two escaped convicts. By the time I got the boat back and tracked them down, they were inside Lemmon's house, cowering in a corner. I put on their leads and walked them to the pig barn, thanking my lucky stars that Lemmon hadn't been home and no damage had been done.

Later that night I got a phone call from Lemmon. He was in a rage.

"Who shit in my bed?" he screamed.

NOW THAT I HAD ESTABLISHED something of a routine for the chimps, I was anxious to explore one of the most tantalizing scientific questions the Gardners had left unresolved. And the chimpanzee island nursery gave me the perfect opportunity to do so. I wanted to determine whether Washoe was simply an exceptionally bright chimpanzee or whether *all* chimps could learn signs. Many linguists who acknowledged that Washoe was using American Sign Language at the level of a two- or three-year-old claimed she was a kind of "mutant genius." They doubted that any other chimpanzee would learn to sign. I thought they were wrong.

At stake was Darwin's theory that human language originated in the cognition of our apelike ancestors. If only one chimpanzee can sign, it may only be the unique behavior of that one chimpanzee. But if *many* chimpanzees are signing, then it is much more likely that signing, or gesturing, has a biological basis in evolution. And if *all* chimpanzees can learn to sign, then it's a pretty good bet that there is a link between ape and human cognition and between ape communication and human language.

With that in mind, I began teaching ASL to Thelma, Cindy, Bruno, and Booee. Their language education obviously would have to be different from Washoe's. Washoe had learned language like a child does, in the course of daily family life. But Washoe's new playmates had no regular family life. Their few hours a day with Debbi, me, or the student volunteers hardly qualified as cross-fostering. I had to figure out how to demonstrate new signs for these youngsters who didn't have the daily routines of getting dressed, eating breakfast, reading books, and using the potty. On top of that, we had to find a way for one or two adults to hold the attention of four hyperactive chimps.

I decided to set up a kind of mobile classroom made out of a cage that was six feet long by three feet wide. On both ends of the cage was a metal bench, so that the chimpanzee student could sit facing me or one of my volunteers from the university. The teaching sessions lasted for thirty minutes, and there were up to three of them a day, five days a week. It was a lot like school, something Washoe never experienced. No three-year-old chimp or child wants to sit still for thirty minutes. So whenever one of the chimps felt a burst of energy he would interrupt the teacher for a quick game of tickle or a swing around the cage. If it was clear he was too hyperexcited to learn, we'd give up and let him out to play with his friends.

Despite its drawbacks, these controlled conditions let me compare how quickly these four different chimpanzees could acquire signs. I decided to teach each of them ten different signs from Washoe's vocabulary list: HAT, SHOE, FRUIT, DRINK, MORE, LOOK, KEY, LISTEN, STRING, and FOOD. While Washoe had learned language by many different methods—guidance, observation, and imitation—I used only guidance with Thelma, Cindy, Bruno, and Booee. I molded their hands into the proper sign and gradually stopped the molding when they learned to make the sign themselves. To help them along, I used raisins as rewards, something my years in Project Washoe should have taught me not to do. Rewards are irrelevant at best and destructive at worst. But I was very tempted by Skinner's methods of reinforcement and their promise of control. I was a new, know-it-all Ph.D., and I thought I could get away with anything.

Bruno, Booee, Thelma, and Cindy straightened me out in a hurry. Only Booee cared a thing about rewards. He would do anything for food, which made him the perfect Skinnerian subject. He figured out how to sign at frenzied speed just so he could get his raisins. But Booee sacrificed quality for quantity.

His signing was sloppy and often degenerated, like Washoe's at dinnertime, into a frenzy of begging: FEED BOOEE, FEED BOOEE, FEED BOOEE.

Bruno, on the other hand, didn't care a hoot about raisins—or about signing for that matter. He would just sit there on the bench, staring at me, as if to say, "I have absolutely no idea what you want me to do." When I tried to teach him the sign for HAT—made by patting the top of the head—I would place his limp hand on his head and then let go. Bruno would let his hand slide off. I'd put his hand back on his head, and he'd let it slide off again. On and on it went. Eventually, I gave up on raisins and offered Bruno apples, bananas, and the ultimate treat for a chimp, Coca-Cola. By this point, Booee would have been beside himself, but Bruno wouldn't budge.

I knew that Bruno was extremely bright because he comprehended English better than any chimp on the island. I was pretty sure he was just playing me for a chump, and one day I decided to test my theory. I reached down and triggered the cattle prod that Lemmon had ordered us to carry at all times. It started buzzing loudly, and Bruno immediately started patting his head wildly: HAT HAT HAT HAT. The game was up. Now Bruno knew that I knew he could learn, and he quickly became one of the best signers on the island.

Cindy adopted signs nearly as fast as Booee did but it had nothing to do with raisins. Poor Pitiful Pearl was so desperate for approval she would do anything to please her teacher. When I first placed her hand on her head to teach her HAT, she left it there for a full minute, like a chimpanzee statue. And after her first few sessions of school, Cindy would enter the cage, sit down, and thrust her two hands out, as if to say, "Do with me what you will."

Cindy always demanded that a bowl of raisins be set in front of her, not because she ate them, but because they proved that we loved her. Every time she signed correctly, her teacher would have to praise her profusely in English. "What a good girl. You're so smart, Cindy." If you hesitated in your encouragement—or, God forbid, forgot altogether—she would give you her most defeated look until you heaped praise on her. Whenever possible I tried to have other people outside the cage urging Cindy on. She just couldn't get enough.

Thelma, like Bruno, was stubborn in all things, including school. She was unmotivated by raisins, praise, or anything else. She let us know that she would sign in her own good time, and only when there wasn't anything better to do. A fly in the cage was enough to divert Thelma's attention for a good five minutes. If a car drove by she acted as though she'd never seen such a strange thing before. Raising my voice to get her attention was out of the question because she would begin crying at the first hint of disapproval. Depending on one's educational philosophy, Thelma was either a creative dreamer lost in imaginative reverie, or she was a good example of attention deficit disorder.

After a few months, all four chimps had acquired their first ten signs. But their speed of learning varied considerably from chimp to chimp. I compared their performances by keeping track of the number of minutes each of them took to learn each of the ten signs. I didn't consider a sign reliable until the chimp could use it correctly five times in a row without any prompting. Here are the results as they appeared in the journal *Science* in June 1973:

Booee: 54 minutes on average to learn a new sign
Cindy: 80 minutes on average to learn a new sign
Bruno: 136 minutes on average to learn a new sign
Thelma: 159 minutes on average to learn a new sign

Anyone who scanned these results without reading the article might conclude that Booee must have been the most intelligent of the group. But of course that would be wrong because chimpanzees, like human children, are individuals, and their learning is powerfully shaped by their personalities and by how they react to different educational environments. Booee learned signs quickly because he loved raisins. Cindy also picked

up signs quickly thanks to all the attention and praise she received. But Booee and Cindy were in no sense "smarter" than
Bruno and Thelma.

That became clear when I tested the chimps using the
double-blind conditions—identifying objects in a box—that the
Gardners had pioneered with Washoe. In this testing situation,
Booee and Cindy scored the lowest, and Thelma and Bruno
scored the highest. Booee was suddenly at a loss without his
beloved raisins. And with no praise at all, Cindy's attentiveness
dropped in a hurry. Bruno and Thelma may have learned more
slowly, but they had retained what they learned.

The publication of this data in *Science* went a long way
toward proving that Washoe was not some kind of freak genius
of her species. Five chimps had now been exposed to ASL and
all five had begun learning it. It now appeared likely that *all*
members of Washoe's species had an ability to think symbolically and learn a gestural system of communication. This, of
course, meant that Darwin was right when he said that language
in hominids had emerged from the cognition of our common
ape ancestor. And this data also confirmed that ape gesture, not
vocalization, was the most likely evolutionary pathway for language.

But it was the variety of chimpanzee learning styles that most
profoundly shaped my own view of how human children acquire
language. Genetic variation is a central tenet of Darwinian biology: a species is a collection of individuals, no two of which
are the same. This, of course, was the first lesson I learned as a
young boy on the farm. There were a variety of cows, a variety
of pigs, and a variety of horses.

When it comes to intelligence, language, or learning, it is
not enough to know the species. One has to know the individual. B. F. Skinner was clearly wrong that all species learn in the
same way, namely through reward and punishment. But linguists
were equally wrong in the 1960s when they assumed that all
individuals within the same species develop language in exactly
the same way, according to some genetically encoded plan. This

approach discounted any and all differences among children—differences of neurology, personality, cognitive development, and family environment.

Children are not born, nor do they develop, according to an identical plan. A class of thirty schoolchildren contains as many different brains as it does faces. The human brain is a very plastic organ, and the development of language is as malleable as the brain itself. In the most extreme cases, a child can recover from head injuries that would cause irreversible brain damage in an adult and still go on to develop normal language skills.

This resiliency was most dramatically demonstrated in the recent case of Alex, an English boy who was born brain damaged and suffered from a relentless series of epileptic seizures in the left hemisphere of his brain. Alex failed to develop any language abilities at all and, when he was eight years old, surgeons removed the entire left hemisphere of his brain. Several months after the operation, Alex began speaking, and he has since progressed to almost normal language abilities. Like most children, Alex was probably predisposed to handling language in the left hemisphere of his brain, but the healthy right hemisphere proved capable of taking over the job once it was given the chance. The case of Alex not only undermines Chomsky's contention of a language organ located somewhere in the left hemisphere, it overturns the theory that children must acquire language during a so-called "critical period" before the age of six—in other words, that they must "use it or lose it." Evidently, a child's brain is quite capable of reorganizing its neural pathways to compensate for neurologic deficits until at least the age of nine, making even the late acquisition of language possible.

In the same way that every brain handles language distinctively, every child *learns* in an individualistic manner. My own findings about Booee, Bruno, Thelma, and Cindy were mirrored in the 1970s by studies of human children. When linguists began focusing on what children actually did when they acquired language, they discovered that each child "cracked the code" of language in his or her own way. Some children—so-called "concept learners"—focus on single words and content. Others who focused on emotional meaning were called "expressive learners."

Further studies revealed that language skills do evolve together with other cognitive skills. For example, the burst of language that usually happens between sixteen and twenty months—when the child begins combining single words into phrases—coincides with the child's increasingly complex play with dolls and blocks. Almost all children play with dolls or blocks, and all of them develop complex sequences of play. But as any parent knows, no two children play in exactly the same way. The unique way that each child patterns his play—and creates meaning out of that order—seems to be a function of personality.

The same is true with language. Over the past two decades, dozens of linguists have proposed dozens of learning strategies to explain how children learn the rules of language. Then, in 1987, the psychologist Melissa Bowerman made the most radical proposal: *all* the theories are right. After studying children who were acquiring a variety of languages, Bowerman concluded that different children apply different learning strategies. And though a particular strategy that works for one child may not suit another child, *every child employs a strategy that works.*

Nature, it seems, has made the process of acquiring language every bit as flexible as the infant brain itself. All these neurological backup systems and different strategies for learning are a matter of survival. No matter what injuries there are to the brain, no matter what negative pressures arise in the environment, human children will almost always find a way to learn language.

In our haste to find one-size-fits-all theories, we have too often wiped out the bell curve and acted as if there were a single way to learn language and that all children should follow it. As a result, in the past many schools refused to teach ASL, *forcing* their deaf children to learn spoken language. Dyslexic children, who have trouble organizing visual images, were *forced* to read. Today we don't hesitate to teach sign language. And when I teach dyslexic college students, I encourage them to use books on tape and auditory tests instead of written ones.

It was Booee, Bruno, Thelma, and Cindy who taught me the evolutionary principle of individual variation. Thanks to them I started taking every child or student I worked with on his or her own terms. I made an effort to locate each child's singular way of learning. And several years later, when I began working with autistic human children, the lessons I learned from the chimps led to the biggest breakthrough of my career.

SEVEN

HOUSE CALLS

NORMAN, OKLAHOMA, must have held some kind of world record as "the city with the most chimpanzees being raised as human children," thanks to Dr. William Lemmon's habit of prescribing chimpanzee rearing to his psychotherapy patients. And like a piano teacher making his rounds, I would visit these young chimpanzee children in their homes to give them lessons in American Sign Language.

The first home on my route each morning belonged to Jane and Maury Temerlin. Maury Temerlin, a psychotherapist and psychology professor at the university, was one of Lemmon's former students and longtime patients. At 8:30 A.M. the Temerlins' chimpanzee foster daughter, Lucy, would greet me at the front door, give me a hug, and show me into the house. While I sat in the kitchen, six-year-old Lucy would go to the stove, grab the teakettle, and fill it with water from the kitchen sink. She did all this chimpanzee-style, by jumping from counter to counter. After getting two cups and tea bags out of the cupboard, she would brew the tea and serve it like a perfect hostess. Then her ASL lesson would begin. It was all very civilized.

Calling Lucy "cross-fostered" doesn't quite do justice to her privileged upbringing. She was born into a colony of carnival chimpanzees in 1964 and was sold, at two days old, to Lemmon, who gave her to the Temerlins to raise. The Temerlins embraced Lucy as their own flesh and blood to an extent that went significantly beyond the Gardners' more arm's-length parenting of Washoe. As Maury Temerlin later wrote in *Lucy: Growing Up Human*, "Shortly after we adopted Lucy I began to love her without reservation. I do not remember how long it took—I would guess not more than a week or so—before I failed to make human-animal distinctions with Lucy. She was my daughter and that was that!"

Lucy slept in a bed between her parents, ate from a silver cup and spoon, and grew very attached to her human brother, Steve. When she was three years old Lucy, like all young chimps, began ransacking the house almost daily, but the Temerlins never considered giving her away. Instead, they built a chimp-proof house with concrete construction, steel exterior doors, locking interior doors, and a central courtyard with floor drains. During the day, when her parents were at work, Lucy was confined to her steel-reinforced concrete bedroom and a spacious "penthouse" on the roof that she could get to through a trapdoor in her ceiling. All in all she had nearly one thousand square feet of safe play area. But even these elaborate defenses couldn't keep Lucy from escaping and raiding the kitchen. Her parents couldn't figure out how she was doing it, until one day they caught her red-handed, letting herself out of her locked bedroom with a stolen key she hid in her mouth every morning.

Whenever Washoe's antics tested my patience, I used to conjure up the imaginary BLACK DOG to scare her into cooperating. Maury Temerlin, ever the psychotherapist, would manipulate Lucy with guilt. This was remarkably effective. If Lucy was refusing to eat her dinner, Maury would plead, "For God's sake, Lucy, think of the starving chimps in Africa." She'd then take just a bite or two. Unsatisfied, he'd beg, "Take at least three more bites for your poor suffering father who loves you." Lucy would eat with a little more enthusiasm. Finally, when Maury whined, "Lucy, how could you do this to me?" she became putty in his hands. After a few years, Lucy developed a guilty expression that immediately gave her away whenever she was hiding a key, smuggling a cigarette lighter, or committing some other household crime.

Lucy was no slouch herself when it came to guilt-tripping her parents. Sometimes we'd be in the middle of an ASL lesson on the couch when Lucy would suddenly realize that her mother was about to leave for work. She would immediately assume the fetal position and, while Jane stood there fretting over her distraught daughter, would begin rocking back and forth like an autistic child or a depressed chimpanzee. But as soon as Jane was out the door, Lucy would turn off her act like a light switch and go right back to her signing lesson. When she heard Jane returning later in the day, Lucy would jump back on the couch and repeat her award-winning performance.

Lucy's presence seemed to have little inhibiting effect on the Temerlins' sexual relations, and Lucy frequently witnessed what Maury called "the primal scene." In the wild an infant chimpanzee will throw tantrums when she sees her mother copulating and will literally attack the amorous male, especially his face. Lucy had a more American version of this: she would run around the room trying to distract her entwined parents by turning somersaults, spilling water, turning on the TV, and flicking the lights on and off. When all else failed, she'd grab her father by the feet and drag him off the bed.

Lucy's own sexual coming of age was allowed to proceed without inhibition or judgment. At age three Lucy began exploring her own anatomy with the help of household tools. She would squat over a hand mirror, pull her labia apart with plier handles, and rub her clitoris with a pencil. When she was eight years old, she began ovulating and menstruating, and her methods of self-gratification became more creative. One afternoon, while sitting on the sofa leafing through *National Geographic* and drinking straight gin, which she used to serve herself from the liquor cabinet, Lucy sat straight up as if struck by a great idea. She put down her magazine and drink, ran down the hall to the utility closet, and pulled out the Montgomery Ward canister vacuum. She brought it back to the living room, plugged it in, removed the end brush from the long hose, turned it on, and applied the suction to her genitals until she reached what Temerlin surmised was orgasm—"she laughed, looked happy, and stopped suddenly." Then Lucy turned off the machine and went back to her gin and *National Geo.* After that, the vacuum became one of Lucy's favorite toys.

As for magazines, when Lucy was in estrus she put aside her *National Geographics* in favor of *Playgirl.* She would squat over the male nudes and rub back and forth against the photos of their genitals. Lucy's love of *Playgirl* began turning up in the scientific literature as the most dramatic, and certainly the most unusual, evidence to date that a chimpanzee could respond to photographic representations.

Lucy and the Temerlins communicated, as most crossfostering families did, through a combination of chimpanzee vocalization, gesture, and English. Lucy had been with the Temerlins for six years before she ever learned ASL, and they all read one another's emotions and intentions with ease. Lucy's grasp of English was also quite good. If I said, "Please make me some tea," she would rush off to do it.

Lucy took to ASL like a duck to water. Our typical session would begin with tea, followed by some warm-up tickling on the floor, and then a signing lesson on the couch. Lucy picked up many signs after only one or two prompts, and she loved to learn by playing games, especially ones that involved ordering me around. WHERE BRUSH? I'd sign, and Lucy would answer, BRUSH LUCY BRUSH LUCY until I brushed her all over. Or I'd sign WHO EAT? and she'd sign ROGER EAT, stuff an apricot in my mouth, and then get right up in my face and watch me chew. (Trying to condition Lucy with food rewards was not only futile, it was a surefire way for *me* to gain weight. Like her father, Lucy was obsessed with feeding people and making sure they cleaned their plate.)

When I'd sign WHAT YOU WANT? Lucy loved to respond WANT ME HUG until I hugged her over and over. Her hugs and tickles had a way of getting rough—though they never equaled Washoe's shirt-tearing frenzies—and when I needed to calm Lucy down I would groom her until she fell asleep right there on the couch. Another strategy I used when she started rough-housing was to distract her by talking. When I spotted her charging me like a freight train from across the room I'd sign WHAT YOU WANT? She'd stop dead in her tracks and sign back TICKLE! Then we'd sign back and forth until she would ask me nicely: PLEASE TICKLE LUCY. By that time she'd be calmed down, and we'd have a friendly tussle.

Sometimes I took Lucy out for a drive in my Volkswagen bus. She had a great sense of direction. WHERE YOU WANT GO? I'd ask her. GO THERE, she'd sign, pointing left, then GO THERE, pointing to the right. After a few minutes and many turns, we'd be in front of Jane Temerlin's office. No matter where we started, Lucy could always find her mother's workplace.

Lucy treated her pet cat the same way Washoe treated her favorite doll—as a baby that should never be put down. But unlike Washoe's doll, Lucy's cat was always trying to escape the clutches of its smothering chimpanzee mother. The cat would run up the nearest tree, which was an effective strategy for evading humans but not chimpanzees. After retrieving her runaway infant, Lucy would scold it—sometimes in sign—and then cradle her baby as wild chimpanzee mothers do, in her arms or pelvis. When I'd tell Lucy to let the cat walk, she would carry it just above the ground so that the cat appeared to be skimming across the floor. It didn't take long for the poor cat to realize it didn't have a life of its own, and whenever Lucy entered the room the cat would immediately go limp and fall to the ground in a heap.

Lucy's cross-fostering of her feline baby was very thorough. One day she sat on the floor, placed the cat between her legs, and held up a book so that the cat could see it. Then, while pointing to the book, Lucy gave the cat a lesson in the sign for BOOK. On another occasion, Lucy and I saw the cat as it was pooping in the litter box. This upset Lucy, and she hauled the cat out of the litter box and carried it twenty feet down the hall to the bathroom. When I entered the bathroom, Lucy was dangling the cat over the toilet, encouraging her to finish her business. Satisfied that her baby was done, Lucy put her down and flushed the toilet.

Even before learning ASL Lucy had an extraordinary ability to communicate her own feelings and to read the moods of others. When she sensed that someone was distressed, she would put her arm around them and kiss them. When she detected anger between two people she would separate them by distracting one of them. When she met someone new, Lucy would go right up to the stranger, smell him or her, and then size them up in a rather unnerving way that Jane Goodall remembers from her first meeting with Lucy: "Lucy came and sat close beside me on the sofa, and simply stared into my eyes for a long, long time. It gave me a strange feeling. . . . What, I kept wondering, is she thinking about me?" (Lucy must have liked what she saw, because she then gave Goodall a big wet kiss, got up, poured herself a gin and tonic, and turned on the television.)

I was not surprised that Lucy used her sign language from the very beginning to express her varied emotions and deep sensitivity. One day, in the middle of Lucy's lesson, Jane Temerlin drove up to the house. Lucy jumped up and wanted to end her lesson, but Jane was in the house only for a minute before she left again. Lucy pulled her chair to the window, watched her mother drive off, and then signed to me CRY ME, ME CRY. Another time, when Lucy was told that she had injured her cat's paws—she had tried to pull the cat away from a fence it was clinging to—she cradled her baby and signed HURT HURT. Whenever she met someone new she regularly inspected them for bandages or scabs and signed HURT HURT very sympathetically.

Lucy also found language useful for deceiving others. Deceit is one of those traits that was long believed to distinguish humans from nonhumans; people are capable of it, but animals aren't. Of course Washoe, and more recently, Thelma and

Cindy, had suckered me on more occasions than I cared to remember or admit. But Lucy was the first chimp who ever tried to put one over on me using sign language. One day she defecated in the living room when I wasn't looking.

Roger: WHAT THAT?
Lucy: WHAT THAT?
Roger: YOU KNOW. WHAT THAT?
Lucy: DIRTY DIRTY.
Roger: WHOSE DIRTY DIRTY?
Lucy: SUE (*a graduate student*).
Roger: IT NOT SUE. WHOSE THAT?
Lucy: ROGER!
Roger: NO! NOT MINE. WHOSE?
Lucy: LUCY DIRTY DIRTY. SORRY LUCY.

Chimps, like children, have minds of their own. Having provided Lucy with a few dozen signs—she learned seventy-five signs in her first two years of lessons—there was no telling what she would do with them. Asking her to use her signs according to my lesson plan was like asking her to use the vacuum only to clean floors. Lucy's most creative use of language occurred serendipitously when I was trying to study something else. In the most memorable case, I wanted to know whether Lucy could conceptualize categories the way Washoe did—that is, by grasping the "treeness" of TREE and generalizing that sign to all trees. So, over several days, I put out the same twenty-four fruits and vegetables and asked Lucy to name them. Her food-related signs at that time were limited to FOOD, FRUIT, DRINK, CANDY, and BANANA. She showed that she had a knowledge of categories; she called apples, oranges, and peaches FRUIT but referred to corn, peas, and celery as FOOD (she had no sign for vegetable).

But even more interesting was what happened when Lucy used her limited vocabulary to *describe* some of the new foods. When she tasted a watermelon, she called it CANDY DRINK or

DRINK FRUIT. That is about as close to "watermelon" as you can get if you don't have the signs for WATER and MELON. When she tasted a radish for the first time she called it CRY HURT FOOD. She referred to citrus fruit as SMELL FRUIT, probably because of the aroma it gave off when she bit into the skin. Lucy also called a stalk of celery a FOOD PIPE; she knew PIPE because I smoked one. And she called a sweet pickle PIPE CANDY.

Lucy was also spontaneously combining words to create new meanings, just as Washoe had recently done when she called a Brazil nut ROCK BERRY and a swan WATER BIRD. Lucy also took to swearing at a local tomcat she disliked, calling it DIRTY CAT. Previously she had used DIRTY only for her activities on the toilet. DIRTY soon became Lucy's all-purpose derogatory sign; when we prepared to go for a walk she referred to her lead as the DIRTY LEASH.

By a strange coincidence, around this same time, Washoe *also* began referring to her adversaries in scatological terms. (I guess this is not uniquely human.) Her cursing started when Lemmon brought in some new, rather territorial, residents to the pig barn: monkeys. One monkey, a rhesus, always greeted Washoe and me by baring his teeth and threat-barking in defense of his space, causing Washoe to swagger back at him. To ease the strained relations I decided to teach Washoe the sign for MONKEY. I pointed at the rhesus and signed MONKEY. Washoe promptly stormed over to the angry rhesus and started signing DIRTY MONKEY. After this, Washoe took to using DIRTY as an adjective to describe any bad person who didn't give her what she wanted. For instance, if she signed ROGER OUT ME, wanting to leave the island, and I replied SORRY, YOU MUST STAY THERE, she would respond with DIRTY ROGER over and over again as she walked away.

I got plenty of laughs from Lucy's and Washoe's creative use of the DIRTY sign. But all this cursing was significant, linguistically speaking. When Lucy called her lead a DIRTY LEASH or a radish CRY HURT FOOD, she was demonstrating a feature of language called *productivity*—the ability to produce an infinite number of new meanings by recombining a finite number of words or signs.

Washoe had been recombining signs to form novel sentences for several years at this point. But Lucy's combinations of signs—DIRTY LEASH, DRINK FRUIT, SMELL FRUIT—was even more dramatic evidence that a chimpanzee could manipulate symbols in an open and creative way. She was using language to convey her own sensory experience, and that gave us a window into one chimpanzee's world of perception. She chose to describe an orange not by its color or by its taste but by its smell; she chose to describe celery not by how it tasted but by how it was shaped. Lucy's idiosyncratic descriptions were very compelling. In a human child these descriptive powers are seen as the essence of language, even as a primitive form of metaphoric poetry.

Lucy's productivity with signs challenged the prevailing wisdom about the extent of a chimpanzee's capacity for language. After Project Washoe, linguists grudgingly acknowledged that chimpanzees could learn a human vocabulary—that they could associate the symbol for HAT with the real thing. But they resisted the notion that chimps could recombine their vocabulary to invent new meanings. For example, Ursula Bellugi and Jacob Bronowski, two critics of Washoe, said in 1970 that the ability to rearrange words into new messages was the "evolutionary hallmark of the human mind." They said that there was no evidence a nonhuman primate was capable of this, "even when he is given the vocabulary ready made."

But Lucy undercut that claim, and in so doing she threatened the bold line between human communication and all other forms of animal communication. According to the prevailing theory at that time, it was the flexibility of meaning that made human language an "open" system of communication; all modes of animal communication were supposedly inflexible and "closed."

According to this view, nonhuman communication takes two main forms. In the first, an animal employs a finite repertoire of signals, each carrying a fixed message. For example, a vervet monkey will use different alarm calls, one to warn of a snake, another to warn of a leopard. In the second case, an animal employs a variable signal. For example, whales sing complex songs in which notes or phrases are combined in new ways to produce variations on a theme. Both forms have superficial characteristics in common with human language—vocabulary in the first case and recombination of signals in the second case—but neither of them is capable of creating new meanings to describe novel events—or so many scientists thought.

The problem with either-or categories like "closed" and "open" is that new evidence keeps cropping up that blurs such neat distinctions. Scientists assumed that chimpanzee communication in the wild was closed because they were focusing on the wrong channel—vocalization. Once ethologists began looking as well as listening, they saw that chimpanzees were capable of varying the meanings of their gestures.

IN 1971 THE POPULAR PRESS began beating a path to the door of the Institute for Primate Studies to get a look at the "talking chimpanzees." They were welcomed by William Lemmon with open arms, which was quite a change for me from the impenetrable wall of privacy that surrounded Washoe when she lived in Reno. The Gardners were that rare breed of serious scientists who abhorred publicity. They published their findings in peer-reviewed scientific journals only, and they trusted that their research data would speak for itself. They had no use for sensation-seeking journalists who might distort the record by exaggerating or trivializing Washoe's achievements.

When Life magazine contacted Lemmon in late 1971, asking permission to run a photo spread on Washoe and the other chimps, I decided I'd better alert Allen Gardner. He was, not surprisingly, aghast at the idea of Washoe appearing in the world's most popular magazine. Out of respect for his wishes, I told Lemmon that Washoe would not appear in Life. That suited him just fine because he resented Washoe anyway; he preferred to promote one of his Institute's own chimpanzees.

The February 11, 1972, issue of Life featured a photo spread entitled "Conversations with a Chimp." It captured Lucy and me chatting during her daily language lesson in the Temerlins' living room. In a frame-by-frame treatment, Lucy was shown answering questions (WHO ARE YOU? WHAT DO YOU WANT?) and begging to be tickled—with me obliging, of course.

The whole world now knew about the talking chimps, and Lucy became a staple of the popular press—Psychology Today, Parade, Science Digest, The New York Times, and so on. But the publicity did not change her life at all. Lucy still lived the same secluded life in the Temerlin household, oblivious to her stardom.

There was something poignant about Lucy's few years of celebrity. The world's most famous chimpanzee didn't even know she was a chimpanzee. In Lucy's own mind she was human. Having never met another chimpanzee, her self-image was just like Washoe's before she came to Oklahoma. Washoe had divided the world neatly into "us" (PEOPLE) and "them" (DOGS, CATS, BLACK BUGS). The fact that she looked so different from her own human family never seemed to faze Washoe. When she looked in the mirror she saw a human. And so did Lucy. One time Lucy was sitting on the floor looking at some photographs. She flipped through the pile casually until she came to a photo that stopped her cold. Staring at it in confusion she asked WHAT THAT? It was a picture of a chimpanzee.

This identity confusion seemed to be true of all cross-fostered chimps, including Viki Hayes, the baby chimp raised in species isolation by Keith and Cathy Hayes in the late 1940s. Viki loved to sort things, and one day she was sorting photographs into two piles: humans and animals. When she came to a picture of herself, Viki put it in the same pile with Dwight Eisenhower and Eleanor Roosevelt. But when she came to a picture of her chimpanzee father, Bokar, she put him in with the cats, dogs, and horses.

ALLY WAS ONE YEAR OLD when I first visited his house in October 1970. He'd been born at the Institute—his parents were Pan and Caroline—but when he was six weeks old he was sent to be raised by Sheri Roush, another one of Lemmon's therapy patients. Ally, like Washoe, began learning sign language around his first birthday, and he progressed very rapidly, learning 130 reliable signs over the next few years.

Ally had one sign that I never included on his vocabulary list: the sign of the cross, which he made over his chest. Ally was a Catholic, or at least that's what his mother claimed, and she had him baptized when he was two years old, and proudly showed off the pictures of his baptism party to visitors. "Why hasn't my baby a right to be saved, like anybody else?" she would ask.

Ally always reminded me of a pinball in play—ricocheting off walls, bounding off furniture, cartwheeling over people. Signing to Ally was like talking to a whirlwind; he'd pause long enough to flash a sign, but he was gone before you could sign back. Ally's signs were big, bold, and expressive, the ASL equivalent of a very loud child. They were also emphatic to the point of violence. When Ally signed HAT he would slap his own head so hard you thought he'd fall down. It was like an Abbott and Costello routine.

Ally painted in the same explosive way he talked, and his oil paintings bore an uncanny resemblance to the action-painting style of the 1950s. We found this rather amusing because critics of abstract expressionism used to joke that if you locked a chimp in a room with canvas and some paint he would produce the kind of avant-garde creations that were hanging on New York gallery walls. One of my students, an art history major named Polly Murphy, decided to get a more formal appraisal of Ally's work, and she brought some of Ally's canvases to an art historian, telling him only that they were done by a young painter friend of hers. The expert could barely contain his excitement. "I knew Pollack was coming back!" he effused.

Ally was not always in motion. He could actually sit still for a long time when tackling a project that was his own idea. One time Sheri Roush forgot to lock him in his cage when she went off to work, and Ally spent the entire day prying off every last one of the grouted tiles that covered the bathroom floor and walls. (Home renovation is a favorite chimp activity. Viki Hayes once tore apart an entire wall so she could find out where the termites were coming from.)

Ally was always a pleasure to be around because he never seemed to get sad or depressed in the way Lucy did when her mother left, or the way Cindy and Thelma did when I scolded them. He was always up, always ready for action, always enthusiastic about one more tickle session, even when I was exhausted. He was as sweet as Booee but more gullible. I could always get Ally to fall for the oldest trick in the book: I would point at the ceiling and he'd look up; then I would tickle him under the chin. It bothered him that he always got suckered in this way—he looked dumbfounded—but each time I did it he would fall for it again.

WHEN ALLY WAS THREE YEARS OLD I chose him for a series of studies that would address the major unresolved question of whether a chimpanzee could master a simple rule of grammar. Although Washoe and Lucy had shown that they could create novel combinations of signs—DIRTY LEASH, SMELL FRUIT, and so on—no one had shown that a chimpanzee could understand and create novel sentences. To do so, a chimp, like a child, would have to apply a rule of grammar.

Grammar is what enables a child to link words in his vocabulary in new ways. For example, a child who has heard and understood "Give me the plate" a number of times can also understand "Give Daddy the ball" the very first time he hears it. He does this by ignoring the surface differences in the two sentences and ferreting out the identical word order (verb-subject-object) that underlies both. He substitutes "Daddy" for "me" and "ball" for "plate" and responds appropriately. By applying a rule of word order, he is able to link two words in a completely new relationship.

In our first study I wanted to see if Ally could understand sentences he had never seen signed before. So Ally and my students played a game in which a student would ask Ally to pick one of five items—FLOWER, BALL, DOLL, BRUSH, HAT—from a box. Then Ally was asked to put the item in one of two locations in the room or to hand it to another student. For example, we might say, ALLY PUT BALL IN PURSE or GIVE BILL TOOTHBRUSH. As Ally caught on to the game, I added new items, and pretty soon he was responding to thirty-three different requests.

To test Ally's comprehension, we added new objects, locations, and students, so that he'd have to respond to signed requests he'd never seen before. Also, the box was moved away from the person signing to Ally to ensure that the person couldn't see the items and cue Ally in any way. Finally, we set up a partition between the person signing and the locations, again to prevent any cueing. In a round of testing with a combination of old and new requests, Ally took the correct item to the correct location 61 percent of the time (guessing would have produced a score of 7 percent). In four rounds of testing with all *new* commands, Ally scored 31 percent (again, with guessing producing 7 percent).

Ally would have scored much higher but he paid a price for two interesting personality quirks: when told to PUT SPOON ON CHAIR, he would go to the chair and sit in it, still holding the spoon. He seemed to think that chairs are for sitting, not for storage. If there had been no chairs involved in the test Ally would have scored 50 percent. The test also penalized Ally for being so hyperactive. He almost always chose the right object— about 90 percent of the time—but he often grabbed it and ran behind the screen before we could tell him where to put it.

(Viewers of a 1973 *Nova* television program saw Ally racing around with the various objects.) Ally, like many human children, was a little too energetic for testing conditions. Still, the study proved that a chimpanzee can comprehend differences in meaning generated by a grammatical rule.

Next Ally proved not only that he could understand simple grammar but that he could also use a rule to tell us where something was. When asked, he could tell us that FLOWER ON PILLOW or BALL IN BOX. In double-blind testing, where the experimenter didn't know the answers, Ally was correct 77 percent of the time.

To this day, those who misconstrue ape language research say that chimpanzees were merely trained like dogs to "roll over" or "get the newspaper"; in other words, they are making a simple Pavlovian association between one or more words and a proper response. But Ally could not have been relying on a simple one-to-one association because he was describing relationships between objects and locations he'd never seen before. He was picking two signs from his vocabulary and using a grammatical rule to link those signs in a new way. This kind of linguistic openness is exactly what Chomsky said freed human children from Skinner's laws of reinforcement.

But how was Ally *able* to learn a rule of grammar? This question puzzled me, and it went to the very heart of the controversy about a chimpanzee's capacity for language. While most linguists admitted that a chimp could learn vocabulary, few would say the same about grammar. Chomsky and his followers claimed that the complex rules of word order were genetically encoded and unique to humans. They said these rules could *never* be learned. On the other hand, linguists who broke with Chomsky believed that children did in fact learn these rules by imitating adults, but they still regarded grammar as the hallmark of human intelligence, something that could never be mastered by apes.

Ally's performance was bad news for both of these positions. Ally clearly didn't have specific rules of grammar wired into his brain; a chimpanzee would have no use for these rules in the

wild. So Ally must have somehow *learned* the grammatical rule
for subject-preposition-location (BALL IN BOX). That supported
the non-Chomskyan position that children did in fact learn
grammatical rules, but it also proved that this ability is not
unique to humans but was shared, at least to some extent, by
chimpanzees.

I had always believed that this ability to learn a rule of grammar must be very advanced indeed because all linguists described
grammar as the most complex and mysterious aspect of language
acquisition. But one aspect of Ally's performance made me think
we might all be wrong in that assumption. Ally did more than
learn and apply a rule of grammar. *He never made a grammatical
error.* This stood linguistic theory on its head by suggesting that
grammar might not be so difficult after all. Ally sometimes confused his locations, but he always signed subject-preposition-
location in their proper order. He got this right even when we
changed subjects into locations and locations into subjects. In
other words, Ally always knew the crucial difference between
TOOTHBRUSH ON BLANKET and BLANKET ON TOOTHBRUSH. His
facility for word order suggested that whatever cognitive skill he
was using had its evolutionary roots in some primitive animal
behavior. Otherwise he'd be struggling and making mistakes.

Every behavior, even the most complex, must have some
biological basis or else it cannot exist. Once I stopped focusing
on the mystery and complexity of grammar and began to consider the possible biological basis for Ally's performance, it suddenly made perfect sense to me. All animals impose order on
the world by perceiving and then following rules in nature. So
it was not all that remarkable that humans or chimps would also
impose order on words.

There is one overriding principle that governs the workings
of every brain, from the most primitive to the most complex:
perceive a pattern beneath constantly changing stimuli and apply that pattern to new situations. In other words: follow a rule.
This rule-following behavior has been famously documented in
the case of herring gulls. Researchers removed eggs from a her-

ring gull's nest and placed them outside the nest, right next to an even bigger egg. When the mother gull returned she would roll the *biggest* egg back into the nest, even though it was not her egg. The mother does not recognize eggs individually; she simply applies a rule: "Bigger is better."

Applying general rules instead of creating a new rule for each new situation is a smart way to run a nervous system. After all, a robin catching a worm must learn something in general about worms, because tomorrow it must catch a different worm. A chimpanzee can't learn to climb just one tree; it must learn how to climb *all* trees.

Consider the alternative. If a robin needed to learn worm catching all over again every morning, it would soon be a very dead robin. Even if it did survive, the robin's brain would need a separate circuit for each new worm. Under this scenario—and it was a fashionable theory at one time—the brain would function like a giant switchboard, hooking up stimulus number 2,458 to response number 2,458, just like an operator patching through a specific call to a specific hotel room. The problem with this design is that the brain would quickly run out of circuits. Besides, after the robin eats worm number one it is never going to encounter it again, so dedicating a whole neuron to that worm is a waste of precious brain matter.

The brain does *not* operate like a switchboard. Instead, it seeks out patterns in related stimuli, applies a general rule, and follows that rule the next time it encounters a similar stimulus. A robin can catch *any* worm even though every worm is a little different. A young chimpanzee can climb *any* tree by applying certain rules—he always goes up headfirst, for example—even though he must organize his muscle movements to match that specific tree.

This same rule-following mechanism governs the human mind. We see this in the way children acquire aspects of language that have nothing to do with grammar. For example, it is common for a child to learn the word "dog" and then to call *all* four-legged creatures "dog." (Washoe did this, too, when she called all small objects BABY.) Children also apply rules when they learn to count. I know a four-year-old who is now learning to count past ten. He says: "One-teen, two-teen, three-teen, four-teen, five-teen," and so on. He has created a logical rule—combine a number with "teen"—and he will keep using that rule until he learns the four exceptions: eleven, twelve, thirteen, and fifteen. Adults do the same thing, but more consciously. For example, it's a lot easier to remember "i before e except after c" than to memorize every single word that falls in that category.

Do children learn the rules of grammar in this same way, by ignoring the flux of adult language and perceiving the patterns? There is a lot of evidence to suggest that they do. For instance, children insist on applying rules to their speech, no matter what they hear in adult speech. We have all heard children make charming linguistic errors such as, "I holded the rabbit." The child learns the "-ed" suffix for the past tense and generalizes it to *all* verbs. Only later will he or she learn the *exceptions* to this rule—the irregular verbs like "held."

The fact that children automatically apply grammatical rules—by saying "holded," for example—is often cited by Chomskyans as proof that these rules must be innate and genetically encoded in a language organ. But children don't need to have such rules encoded if they can naturally pick them out of adult speech or signing and then use them. If language does have a biological basis in the brain mechanisms of our mammalian ancestors, as I believe it does, then grammar is just a complex form of rule-following behavior—something that characterizes every aspect of a child's cognitive development. It also explains why an ape like Ally could tell the difference between TOOTHBRUSH ON BLANKET and BLANKET ON TOOTHBRUSH. He simply ignored the variance in stimuli and applied a rule.

LUCY AND ALLY WERE LINGUISTIC SUCCESS STORIES, but two of my other home-reared students were not so fortunate.

I had been teaching Maybelle for about nine months when

her foster mother, Vera Gatch, decided to leave her chimpanzee daughter for the very first time. Vera was one of Lemmon's students and a psychotherapist with her own private practice and a teaching post at the university. She had raised Maybelle from infancy and had never left her daughter alone even for one night. Now that Maybelle was four, Vera felt the time was right to attend a conference out of town, and she arranged for someone Maybelle knew to stay with her in her home.

As soon as Vera was gone a full day, Maybelle went to pieces. She developed terrible diarrhea and a respiratory infection. Those of us who knew Maybelle set up shifts to care for her around the clock. Day after day we sat at her bedside administering fluids and trying to get her fever down, but poor Maybelle was wasting away before my very eyes and I felt utterly powerless to save her. Her diarrhea became dysentery and her lung infection turned to full-blown pneumonia. The doctor came but there was nothing he could do. By the time her mother returned home, Maybelle was dead.

Nearly two years later I watched my youngest pupil, barely older than a baby, also shrivel up and die in the absence of her human mother. Salomé began learning signs at four months of age, about the same age when deaf children begin signing. Thanks to her precociousness she appeared in the 1972 *Life* magazine spread with Lucy. Salomé was raised by Susie Blakey and her husband, Church Blakey, a well-to-do businessman and patient of Lemmon's. Just when Salomé was out of infancy, Susie became pregnant. After the baby was born, the Blakeys decided to take a vacation with their new child, and immediately Salomé lapsed into pneumonia and was close to death. The Blakeys rushed home and Salomé recovered from her grief-induced illness. Shortly thereafter, the Blakeys decided to try another vacation. But this time Salomé didn't make it. She died within a few days.

The way that Maybelle and Salomé died in the absence of their mothers was reminiscent of a death that Jane Goodall observed among the wild chimpanzees of Gombe around this very same time. She told the story of a young male chimpanzee named Flint who was unusually attached to his aging mother, Flo. Flint continued to sleep with his mother and ride on her back like an infant until he was eight years old. When Flo died in 1972, Flint fell into a deep depression, wasted away, and died.

I DON'T KNOW WHY Washoe did not suffer mother loss as dramatically as Maybelle and Salomé. Certainly Washoe had not been as emotionally dependent on Trixie Gardner as the two other chimps, who had no other source of comfort than their own human mothers. Maybe, in the end, Washoe was fortunate not to grow up in a strictly nuclear family. By sharing the parenting, the Gardners had given Washoe the chance to form other attachments. Maybe her bond with me saved her from a tragic end like Maybelle's and Salomé's. Or maybe not. Maybe Washoe was just a born survivor, a headstrong girl with an unbreakable spirit and a talent for landing on her feet.

I often wondered whether Washoe thought about the Gardners and the world of comfort that she had left behind in Reno. (In the sixties I also used to speculate about Washoe's chimpanzee mother, but that seemed like ancient history now.) I never mentioned MRS. G or DR. G to Washoe, and I never saw her sign their names. After our first rocky year at the Institute, our life in Reno was receding further and further into the past. Perhaps Washoe felt the same way.

Then in the spring of 1972, a year and half after we arrived in Oklahoma, I got a call from the Gardners, saying they wanted to come visit Washoe. I gave them the go-ahead, but I had misgivings. I had no idea if Washoe would greet them with open arms or outright anger. There were stories from Africa about chimps attacking the people who had raised and then abandoned them. I didn't want to take any chance that this reunion would get out of control. I decided not to tell Washoe that the Gardners were coming, and I set up their meeting on the island in order to contain it.

The day the Gardners arrived I removed all the chimps from the island, including Washoe. Then I asked Allen and Trixie to hide behind the rundevaal (the new concrete hut in the middle of the island). I brought Washoe over in the rowboat, and she and I stepped onto the island and headed toward the rundevaal. As we walked around it, the Gardners were sitting on the ground, not six feet away, with a pile of wrapped presents stacked in front of them. Washoe stopped short, as if in shock, and let out a piercing scream that scared the hell out of me. Then she sat down right on the ground where she'd been standing and ignored the Gardners completely, as if they were invisible. She obviously remembered them; if they were strangers she would have investigated them thoroughly. She did not ignore the presents, however. She opened all of them like a child at her own birthday party.

For the next two days Washoe continued to pretend that the Gardners were not there. She wouldn't return their big smiles or their signs, and when they tried to initiate Washoe's favorite games, she showed no interest or played dumb. Finally, on the third and last day of their visit, Washoe began warming up to her parents and even agreed to play the old games. But that very same evening the Gardners headed back to Reno and disappeared from Washoe's life once again.

Washoe's snubbing of her parents—her refusal, literally, to talk to them—was reminiscent of separation behavior Jane Goodall has described in the wild. When a young chimp is separated from her mother in the forest, she will cry and scream loudly for as long as they are apart. But once they are reunited, Goodall says, "there is none of the frenzied greeting, none of the hugging and kissing, that one would expect. Instead, the child wanders nonchalantly toward the mother, and may even ignore her. She seems to be conveying a message, 'You're bad. You shouldn't have left me.' "

I wondered how the Gardners felt about the way Washoe treated them. Did they feel guilt? Regret? Sadness? They said nothing. And I wasn't about to ask any questions. Moving

Washoe to Oklahoma was one of those forbidden family topics. It was as if their foster daughter's abrupt departure from Reno was the most natural thing in the world and had no emotional repercussions whatsoever for any of us.

DESPITE THE AVALANCHE of scientific and popular attention that Washoe, Lucy, and Ally brought to the University of Oklahoma, after three years of hard work I was still just a Research Associate and Visiting Assistant Professor, with minimal salary and no immediate prospects for a permanent position on the faculty. The university happily acknowledged my value to the psychology department—it suddenly had a national reputation and graduate applications were flooding in—but they figured I had no alternatives. Where else but Lemmon's Institute could I conduct sign language studies with chimpanzees?

That changed in a hurry when Yale University got into the act in 1973. Yale flew Debbi and me to New Haven so that I could interview for an Assistant Professorship in their psychology department. We were given the royal treatment by Alan Wagner, the very charming head of the search committee. I had never been behind the walls of an Ivy League university before, and I was appropriately awed by the weighty tradition surrounding the place. As we were wined and dined at the Graduate Club I gazed up at the aging crew oars from winning teams past. As we strolled through the corridors of the psychology building I saw the imposing portraits of illustrious and departed professors, each of whom, according to the attached inscriptions, had earned an honorary master's degree at Yale. I asked Alan Wagner about this strange coincidence of the honorary degrees. It was inconceivable, he explained, that someone would become a full professor at Yale without having a Yale degree. To remedy that, all professors were granted honorary degrees.

I certainly needed help in the pedigree department. I still saw myself as a fortunate farm boy from California with less-than-illustrious diplomas from Long Beach State and the University of Nevada. But unlike Oklahoma, Yale not only saw me as a rising star, they *treated* me that way. They were willing to do whatever it took. Moreover, their primate facilities were highly regarded. I began to make plans in my mind. I would move the entire chimpanzee signing project east—Washoe, Booee, Bruno, Thelma, and Cindy would all come, and the research would continue. Three years out of graduate school and I was suddenly on the brink of every academic's dream, a choice teaching and research position at one of the world's leading universities. It was very heady stuff.

All of this was before I toured Yale's primate facility, which was located three stories underground. The first thing I noticed when we descended into the labs was that no natural light ever penetrated this bunker. My tour guide, the head of the animal care department, told me that the animal care techs were in complete charge down here. I would have no authority or responsibility for Washoe. Apparently, most researchers were happy to hand over responsibility for their research subjects.

My guide then showed me the kind of room where Washoe would live. It was a tiny concrete cell, about seven feet square, with a steel door containing a small opening that was the room's only window. Along one side of the floor ran a trench drain for efficient cleaning. All in all it was the closest thing to a jar that you could build to house a primate. A chimpanzee couldn't even climb the walls.

"Do you allow toys in the rooms?" I asked him.

"Never," he answered. "They just collect feces and make cleaning harder."

"What about blankets?" I asked.

"Against the rules. They clog the drains."

Compared to this dungeon, the chimpanzee island in Oklahoma was looking pretty damned good. I may have resented Lemmon's cruelty, but at least the chimps had each other and some room to play.

I climbed up out of the bowels of the primate facility, and by the time I emerged from the building and took a deep breath of fresh air, I knew two things. First, I couldn't work at Yale, and second, my fantasy of how other universities conducted primate research was just that, a fantasy.

My illusions about academia sprang, I suppose, from my early experience in the Gardners' backyard, a place where science was conducted with compassion and respect for the research subject. It was an atmosphere, I now realized, that I had desperately tried to re-create in Oklahoma by defending our island of sanity against the surrounding insanity. Thanks to Lemmon I was no longer a stranger to the dark side of human power mongering. But all along, I had kept telling myself that Lemmon was an aberrant personality and a fringe scientist and that things were surely different elsewhere. I really believed that leading researchers in prestigious universities treated our closest primate relatives with the respect befitting such intelligent and social creatures. One of those universities was going to rescue me from my plight in Oklahoma—or so my fantasy went.

After visiting Yale's primate lab I felt like a child who has just been told there is no Santa Claus. Yale, I now realized, wasn't going to save me from anything. They would happily put Washoe, an eight-year-old social butterfly, in a cell with no toys, no blankets, and no friends. They obviously had no clue what Project Washoe was about, and it was a pretty good bet that no other university did either.

That evening, when Alan Wagner offered me the coveted teaching position, I asked for a week to think it over. I was trying to be diplomatic. He called me a few days later.

"You're not taking the job, are you?" he asked.

"No, to be honest, I've made up my mind not to."

"I knew it," he said with absolute certainty.

"How did you know?" I asked him.

"Nobody asks Yale to wait a week," he explained.

My rejection of Yale seemed to work out for the best. It gave me a new appreciation for the chimpanzee island and the youngsters' freedom to socialize. And after Yale's offer, the University of Oklahoma stopped taking me for granted. They put me on the "wunderkind" track and within four years I was a full professor with tenure.

Although I never once regretted my decision to stay in Oklahoma, many colleagues questioned my judgment. They couldn't imagine why any scientist in his right mind would turn his back on academic prestige. I guess they never understood my relationship with Washoe. She was a member of my family. It was never a possibility for me to lock her away in an underground vault at Yale or any other school, no matter how prestigious. Looking back, this was only the first of many times I would be forced to choose between Washoe's welfare and my own academic ambition.

LIFE FOR THE CHIMPS ON THE ISLAND was infinitely better than solitary confinement in a university basement, but it had its own share of challenges and traumas. The first came in June 1974 when Ally's mother, Sheri Roush, announced that she was getting married and that there was no room for Ally in her new family. The chimp would have to go. Bill Lemmon decided that Ally would come back to the Institute and live on the island with Washoe and the others. After growing up human, Ally was about to discover chimpanzees. I wanted to avoid a replay of Washoe's scary and violent entry into chimpanzee society, so I prevailed upon Lemmon to let me ease Ally into his new life with a few visits before the big move.

I will never forget Ally's first trip to the chimpanzee island. Like a kid visiting the zoo for the very first time, he was full of excitement and surprise. Hand in hand, we walked around as he pointed and gawked at Washoe and the other funny animals on the island. The other chimps were more curious than jealous about my new friend, and they checked him out thoroughly. Booee and Bruno swaggered and displayed at the new kid, and Ally seemed to find it all quite entertaining.

A few days later I took Ally back for another visit. Once again he was excited about seeing the strange animals. When we stepped ashore on the island, Washoe, who now recognized Ally, came over and signed a greeting to him. *This is great*, I thought, *maybe these two will talk and be friends*.

But Ally did not sign back. He froze like a statue, staring at Washoe the way you might if you encountered a talking dog. It was a long and terrible moment. I can only guess that a light went off in Ally's mind and he finally put two and two together: *I am one of them*. Ally had an instant identity crisis. He let out a bloodcurdling scream, then lapsed into a full-blown panic attack.

The scream brought Lemmon running down to the shoreline. He sized up the situation and proclaimed Ally's fate with all the compassion of a hanging judge: "Leave him on the island. Tell his mother he's not coming back."

I didn't know what lay in store for Ally. But Lemmon did. He had written about home-raised chimpanzees who were suddenly taken from their human mothers. Lemmon described their "anaclitic depressions and atypical neurological states which in a human infant would have been indicative of severe central nervous pathology." In other words, they went crazy.

Ally was no exception. He developed a hysterical paralysis, losing the use of his right arm. And he lapsed into a deep depression that was absolutely shocking to see in a chimpanzee who had been a bundle of energy his entire life. Ally stopped eating. He stopped signing. He pulled out his hair. He was unreachable and inconsolable.

The only thing I could think to do was to hold Ally and hope that the physical contact—like a family member's in the wild—would comfort him and get him through the terrible loss of his mother. One of my assistants, Bill Chown, and I began carrying Ally on the island, off the island, in the woods, just about everywhere. Ally probably weighed seventy pounds, but we held him on our chests for every waking minute of every day. He was never alone.

Finally, after two full months of constant loving care, Ally emerged from his terrible darkness. He began eating, and he started signing again. Eventually, we had him start spending more and more time each day on the island with the other young chimps. After establishing their dominance, Booee and Bruno, who were then seven and six, respectively, adopted five-year-old Ally as "one of the boys," like a younger brother who is allowed to tag along. They encouraged Ally's clownish antics. Even when they played pranks on Ally, which was quite often, he always ended up laughing at himself. It was impossible not to love Ally.

By this time, in late 1974, Bruno was no longer the kingpin on the island. Booee, who was bigger and older, had become the dominant male, and his new power came into play in a very interesting way. I was curious to see if the three boys would use ASL to communicate with one another, especially because Ally was so spontaneous and talkative around humans. But it turned out that Ally was extremely hesitant to sign to the more dominant chimps. When Booee wanted something from Ally he would look him in the eye, sign to him, and then poke him in the chest with the YOU sign, like a football coach giving commands to his quarterback. But Ally would rarely make eye contact with Booee, much less sign to him or touch him.

It got to the point where even Booee grew tired of Ally's downcast gaze and reticence. And if Ally wanted something from Booee—usually a piece of food—Booee would wait for Ally to ask for it properly before giving it to him. If Ally was too shy to ask, which was usually the case, then Booee would poke Ally to get his attention and then sign YOU GIVE ME FOOD. Booee would continue this prompting until Ally looked up momentarily and signed YOU GIVE ME FOOD. Satisfied, Booee would hand over the food.

We quickly discovered that dominance played a major role in the chimps' communication; who was talking to whom was very important. This made sense to me. After all, effective communication is the art of knowing your audience, and nothing is more important in chimpanzee society than knowing where you stand vis-à-vis every other member of the community. Today, we know that *all* species of primates adjust their communication to fit the social situation. The baby mouse lemur, for example, produces a distinctive array of vocal calls only when interacting with its mother; likewise, an adult lemur adjusts its vocal calls to suit a dominant or subordinate listener.

Human primates are no different. The "motherese" studies of the 1970s established that both speaking and signing mothers modify their language when addressing their infants. Later studies showed that children altered their *own* speech to suit different listeners. For instance, they spoke one way to younger children, another way to their mothers, and still another to their peers. Even two-year-olds speak one way to familiar adults and another way to unfamiliar adults.

The fact that children speak differently to their teachers than they do to their friends may seem like common sense today. But back in 1973 the idea that someone might shape their language to suit the listener went against the prevailing Chomskyan theory that language was an abstract system of logic that the child applied in one innate way, no matter whom he or she was addressing.

My observations on the island led me to undertake, with the help of two speech pathology students, the first comparative study of how chimpanzees and deaf children communicate with American Sign Language. In addition to Booee, Bruno, and Ally we studied three deaf children—Gwen, Jeff, and Sharon—in a Special Services School in Oklahoma. The three children, who were also six to seven years old, were as sensitive to social hierarchy as the chimps, but for different reasons. Gwen, who had partial hearing, dominated the other two kids, both of whom had profound hearing loss; Jeff was the next most dominant, followed by Sharon, who was the most submissive.

The chimpanzee and human youngsters mirrored one another in most of their dominance behavior. In both groups, virtually all of the touching behavior followed a single rule: the less dominant the youngster, the more that youngster was touched and the less that youngster touched the more dominant individuals. In a typical two-minute video segment, Booee touched Bruno fourteen times and Ally thirty times. Ally, the most submissive, touched Bruno once and Booee not at all. Likewise, Gwen, the most dominant child, touched Sharon, who was the most submissive, four times, but Sharon *never* touched Gwen.

Both the chimps and the deaf children showed similar respect for authority by rarely touching their teachers. And both the chimps and the children tended to touch one another more when the teacher was absent and the setting was less formal. Dominance also determined the amount of sustained eye contact—who was looking at whom. But here the children and chimps followed opposite rules. For the chimpanzees, respect for authority meant that they avoided eye contact, in the same way they avoided touching those more dominant. This is also the case in many human cultures where people signal submission by avoiding eye contact. But, of course, in American culture children are taught to "look up to authority" and to look others "in the eye." In keeping with this, the deaf children paid close attention to their teacher.

When it came to using American Sign Language, we were not surprised to find that the children, like the chimps, obeyed their social hierarchy, and the most dominant kids talked the most. But both groups also shaped their signing to fit the social situation. When the chimps and children were interacting with their teachers, their signing became more formal and precise. But when the teachers went away, the child-to-child and chimp-to-chimp signing became less perfect and much more relaxed.

Booee and Bruno had already taught me that every chimpanzee, like every child, has a distinctive style of learning language. Now they showed me that language couldn't be separated from the social relations in which it emerges. This insight, more than any other, would shape my work for the next twenty years. Other ape language researchers studied one chimpanzee at a time; for example, they observed a chimp interacting with a computer or plastic tokens or movable symbols or a human researcher. It always seemed to me that this approach repeated the error of linguists in the 1960s who assumed that language could be reduced to a system of logical interactions outside of any social context. I wanted to know how chimpanzees talked among themselves.

NOT ALL NEWCOMERS TO THE CHIMPANZEE ISLAND found their place in the primate pecking order as easily as Ally. One time we were attempting to introduce a recently arrived Peace Corps chimp named Candy, but, like a new kid in the school yard, she was mercilessly teased and harassed by Bruno and Booee. After many weeks of hazing, Candy was finally taken under Washoe's watchful eye, and things settled down.

But Washoe couldn't protect Candy twenty-four hours a day. One morning I couldn't find Candy and I became worried that she had attempted to jump the moat and had drowned. My students and I waded into the water up to our chests and began dredging the muddy bottom with poles. After an hour or so, I felt her small body under my feet, and went down to retrieve it. As I emerged from the moat, cradling Candy's lifeless body in my arms, the other chimps kept their distance, eyeing me with the morbid interest of passersby at a traffic accident. I had never seen, much less held, a dead chimpanzee before. It was heartbreaking. Chimpanzees, like human children, are so animated in every expression, so vibrant in every leap, that their spirit seems like the very essence of life itself. Drained of Candy's spirit, the stiff body I held was just an empty vessel.

After Candy drowned, Lemmon installed an electric fence around the perimeter of the island to prevent any more such deaths. But in the summer of 1974, we discovered that the fence was not foolproof. One morning we introduced a new female, named Penny, onto the island. That same afternoon, while playing by the shoreline, I suddenly heard Penny screaming in terror from the other side of the island. She must have panicked at being left alone with the other chimps. The next thing I heard was a loud splash, the sound of Penny hitting the water in the moat. She had taken a running start and vaulted over the electric fence.

I took out my wallet, threw it on the ground, and started racing toward the pond, intent on diving in to save her. As I ran it occurred to me that this rescue attempt could turn into a double disaster. Trying to save a panicked chimp in deep water is dangerous business. She could easily pull me down with her.

As I neared the fence I was surprised to see Washoe sprint ahead of me and leap over the two electric wires. She landed, thank heaven, on the narrow dirt ledge that dropped sharply off into the pond. After sinking like a stone, Penny had now surfaced near the island's shore and was thrashing about wildly. Then she submerged again. With one hand grasping the bottom of an electric fence post, Washoe stepped out onto the slippery mud at the water's edge. She reached out her other long arm, grabbed one of Penny's flailing arms, and pulled her to the safety of the bank. I ran to get the boat and rowed as fast as I could to where the two girls were huddled outside the fence. Penny was in shock, shivering and terror stricken. I got the two of them back onto the island, and Washoe and I sat grooming Penny for a long time.

While Penny was calming down, I had time to gather my wits, and to let the enormity of what I'd just seen sink in. Washoe had risked her own life to save another chimpanzee— one she had known for only a few hours.

THOUGH THE MOAT was an effective, and sometimes deadly, way of keeping the younger chimps from escaping, it was a different story when it came to containing the adult chimps. Despite the fortified main colony and Lemmon's elaborate security procedures, there were the inevitable breakouts. The male chimps, especially, probed their building's defenses like a team of soldiers in a prisoner of war camp.

Their method was quite systematic. One of the chimps would

begin twisting the exposed end piece of heavy chain link that enclosed the main colony. When the first chimp grew tired another would take over. This project would continue for days and days until the chain link cracked from metal fatigue, and the chimps could begin unraveling the entire wall of mesh. The most remarkable thing about this bit of sabotage was that they were able to keep it invisible. We never actually saw them working on the chain link, because they would stop when any of us entered the colony. It was only by peeking in through the windows from outside that we caught them in the act.

Once they escaped from the main colony, the males and females reacted very differently. The females would usually wander off into the woods and go about a quarter mile or so before giving up. But the males would rush out and then stand there stock-still, as if they had no idea what to do next. If they had anything in mind it was usually food. One time when Pan escaped, he raided the food stocks and brought back a supply of Coke syrup to share with his grateful cell mates.

One weekday afternoon I got a call at home from Steve Temerlin, Maury and Jane's teenage son, who was working as a tech for Lemmon. "Come over right away," he said, "I can't get Burris in his cage." Burris was the sorriest chimp I'd ever known. He had been raised by cowboys who kept him chained to a doghouse. When he was twelve years old, Buddy, as the cowboys called him, was dumped at the Institute. Lemmon changed Buddy's name to Burris, after Burris Frederick (B. F.) Skinner. Lemmon hated Skinner.

Raised like a dog, Burris lacked some rather basic chimp behaviors. He didn't know how to groom, and he couldn't climb. He'd come to a fence and just stop. Most of the time, he sat alone in a cage in the pig barn and masturbated. The day came of course when Lemmon socialized Burris by dumping him in the main colony. The adult males promptly beat the hell out of him.

Although Lemmon derided my attempts to relate to chimps, he thought my friendly approach might be useful with Burris and he asked me to try to rehabilitate the newcomer. I put Burris back in his own cage and took him out on long walks. I tried to teach him to groom, and I showed him how to climb trees. Things were slowly improving—until Steve's call.

For some reason Steve had put Burris back in with the adults, and of course they had attacked him. Now Burris was in a holding cage in the main colony, and he adamantly refused to go back to his own solitary cage via the tunnel. When I arrived, Steve was just opening the cage door to try to put a lead on Burris. In a flash, the big male burst out of the cage and fled the building in a rage. He made a beeline for Lemmon's house. This was bad news. The only person home was Lemmon's housekeeper, a sweet old lady named Mrs. Daniels. I sprinted toward the house and somehow beat Burris to the front door, where I planted myself directly in his path. Burris stopped and pounded on the glass window next to the door until he smashed it to bits.

"Who's there?" Mrs. Daniels called out in her cheeriest voice.

"LOCK YOURSELF IN A CLOSET!" I screamed to her.

Burris was now completely freaked out. He looked possessed, and he started circling the house like an armored tank, looking for weak points in the defenses. I ran alongside him, pathetically trying to groom his shoulder. Steve reappeared with a pellet gun and shot a round into Burris's back. There was no reaction at all. Burris now turned and headed for the pig barn, his old home, with me in pursuit.

Once inside the barn, Burris wheeled and displayed in preparation for an attack. I stood my ground, ten yards away. He came at me like a charging bull, and I jumped aside, matador-like, at the very last moment. After missing me, he screeched to a halt, turned around, and began displaying again. Eyeing a bag of raisins on the shelf, I grabbed and opened it, then threw the whole thing at Burris. It bounced off him. He paid no attention to it at all and began his next charge. Once again, I leapt out of his way—this time right into an alcove enclosed on three sides by chain link.

I knew right away that I'd made a big mistake. Burris doubled back and blocked my only escape route. I was trapped. This is it, I thought—I'm dead. To stop Burris from biting me, I grabbed his head with both hands and held on for dear life. He grabbed me around the legs, picked me up, and began slamming me into the concrete floor—BAM BAM BAM—like a pile driver hammering away at a sidewalk.

At that moment Steve reappeared, aimed his gun at Burris, and blasted him with another round. Enraged, Burris dropped me, mid-slam, and turned to Steve, who now fled the barn, closing and bolting the door behind him. Great, I thought. Now I'm *locked* in a room with a murderous chimp—I'd have better luck in a ring with Muhammad Ali. Burris was heading back to finish me off. I started praying.

But just then Burris stopped and looked down. I followed his gaze to the floor and saw that giant pile of raisins. Burris couldn't believe his eyes! He snapped out of the trance he'd been in for the last twenty minutes, sat down, and started stuffing his face like a kid in a candy store. I edged out of my death trap, grabbed a lead, and gently attached it to his neck chain. Burris and I went for a little stroll, sharing our raisins and grooming one another, just like it was a lovely day in the park and he hadn't come within an inch of killing me. After I locked him in his cage, I lay down on the grass and fell into a deep sleep.

Twenty minutes later I woke up to the sound of Lemmon's Mercedes roaring up the driveway and screeching to a halt. Evidently, Steve had called and pulled him out of a therapy session, yelling: "Burris is on the loose! Burris is on the loose!" The raisin ploy may or may not have saved *my* life, but there's no question it saved Burris. Had he still been running amok when Lemmon showed up, Burris would have been taking real bullets, not pellets.

EIGHT

AUTISM AND THE ORIGINS
OF LANGUAGE

ONE DAY IN LATE 1971 I left the chimps on the island and
drove to the University of Oklahoma Medical School in
Oklahoma City. Accompanied by an old friend and clinical psychologist named George Prigatano, I passed through the double
doors of its hospital, proceeded to a small room on the second
floor, and met a nine-year-old boy named David.

David was a classic case of infantile autism. Autism is a developmental disorder characterized by lack of speech and eye
contact, obsessive and repetitive body movements, and an ina-
bility to acknowledge the existence or feelings of other people.
The autistic child lives in a kind of glass bowl, inhabiting a
separate reality from those around him. Autism had intrigued
me ever since I studied it in college, so when George began
telling me about his young patient David, I listened intently.
Soon, a rather unorthodox idea for treatment began to take
shape in my mind.

There was no shortage of theories about autism at the time.
Bruno Bettelheim, the renowned psychologist who ran a school
for emotionally disturbed children, blamed autism on cold, unfeeling mothers and ordered that they not visit their children at
his school. Dr. Ivar Lovass took a more Skinnerian approach,
treating autistic children with reward and punishment therapy
that included shocks from a cattle prod to discourage aberrant
behavior. And another physician, at our medical school's

hospital, claimed that autistic children were overstimulated; he recommended depriving them of stimuli by locking them in a padded room. All of these approaches stressed the use of speech therapy to one degree or another—with discouraging results—and none of them had helped David.

I came on the scene right after David's first and last day of deprivation therapy. When they'd locked him in the padded room, David had screamed so loudly that his mother demanded he be let out immediately. The hospital then wrote David off as a hopeless case. He was now too old, the doctors said, to respond to psychological therapy or speech therapy. Out of desperation, David's mother agreed to give my new idea a go.

I had a hunch that David could learn American Sign Language. I was no expert in autism, but certain aspects of autistic behavior begged for a visual approach to language. For one thing, most autistic children seemed to have problems with processing and responding to auditory stimuli—sounds. So when doctors said that autistic kids had "language problems," what they really meant was that these children had problems with *spoken* language. Meanwhile, I had turned up several studies from the late 1960s that showed that many autistic children responded to facial expression, gestures, and being touched. It seemed to me that modern psychologists were focusing on the wrong channel of communication, just as the early ape language experimenters had done. Forcing vocal speech on an autistic child made about as little sense as forcing it on a chimpanzee.

It wasn't all that surprising that no one was trying sign language with autistic kids. Most psychologists, like most linguists, were from the "speech is special" school. Until the 1970s deafness itself was viewed as a kind of pathology that was "treated" by forcing deaf children to lip-read and speak. Similarly, doctors expected autistic children to use language the "real" way, either to speak it or not to bother.

My sign language theory seemed reasonable enough. But Booee, Bruno, Thelma, and Cindy had taught me that there is no one approach that works for everyone; every child learns differently. So I asked David's mother to let me visit her son one or two times to observe his behavior. I wanted to approach him like an ethologist who respectfully studies another species. I assumed that David was seeking out and processing information in some meaningful way. My job was to figure out how.

When I came into the hospital room that first day, David was sitting in a chair, staring up at the ceiling light and waving his right hand back and forth rapidly in front of his eyes. Then he began rocking back and forth, reaching behind himself to rub his thumb along the back of the chair. After a few minutes of this, David got up, went over to the desk, and began quickly leafing through a book, over and over again. Then he turned toward me and approached me, never making eye contact, as if I were just another piece of furniture. He reached into my shirt pocket, grabbed my pipe, and began playing with it. Toward the end of the session, David walked over to the wall, planted his face against it, and let out a piercing scream while nodding his head up and down.

These were all examples of classic "stereotypies"—meaningless, repetitive behavior. But while that behavior may have seemed meaningless to us, it must have been meaningful to David, so I began searching for clues that might explain how he processed information. For starters, it was obvious that David sought out visual stimulation (staring at the fluorescent light and the pages in the book). In addition, he didn't have any problem responding to these visual stimuli with motor activity (waving his hand in front of his face, flipping the page, playing with my pipe and keys). He could also make motor-motor connections—rocking his body while rubbing his thumb. But David seemed unable to process auditory stimuli and visual stimuli at the same time. That's why he hid his face in the wall when he screamed; he was shutting out his visual field while he made noise.

David evidently had a problem connecting information from the auditory channel with information from the visual channel. This is a brain function known as cross-modal transfer. He could associate visual with visual, visual with motor, and motor with motor, but he could not integrate visual and auditory information. And any person who doesn't have this auditory-visual link is going to have a very hard time learning spoken language. If I hold up a pen and say, "This is a pen," most children immediately make the connection. But for someone like David, who had two separate sensory systems, it was like watching a movie with very bad dubbing; the sound was confusing at best and terrifying at worst. No wonder he avoided his mother and everyone else. Their speech was disorienting. (I got a better feeling for this distorted auditory reality while working with another autistic child. She wouldn't respond whenever the phone rang, but five seconds later she would start screaming.)

After observing David I was more certain than ever that sign language might work for him. Signing would capitalize on David's two working channels, the visual and the motor. His mother agreed to try, and the next week we met in the hospital again. After David sat and rocked for a while he went to the door and began turning the locked doorknob furiously, as if he wanted to leave. I took both his hands in mine, shaped them into the ASL sign for OPEN (palms down, side by side), and then moved his hands through the sign (opening them up and out, as if opening a book). When we got into the hall, David began to run. I stopped him, took his hands again, and moved them through the sign for RUN (brushing the right-hand palm against the left-hand palm).

A week later, we met again. This time David went over to the door and signed OPEN. When we got into the hallway, he signed RUN, and we ran up and down the halls together. From that moment on, I had no problem getting David to learn signs. He grabbed them from me like he was possessed by a pent-up need to communicate visually, which of course he was. David and I met only once a week for half an hour, but within two months he mastered a small vocabulary of signs and began combining them into phrases like YOU ME RUN.

Armed with these signs, David broke through the "glass

bowl" that had encased him in a separate reality for nine long years. His behavior changed dramatically. There was no more screaming and rocking, and he was actually making eye contact with other human beings, initiating games, and inventing his own gestures to communicate what he wanted. For David's mother, the sight of her autistic son signing MAMA was practically a miracle. The doctors and nurses at the University Medical Center were equally flabbergasted. It was the eye contact, more than anything, that stunned them. They couldn't believe that this was the same child who had never acknowledged their existence.

David's signing and the radical change in his demeanor were impressive, and my approach would certainly qualify as a breakthrough if it could work with other autistic children. But that was not the end of the story. A few weeks after David began signing, something quite extraordinary and unexpected happened: David began *speaking*. First it was only one word at a time: "open," "mama," "drink." Then, as he began combining his signs into phrases, David also began putting words into phrases: "Gimme drink."

I was completely stumped by this development. Why would his signing, which was visual, lead him to speak, which was auditory and vocal? I had used spoken words with David while signing (an approach called "total communication"), but I never tried to get him to speak back. It seemed that his signing had triggered a capacity for speech. But how?

It could have been a fluke, so I decided to try the same ASL therapy with another autistic child, a five-year-old boy named Mark. Mark was extremely hyperactive. The first time I visited him at home he was spinning wildly, wringing his hands, and making unintelligible sounds. His parents said he often laughed and cried uncontrollably and attacked himself and others. He had been seen by five different doctors, including two pediatric neurologists and a psychiatrist. He had been thrown out of three different schools, including one for learning disabled and emotionally disturbed children.

I began teaching Mark sign language in half-hour sessions, twice a week. Like David, Mark learned his first sign—GIMME—during the very first week. In our second week he signed his first phrase—GIMME KEY. By the fourth week, I counted a cumulative total of one hundred signed responses from Mark.

Then, as if on cue, Mark began speaking. First it was one word in our fifth week. Then more and more words until the tenth week, when he began combining words into phrases. As his signed phrases grew longer (GIMME MORE DRINK), so did his spoken phrases. When I charted Mark's progress in language development, there were four adjacent curves, each beginning a few weeks apart, that traced the same arc. There was one curve for signs, another for signed phrases, a third for words, and a fourth for word phrases. Mark's signing was clearly facilitating his speech. By our twentieth hour together, he created a game by putting keys in my pocket and then removing them while saying, "find a key," "gimme back key," or "back key going out."

Like David, Mark underwent a complete personality change. He became attentive during our sessions, which he always looked forward to. He initiated games with me, signing PLEASE TICKLE or molding my hand into the TICKLE sign, then playfully attacking me. When his parents hugged him he would hug back, which was an amazing development. Mark was still not an average five-year-old, but he was no longer living in his own world.

By the time I reported the results of this study in the *Journal of Autism and Childhood Schizophrenia* in 1976, I had discovered that at least two other teams of researchers were trying sign language with autistic children and had reported results similar to mine: all of the children had learned some signs, all became more interactive with the people around them, and *some* of the children had begun spontaneously speaking. Speech was beginning to look like a pleasant side effect of sign language training in autistic children, although nobody had the slightest idea why. This "side effect" phenomenon happened to disprove the old canard that teaching a child sign language would prevent him from ever learning to speak. Clearly, that was wrong.

As I traveled around the country, talking to dozens of university audiences about ape language study, I made a special plea to speech therapists and psychologists to abandon vocal training with the autistic and try sign language instead. Today, twenty years later, ASL training is regularly used as a language intervention technique with many autistic children.

But the question remained, *why* did signing promote speech? I puzzled over this endlessly until, as so often happens in science, I stumbled on the answer where I least expected to find it. Early in 1977 I was speaking at the University of Western Ontario in London, Canada. One of my hosts was a neurologist, Dr. Doreen Kimura, who had just done some interesting research on aphasics, people who suffered loss or impairment of speech as a result of damage to the left hemisphere of the brain. Kimura discovered that these patients also had difficulty with sequences of fine motor movements that involved the fingers. For example, when told to press a button and then grab a handle, they would press the button but then they would *press* the handle instead of grabbing it.

Kimura explained to me that the region of the brain that controlled speech also appeared to control precise hand movements. Patients with aphasia could understand or produce a word, but they could not combine words into a sentence. Likewise, they could carry out one motor movement—pressing a button—but they couldn't organize a sequence of such movements.

It took about one second for this to sink in. *Speech involves precise and sequential motor movements*. That explained perfectly why the autistic children had progressed from signing to speech. Once they learned to communicate by the fine motor movements of signs, they naturally began expressing themselves through another form of fine motor movements—spoken words. By focusing on the *difference* between signed language and spoken language—you see one and hear the other—I had overlooked the obvious fact that they are both forms of gesture.

Sign language uses gesture of the hands; spoken language is

gesture of the tongue. The tongue makes precise movements, stopping at specific places around the mouth so that we can produce certain sounds. The hands and fingers stop at precise places around the body to produce signs. These precision movements of the tongue and hands are not just related, they are *connected* through the motor regions of the brain. Charles Darwin noted this connection in an activity we are all familiar with: when people move their fingers very precisely—while threading a needle, for example—they often make sympathetic movements with their tongues. And Doreen Kimura noticed that people make certain kinds of free hand movement only when they're moving their tongue while speaking.

Kimura's breakthrough—that precision tongue and hand movements are controlled by the same areas of the brain—supported a theory, put forward several years earlier by an anthropologist named Gordon Hewes, that the origins of language were gestural. Hewes said that early hominids communicated with their hands, which naturally led them to develop other skills, like toolmaking, that also required precise hand movements. Speech evolved later out of this same ability to follow "complex patterned sequences." According to Hewes, the distinguishing characteristic of early man was his evolving capacity for "syntax"—an ability to devise and follow complex programs of action, whether in tools, signs, or words.

Hewes's theory helped explain why modern chimps were able to make and use simple tools; our own toolmaking skills, like theirs, were rooted in the cognition and neuromuscular control of our common ape ancestors. It also explained why Ally could apply a simple rule of grammar. Language, like toolmaking, was grounded in a neuromuscular syntax that arose from the animal kingdom.

But Hewes didn't answer the puzzling question of how speech might have *physically* evolved out of gestures. Words and gestures may both share a "syntax" of complex sequences, but there is a big jump from motions of the hand to streams of words from the mouth. How did our hominid ancestors bridge that gap?

Kimura located that bridge in the neural mechanism that connects movement of the hands with movements of the tongue. But it was two autistic boys, David and Mark, who dramatized this bridge by crossing it in a matter of weeks. In doing so, David and Mark may very well have retraced the evolutionary path of our own ancestors, a six-million-year journey that led from apelike gesture to modern human speech.

FOR THOUSANDS OF YEARS, people have noticed two curious facts about human communication. First, infants begin gesturing—showing, pointing, glancing—before they begin speaking, and second, gesture is a kind of universal language that all of us fall back on when we can't communicate through a common spoken language. Both of these observations led to the very old idea that language may have originated in gesture.

Modern linguists, however, dismissed the gestures of early hominids as unrelated to the spoken language that followed. And they also dismissed the first gestures of modern-day infants as unrelated to the words those children later form. This discounting of gesture was partly due to an inherent bias of speakers—whether they are linguists, child psychologists, or anthropologists—who tend to equate language with speech. The power of words is vested with a great deal of magic and mystique in every human culture. Speech, after all, is the distinguishing trait of humankind in creation myths around the world. It seems difficult for most people to imagine Adam in the Bible naming the animals with signs instead of words. In addition to this bias, gesture looked like an evolutionary dead end to linguists until they discovered, in the 1960s, that sign languages are every bit as complex and grammatical as spoken languages. (Darwin himself did not believe that spoken language evolved from gesture, probably because sign languages were so poorly understood during his time.)

Considering how unpopular and misunderstood gesture has always been, it's not surprising that the two main schools of thought about language origins both focus on speech. The "early origins" school says that language first appeared more than one million years ago, along with stone tools and the enlarging brain of *Homo habilis* or early *Homo erectus*. The "recent origins" school argues that language arose only in the past hundred thousand years among fully modern, big-brained humans who had fully descended vocal tracts that enabled them to produce speech.

Both of these theories of a vocal origin for language run into several major evolutionary obstacles. Pick up almost any book on the origins of language and you will find the author grappling with the first problem for the vocal approach: *How did the grunts of apes evolve into the words of hominids?* This doesn't seem too difficult—perhaps "ugh ugh" became "ma ma"—until you realize that an ape's involuntary grunts, like human screams, are controlled by the limbic system, the most primitive part of the brain. If our power of speech evolved directly out of the limbic system we'd never be able to convey a simple message like "There's a lion standing behind you" without bursting into uncontrollable alarm calls or screams. And in fact, in the human brain, voluntary speech is not controlled by the limbic system.

Fortunately for us, language didn't have to evolve out of the grunts of apes, or we might still be grunting. Washoe's ability to sign showed us that even though our common ape ancestors may not have been able to control their food barks or alarm calls, they could communicate with voluntary and visible gestures. We know that evolution always follows the path of least resistance, so our earliest quadrupedal, hominid ancestors must have communicated with their hands, just like their quadrupedal ape cousins did. Once these early hominids began walking upright, their hands were freed to make even more elaborate gestures, eventually stringing together sequences of gestures to convey more specific information.

Here, in the expansion of the language system, is where the vocal schools run into a second, and even more formidable, evolutionary roadblock. Even if words somehow emerged from ape grunts, and even if humans built up a large vocabulary for naming different objects, this naming ability is still a far cry from a rule-governed language. How did early humans get from "I" and "bear" to "I caught the bear" or "The bear caught me"? How did they make the great leap from individual symbols to a logical system that can create millions of meanings?

This leap is so vast that most linguists say random chance must have been involved. The linguist Derek Bickerton has said that "syntax must have emerged in one piece, at one time—the most likely cause being some kind of mutation that affected the organization of the brain." In other words, we hit the jackpot, biologically speaking, and a universal grammar was the result. Other linguists have suggested an even more unlikely scenario: a *series* of fortuitous mutations that hardwired a universal grammar into the hominid brain over time.

But experts in sign language, who assume a gestural origin for language, can explain the emergence of syntax in a much simpler, more commonsense way. You can test it yourself right now by following this suggestion of David Armstrong, William Stokoe, and Sherman Wilcox from their book, *Gesture and the Nature of Language:*

"If you will, swing your right hand across in front of your body and catch the upraised forefinger of your left hand."

By enacting this gesture, say the authors, you have just illustrated the most primitive form of syntax. "The dominant hand is the agent (it acts), its swinging grasp is the action (verb), and the stationary finger is the patient or object. The grammarians' symbolic notation for this is familiar: SVO [subject-verb-object]."

It is easy to imagine our earliest ancestors using this gesture to communicate, HAWK CAUGHT GOPHER. And they might have modified this sentence with adjectives (two fingers for two gophers) and adverbs (raised eyebrows for expressing disbelief: HAWK *SOMEHOW* CAUGHT GOPHER). These variations on a relationship are the beginnings of language as we know it.

The above example illustrates the essential difference between a primitive vocal system and a primitive gestural system: words symbolize objects; gestures symbolize relationships. Getting from spoken words to spoken grammar involves a huge leap, which is why linguists assume that one or more major brain mutations were necessary. But getting from *gestures* to grammar is no leap at all. Gesture *is* grammar. Early humans didn't need to have the grammatical rule for subject-object-verb encoded in their brains if they could already perceive that relationship in the world and mirror it in a gesture.

Over time, this gestural grammar would naturally become more complex, and the gestures themselves would evolve from gross motor movements to more precise motor movements. Driven by this, the human brain would get better and better at producing long sequences of fine motor movements. And that sequential cognition would produce, as Gordon Hewes theorized, a dividend: the ability to make and use more complex tools.

This is the point where the theory of a gestural origin for language traditionally ran into its own roadblock. It was fairly easy to see that as the gestural system became increasingly precise it would eventually produce today's modern sign languages. But how did a gestural system lead to *spoken* languages? This mystery was solved by my autistic students, Mark and David. Just as their signing triggered their first meaningful vocal sounds, our ancestors' precise gestures and toolmaking triggered precise movements in their tongues. My own guess is that our species began shifting to speech about two hundred thousand years ago. That date coincides with the appearance of markedly improved toolmaking among early *Homo sapiens*. These specialized stone tools were made by a process that required a precision grip, exacting pressure, and the kind of eye-finger-thumb coordination that Doreen Kimura found to be associated with vocal speech. In other words, the early humans who produced these tools possessed the kind of neural mechanisms that would have also let them produce words.

At this point, vocal words became a part of our ancestors' gestural communication. And there were immediate advantages even to primitive speech. Those who spoke could communicate words when their hands were full or when the listener's back was turned. Eventually, evolutionary pressures would bring about the innovations in our anatomy that were necessary for full-blown speech: we would develop a fully descended vocal tract and the ability to speak and comprehend words at increasing rates of speed. Over tens of thousands of years, spoken words would slowly crowd out gestures and become the dominant mode of human communication. In the meantime, humans would blend precise gestures and spoken words into a unified language system.

This prolonged overlap of gesture and speech overcomes the third and final hurdle that arises for any vocal origins theory. Before a spoken language could function on its own, it would require a low enough vocal tract, a minimal number of phonemes, and the ability to transmit those sounds quickly. In the meantime, our ancestors would have been speaking with a limited number of confusing sounds at a very slow rate and with lots of errors in comprehension—much like a modern two-year-old child. These inefficient and arbitrary sounds would probably not have conferred any adaptive advantage on our ancestors—*unless they were able to make their meanings clear by using gestures.* Without a gestural system to supplement speech in its first millennia of use, it probably would not have survived. As Gordon Hewes once wrote, "If all adults were stuck with the kinds of speech deficiencies normal enough in early childhood, we would probably still be using a well-developed sign language."

This gestural scenario explains how language might have evolved along an unbroken continuum over millions of years, without resorting to unlikely mutations or impossible leaps. It's also consistent with Charles Darwin's radical thesis that human language emerged from other forms of animal communication. Language is firmly rooted in the anatomy, cognition, and neuromuscular behavior of our common ape ancestors. Without this

Moja, seen here at age twenty-two, joined Washoe's family in 1979. She was the first chimp to paint representationally and has always loved dress-up clothes.

In the fall of 1967, I first entered the Gardners' backyard laboratory and began communicating with two-year-old Washoe in American Sign Language. TOP: Washoe swings from her favorite willow tree. BOTTOM: Washoe plays on the "monkey bars" with Susan Nichols, another student.

In 1970, I moved with Washoe to Dr. William Lemmon's Institute of Primate Studies at the University of Oklahoma, where I continued my chimpanzee language studies. One of my first students was a defiant young loner named Bruno. Here, Lemmon looks on as four-year-old Bruno signs TREE during a photo shoot for *Life* magazine.

When we moved to Oklahoma, Washoe's new chimpanzee friends became a part of our family. The sweetest of them all was Booee. Born in a biomedical lab, he was raised by a human family before coming to the Institute. ABOVE: Debbi carries our middle child, Rachel, on her back while holding four-year-old Booee. LEFT: Our son, Josh, plays chase with his new pal, Booee.

The "chimpanzee island" was a refuge for the young chimps and for
me as well. Communication on the island was a blend of ASL,
English, and chimpanzee vocalization, like some primate Tower of
Babel. TOP: Booee and Bruno communicating in ASL.
BOTTOM: Ally and myself on our way to the island in the rowboat.
After being raised as a human child for four years, Ally came to the
island in 1974 and met other chimpanzees for the very first time.

TOP: Ally signs MORE during a conversation in the woods.
RIGHT: Washoe and I go for a walk near the pond.

Every day I would make my rounds in Norman, teaching ASL to cross-fostered chimps being raised by human families. Salomé began learning signs at four months of age, about the same age when deaf children begin signing. Here, she and her human "sister," Robin, hug their mama, Susie Blakey.

Lucy was nearly six years old when I met her in 1970. She drank chablis with dinner, was adept with household appliances, and loved *Playgirl* magazine. TOP: Lucy and I share a cup of tea, which she brewed, before starting our daily ASL lesson. MIDDLE: Three-year-old Ally shares his lunch with the family cat, Talbot.
BOTTOM: Washoe signs RIDE.

In the summer of 1978 Washoe became pregnant. She and I began going for long walks in the woods where we would share food and converse. TOP LEFT: Washoe signs FRUIT. TOP RIGHT: I respond FRUIT as she reaches for some. BOTTOM LEFT: After I ask Washoe WHERE BABY? she points to her belly. BOTTOM RIGHT: Washoe is ready to head home, and she signs GO.

TOP: Washoe kisses her newborn son, Sequoyah.
ABOVE LEFT: After adopting ten-month-old Loulis from the Yerkes Regional Primate Center in Atlanta, we brought him back to Oklahoma in the van. ABOVE RIGHT: Loulis arrived home on March 24, 1979, Josh's twelfth birthday. Later that morning, we introduced him to Washoe.

In 1981 Tatu and Dar, the two youngest chimps in the Gardners'
second round of language studies, came to live with Washoe's
family. They had both been signing since infancy and each had a
vocabulary of more than 120 signs. TOP: Tatu with Debbi and
me. BOTTOM: Dar *(left)* and Loulis (signing WANT) became best
friends right away and have remained buddies ever since.

TOP: Tatu and I during a rare outing in Ellensburg, Washington.
BOTTOM: When my federal grant money ran out in 1981, we had to
become resourceful in order to feed Washoe's growing family. Here,
Debbi, six-year-old Hillary, and I sort through leftover-fruit bins at
the Albertsons supermarket.

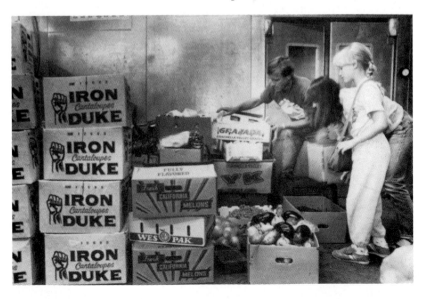

In March 1987, Jane Goodall and I toured the federally funded Sema biomedical lab in Rockville, Maryland. The appalling conditions we witnessed there led us to work together on behalf of all captive chimpanzees. LEFT: rows of "isolettes," the stainless-steel boxes in which many of Sema's chimpanzees were sealed. BOTTOM: An infant research subject alone behind bars.

Jane has visited Washoe's family many times over the years, and she played a crucial role in winning state funds to help build our Chimpanzee and Human Communication Institute in Ellensburg. Here she visits Tatu and me in 1983.

After years of planning, Washoe's family finally moved into its new home in 1993. TOP: The Chimpanzee and Human Communication Institute features a three-story-high, open-air mesh roof that enables the chimps to swing above their living space as they might do in a rain forest. BOTTOM: Washoe, Moja, and Tatu greet us at the door of their new home.

We want Washoe's family to live as naturally as possible within its social community, free from human intrusion. As a result, we do not enter Washoe's home except to clean, make repairs, and provide medical care. But we still visit and communicate with the chimps every day from the outside. BOTTOM: We show Loulis a book.

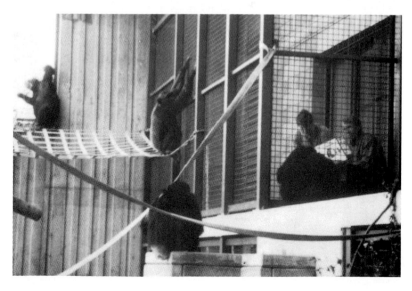

In 1996, after thirty years with Washoe, Debbi and I at last visited Africa and observed chimpanzees in the wild. Here, we visit some lucky chimps who were rescued from the poaching trade and are now living in sanctuaries run by the Jane Goodall Institute. We want to build similar sanctuaries in the United States as refuges for hundreds of chimpanzees who have outlived their usefulness to biomedical experimentation.

evolutionary continuity it is impossible to explain why modern-day chimpanzees are able to manipulate linguistic signs.

The continuity between gesture and speech also explains our own modern-day use of gesture whenever spoken language proves useless. As our species' oldest form of communication, gesture still functions as every culture's "second language." For example, when we are in a foreign country where we don't speak the language, standing near a noisy jet airplane, scuba diving beneath the ocean, or sending signals on the baseball diamond, we automatically revert to gesturing. And when the mechanisms for vocal language break down in any one individual—the deaf, the autistic, the mute, or any number of others—that person naturally adopts an entire *system* of gestural communication: sign language.

Human infants also illustrate the continuity between gesture and speech through that famous dictum of biology: *ontogeny recapitulates phylogeny*. The history of the individual retraces the evolutionary history of the species. Through the development of his body and behavior, each human infant roughly reenacts the multimillion-year journey of our ancestors from gesture of the hands to gesture of the tongue.

The human infant is born with a vocal tract like a chimpanzee's and is incapable of speech. He first communicates by facial expression and simple hand gestures. At five or six months of age, an infant exposed to sign language begins forming his first signs. At this same age, the larynx begins its long descent into the child's throat (it will not achieve the adult position until fourteen years of age), but he will not control his tongue sufficiently to form his first word until he is about one year old. Once he begins speaking he does not suddenly stop gesturing with his hands. Just as our hominid ancestors did, the child strings together words and gestures to make himself understood. Then, sometime between the ages of two and three, the vocal apparatus fully kicks in and there is an explosion of words that add to the gestural signals that we never stop using. Gesture of the hands and gesture of the tongue are forever inseparable.

It is important to point out that human language, whether signed or spoken, is not in any sense "better" than the communication system of wild chimpanzees. Evolution is not a ladder of "improvement" culminating in the human species; it is an ongoing process of adaptation for millions of related species, each on its own evolutionary pathway. Modern human communication and modern chimpanzee communication—like our different ways of walking, eating, and reproducing—are each an ideal product of six million years of adaptation. And both of these specialized products can be traced back to the gestures of our common ape ancestor. As a result, every time we speak or sign, we are displaying our evolutionary kinship with Washoe and other chimpanzees.

BY THE MID-1970S I seemed to have everything I ever wanted, personally and professionally. Debbi gave birth to our third child, Hillary, in 1975, and the five of us lived in a little country house on two acres along with some rabbits, chickens, cats, dogs, and an Appaloosa horse. It was like being back on the farm I loved so much during my childhood.

I was also riding a wave of enthusiasm for ape language research that brought me professional acclaim as well as ample research funding. The University of Oklahoma was not Yale, but it was the center of the universe when it came to chimpanzee signing. My work was being published regularly in the most respected journals, and doctoral candidates were flocking to Oklahoma to work with the chimps and me. I loved teaching and I got a great deal of satisfaction from sharing credit with my assistants and helping them launch their own scientific careers.

In 1974, I was invited to give a presentation on chimpanzee language acquisition at the world's first Conference on the Behavior of the Great Apes, held at Burg Wartenstein, Austria. There I was, just thirty-one years old, being welcomed by the giants of primatology—Jane Goodall, Junichiro Itani, Diane Fossey, Toshisada Nishida, and Biruté Galdikas. Just seven years earlier I had been one chimpanzee hug away from being a plumber. It felt like a dream, especially when I saw myself, along with my wacky friend Ally, in *People* magazine, a brand-new bellwether of American pop culture.

But the most unexpected dividend of my work was that I, an animal behaviorist, had somehow found a way to work with human children, which was the reason I had gone into psychology in the first place. I still can't imagine anything more dramatic and moving than seeing an autistic child stretch out his hands and form his very first sign, or open his mouth and speak his first word. The scientific significance of this was always secondary to me. Seeing "noncommunicative children" communicate was reward enough. And watching their families begin to heal and bond was a memory that would sustain me through the dark days ahead.

And I was heading into some pretty dark days—years of them. I may have been on top of the world, but something had been eating away at me for a long time, and now it threatened to destroy everything I'd achieved. Looking back, I realize that the trouble began when I started working with the autistic children in late 1971. The problem was not the kids. I loved the few hours I spent with them every week. The problem was returning to the Institute. Seeing the chimps confined to cages was much harder to stomach after spending the afternoon in a child's loving home.

What disturbed me most was my own role in these two environments. My work with David and Mark, who were "abnormal" children by any psychological standard, brought about obvious improvements in their lives. But Washoe and the other chimps, who were perfectly normal, were doomed to a life of confinement, far from their native African home. For the children, science brought hope and freedom. For the chimpanzees science meant imprisonment. This disparity was all the more upsetting because the sign language therapy that the children were benefiting from came directly from the research with Washoe.

I was beginning to see that all of my work was made possible only by the incarceration of my chimpanzee research subjects. I had become a jailer. And all the scientific acclaim in the world couldn't change that fact. My day-to-day interactions with the chimps involved cages, locks, keys, leads, cattle prods, and guns—and these tools of control had come to seem absolutely routine.

My awakening was inevitable, I suppose. I tried to make the best of a bad situation back in 1970 by compromising with William Lemmon. I leveraged Washoe's scientific status and my own funding to protect orphaned chimpanzees like Ally, Booee, Bruno, and the rest. I comforted myself in the knowledge that without me, Washoe and the others would be alone in the pig barn, as lonely and distressed as the pigs on the electric grids and the dying, lovelorn siamangs.

I had fought and won the small battles. The juveniles got their rundevaal shelter on the island. I got permission to take them on wonderful walks to forage in the woods. I won for Burris the right to live alone. I carried blanks instead of live ammo in my gun. But all of my victories had ultimately won me only one thing: the right to conduct science inside a prison. I may have been a *nice* jailer, but I was still the keeper of the keys. Every morning I let the chimps out of their cells, put on their leads, and led them, like a chain gang, to the island.

And things were only getting worse. In late 1974 Lemmon told me that he was going to move Washoe, Booee, Bruno, and Ally into the main adult colony permanently (Thelma and Cindy were already living with the adults). There was a new generation of young chimps on the island, and Lemmon wanted Washoe and the other adolescents to join the adults so that they would breed. Although Lemmon's students conducted some studies of maternal and sexual behavior in the main colony, the adult chimps' main function was to produce infants that could be sold or loaned to other researchers.

My truce with Lemmon had bought us four years of peace on the island, but now he was growing more and more obsessed with controlling the chimps. One day Lemmon told me he'd devised the ultimate "chimp containment plan," and he trotted out some Doberman pinschers. His plan was to build two concentric fences around the property. The Dobermans would live between the two fences, creating a death zone that no chimp would even think of crossing. Until he could build the fences, Lemmon chained the Dobermans at different locations each day to add an element of surprise that would discourage chimpanzee escapes.

One afternoon I was out walking with Booee on my back when we suddenly heard a dog chain uncoiling. My heart stopped, and Booee scrambled over my head and onto my chest. As I glanced back over my shoulder, I saw a Doberman already airborne, lunging at us with its fangs bared. He came within eighteen inches of us before the chain jerked him back to earth. Had we woken the Doberman two seconds earlier Booee and I would have been dog food.

DURING THIS SAME PERIOD, I came to the sobering conclusion that the entire field of ape language research was not as beneficent and benign as I had once thought. One by one the talking chimpanzees were growing up, and by the age of seven they were seen as obsolete by behavioral researchers. A full-grown chimpanzee was still learning; he was just too big, too strong, and too unpredictable to work with in a house or even in a caged environment. As a result of this, scientists were facing tough choices about what to do with their outsized chimpanzee subjects.

Project Washoe was first and foremost a cross-fostering experiment, in which a chimpanzee became socially and emotionally attached to a human family. But attachment is not a one-way street. Washoe had imprinted on us, and we had imprinted on her as well. I felt like that old mama cat on my family's farm, the one I tricked into hatching and raising a brood of ducklings. Those newborn ducklings may have attached to the mother cat, but she was hardly disinterested in them. In fact, she began acting strangely ducklike, proving that it takes two to cross-foster.

From the day I began working on Project Washoe, I had to break the first commandment of the behavioral sciences: *Thou shalt not love thy research subject.* I was being *paid* to love my research subject so that she would learn language in a natural family setting. The Gardners had shown me that behavioral research could be conducted humanely and compassionately while still maintaining scientific objectivity. Unfortunately, nobody warned me that I was supposed to stop loving Washoe when the experiment was over. By the time I realized the depth of my own feelings for her it was too late. I was attached.

But as I looked around me, I noticed that other scientists did *not* seem so attached to their chimpanzees. The Gardners obviously loved Washoe, but in the end they let her go. When it came time to choose between keeping the foster family together or advancing science, they chose science. And by 1972 it was clear where their own science was heading. Just months after they visited Washoe in Oklahoma, the Gardners adopted another infant chimpanzee, named Moja. And over the next four years, from 1973 to 1976, they adopted three more newborn chimps: Pili, Tatu, and Dar. Three of them—Moja, Pili, and Dar—were born in biomedical laboratories. The Gardners acquired Tatu from William Lemmon. She was the daughter of Thelma, my own distracted, dreamy pupil.

The Gardners' new experiment was extremely ambitious. They wanted to repeat Project Washoe, except this time the chimps would be exposed to ASL from birth, and they would have one another as playmates to talk with. They would also be surrounded by deaf, native signers. In short, they would have all the linguistic benefits that most human children have from the moment they're born. There was no doubt the Gardners' experiment would reveal, to a fuller extent, the chimpanzee's capacity to learn language. It promised to be a milestone of behavioral science.

But then what? What would happen to Moja, Pili, Tatu, and Dar after they became too big to handle? Where would they go? Would they survive like Washoe? Would they go crazy like Ally? Would they grieve to death like Maybelle? Now that we'd proven that chimpanzee children could bond with human parents, didn't we have some moral obligation to meet their emotional needs?

I'd begun asking myself these questions the day I left Reno with Washoe. In the intervening years I had sat at Maybelle's deathbed, I had watched little Salomé shrivel up and die, and I had clutched Ally's partially paralyzed body to mine day after day, desperately trying to rekindle his lifeless spirit. A person would have to be blind not to see that chimpanzee children were shattered emotionally by the separation from their human mothers. The ultimate proof that cross-fostering between humans and chimpanzees was successful was the fact that the fosterlings were dying when the family bond was broken.

These "adoptions" were beginning to look less like experiments in family attachment and more like nightmares of separation. No one ever seemed to consider the chimpanzee's point of view. And as I agonized over the fate of my chimpanzee friends, I suddenly found myself asking the questions I had never asked my mother as a child: Why did Curious George have to leave his home in the jungle? Why did "the man with the yellow hat" put the "good little monkey" in a zoo?

The answer, I now realized, was curiosity—*not* Curious George's but our own. We were all like the man with the yellow hat. Scientists were so curious about chimpanzees that we rationalized almost any behavior to satisfy that curiosity. We would sanction any use of chimpanzees if it might help answer an interesting scientific question. The space program invoked science to justify kidnapping chimpanzee infants from Africa in the first place. The Gardners invoked science when they sent Washoe away and took Tatu from Thelma. And in the name of psychotherapy Lemmon prescribed chimpanzees to his patients as if they were pills.

Nobody seemed to notice, much less question, that the chimpanzees were suffering. On the contrary, the scientists around me swore that this shuttling back and forth of chimpanzee babies was a *good* thing because it advanced our own human knowledge.

I don't know why I saw things differently, but I did. All the scientific rationalization in the world could no longer drown out the inner voice of my conscience. I was supposed to take comfort from the fact that my language research with chimpanzees was benefiting autistic children, but it only made me feel worse. The disparity between these loved children and the imprisoned chimps was now unbearable. People were constantly asking me, "Why don't you leave the chimps and work full-time with autistic children?"

"Because those kids have families," I answered. "The chimps have no families. All they have is me."

But by late 1974 I'd had enough. I didn't want to be part of a system that was breeding more chimpanzees for more suffering. Most of all, I didn't want to be a scientist if it meant incarcerating Washoe.

"I WANT TO SEND WASHOE BACK to Africa," I said to Debbi one night at home. "That's where she belongs. Things are only going to get worse here."

Debbi knew exactly what I meant. Washoe was now nine years old, and she was about to enter the adult colony. She would spend her lifetime behind bars. Her children, if she had any, might very well be taken from her. Her chimpanzee friends would come and go at the whim of science. She had Debbi and me, but that hardly seemed like enough. In Africa she could live a free existence in the jungle that nature had designed her for. Of course, I would miss Washoe terribly and without her I wouldn't have the heart to continue in ape language research. But at this point I would welcome the opportunity to spend more time working with autistic kids. I'd finally be free of the gloom of prison life.

"You have to be able to live with yourself," Debbi counseled.

That night I wrote a letter to Jane Goodall at her Gombe research station in Tanzania, telling her about my soul-searching and my new determination to return Washoe to the wild and get her as far away from science as possible.

I first met Jane when she visited Oklahoma in 1971, the same year she published her classic field study and widely read book *In the Shadow of Man*. Goodall had a very quiet but powerful presence, and I stood in awe of her scientific scholarship. But I was struck by something else the day Jane met Lucy and Washoe's friends: her sensitivity toward the chimpanzees as individuals. She saw each of them as a person. I felt certain that if anyone would understand my decision it was Jane. I asked her to help me with my plan for Washoe. Then I waited for her reply.

When it came two months later, I was totally blindsided. "That is the worst idea I've ever heard," she wrote. She went on to explain that Washoe could never be integrated into a chimpanzee group in the wild. As an outsider, she'd almost certainly be killed. Furthermore, the African nations were not financially able to support their own people, much less an expatriate chimp. Finally, it would be cruel to Washoe. She had been reared as a human child, diapered, spoon-fed, and privately tutored. How did I expect her to survive for ten minutes in the African jungle? Just to be sure I grasped this last point, Jane compared my proposal to abandoning a ten-year-old American girl in the wilderness, naked and hungry, and telling her to return to her roots in nature. This was romanticism at its most dangerous.

After a moment's reflection, I knew she was right. My saying that Washoe belonged in Africa was nearly as egocentric as Lemmon saying she belonged in a laboratory prison. Washoe may have been born in Africa, but she was psychologically human and culturally American. It would be an act of cruelty to make her conform to my idea of where a chimpanzee should live simply to salve my conscience. She could never go "home" again, nor could Ally, Bruno, Booee, or Thelma. As far as they were concerned, they *were* home.

I was brought up short at my own readiness to desert Washoe. Here was a child who had known nothing but abandonment. Her mother was murdered. She was sold off by the animal dealers who kidnapped her. She was deserted by the Air Force lab techs who cared for her at Holloman. She was sent away by the Gardners, her human foster parents. Debbi and I were the only fixed points in Washoe's young life. As much as I wanted to drop everything and go work with autistic kids, I knew I couldn't. For the second time in my career, I had to accept that my first allegiance was to Washoe.

The sudden collapse of my Africa fantasy sent me into a tailspin. It was as if I had been awaiting parole from prison, only to be hit with a life sentence instead. Washoe and I were not going to be saved by Yale or Africa or anything else.

In my despair, I turned to alcohol. Late one afternoon, after I finished teaching, I drove out to a redneck bar as far from the university as I could get. I ordered a pitcher of beer, then another and another. When I was dead drunk I said to myself, "Roger, you are caught up in something that you cannot handle and it could last a very long time. Washoe might live for another forty years."

I cursed the day I met Washoe on that playground in Reno. That was the day I imprisoned my family for the rest of our lives. I cursed Lemmon and I cursed science. But mostly I cursed myself. I was trapped in a bad dream. "All I ever wanted was to be a child psychologist," I kept saying to myself. "What happened to me?"

I suppose, like many people, I had a predisposition to alco-holism. And once it was triggered I had a hard time turning it off. I drank heavily for the next four years. At first I drank just to blot out the day-to-day reality of cages, guns, Dobermans, and the paranoia surrounding William Lemmon. I always drank in bars where I'd never run into my students. I sought out places where no one would call me "Professor" or remind me of who I was. I was just another guy, someone whose work didn't in-volve imprisonment and terror. When I was drinking I could lose myself in fantasies of becoming an Oklahoma potato farmer or a scientist who studied cockroaches.

But soon I was drinking out of pure self-pity. Forgetting wasn't enough. I wanted to obliterate myself. I'd smoke three packs of Marlboros during the day and then sit at a bar for most of the evening. By the time I got home I felt nothing. Just looking at my kids made me feel guiltier about what I did all day. Their father was a jailer. If I was drinking gin, it brought out a mean streak. I would pick a fight with Debbi, and we would dissect each other in our most tender spots, as only long-married couples know how to do.

"Roger, please be good to yourself," Debbi would finally beg. "Stop the drinking."

"This is my devil," I'd answer. "I want it like this."

Strangely enough, my drinking didn't get in the way of my career. I didn't publish less; I published more than ever. In the years 1975 to 1979 I produced more than twenty papers, ad-dressed dozens of symposia, taught hundreds of students, and worked with a dozen Ph.D. candidates. I seemed to want to prove that I could still play the science game—*and* drink like a demon. But beneath my success I was a hollow man.

I am not proud of those years. But I don't like to bore people with my personal demons either. Suffice it to say that I was an absent father and a lousy husband. I preferred to wallow in self-pity rather than to change my life. I just thank God that I had a wife who believed in me and who had faith that somehow I would emerge from my very dark tunnel.

· · ·

AT THE END OF 1974 Lemmon moved the older juvenile chimps off the island and into the adult colony, and Washoe went with them. I could have insisted that Washoe stay on the island, but I figured she was better off going with her friends. Debbi and I continued to take them out for walks in the woods and to study their progress in sign language, but the bulk of their day-to-day life was now spent behind bars.

Now that she lived in the adult breeding colony, Washoe faced another rite of passage: sexual activity. The year before, Debbi and I had noticed that Washoe's genitalia were slightly pink and swollen, the first signs of sexual maturation. It was hard to believe that the baby girl I had once diapered and bottle-fed was now on the brink of adolescence.

In a wild female chimpanzee these first swellings of the labia gradually grow larger and larger until age ten or eleven, when she goes into her first estrus and a full-sized swelling appears. Over a week's time her external genitalia will grow six times their normal size and her bottom will look like a large pink ball for ten days. This is the time when she is receptive to courtship and mating, and it coincides with ovulation, which usually occurs on the last day of full swelling. After her swelling recedes, she will bleed for about three days, followed by two weeks of normal genital size and color—all in all, a thirty-six-day menstrual cycle.

Although the female chimp in the wild begins menstruating at age ten, nature gives her a kind of grace period by keeping her sterile for the first one to three years of her puberty. Generally she does not become pregnant until she is twelve to fourteen years of age. (For unknown reasons captive chimps often get pregnant at a younger age.) This period of sexual learning gives her a chance to leave her mother and to begin socializing with adult males in her own community or to cross over into neighboring communities. After mating with some or all of the males who court her, she will most likely pair off with one male and go on what Jane Goodall calls a "safari" or "consortship." This is a private tryst in the jungle that can last anywhere from two weeks to three months.

Although it is usually the male that attracts the female's attention with a courtship display, adolescent females are not the types to stand around waiting, and they are known to solicit males of all ages. Washoe was hardly shy and retiring herself. She suddenly became very aggressive with members of the opposite sex, especially humans. Washoe in estrus was a force to be reckoned with. When she developed a crush on one of my male graduate students, she would literally throw herself at him, leaping into his arms. She would wrap her arms around his neck, plant her wide-open mouth right over his, and start thrusting her pelvis into his midsection. Chimps in the wild often greet each other with this kind of wet open-mouth kiss—minus the pelvic action—but the first encounter with a one-hundred-pound ape giving you mouth-to-mouth resuscitation can be rather disconcerting.

Other cross-fostered chimps romanced me like this from time to time, but Washoe never did. She seemed to see me as off-limits, probably because of our close family relationship. In fact, she hardly noticed me at all during her monthly bouts of love-sickness. When Lucy was in estrus she'd go out of her way to avoid physical contact with her foster father, Maury, and brother, Steve. But like Washoe, Lucy was jumping into the arms of total strangers. Washoe's and Lucy's sexual aversion to their older brothers is not surprising. The taboo on incest has deep biological roots in our ape and monkey ancestry. Jane Goodall has noted that it is extremely rare to see chimpanzee females mating with their older brothers in the wild. A chimpanzee brother generally shows very little interest in his sister even when she is copulating with every other mature male in the community. If a brother does become sexually aroused around his sister, she usually does her best to avoid him or else protests his courtship violently.

Considering that Washoe's formative experiences of love,

intimacy, and attachment all unfolded in a human family, it's understandable that she preferred human males over chimpanzee males. Like Lucy and Ally, Washoe thought she was human. Why wouldn't she expect to mate with one? Sooner or later, of course, Washoe was bound to get frustrated and start going after males of her own species.

When she turned nine years old and entered the adult colony, Washoe began "cycling"—menstruating and ovulating regularly—though her cycles were probably sterile. There was no mistaking when Washoe was in estrus. Her pink bottom swelled up as big as a volleyball. The adult males, especially Pan, the alpha-male, became extremely interested in her then, but Pan never seemed to get anywhere with Washoe. She didn't go for domineering types at all, and she certainly didn't throw herself at him or any of the other chimps the way she did with humans.

But sometimes, Washoe would crouch and present her rear at the wire mesh as an invitation to one of the males in the cage next to hers. More often than not, that special someone was five-year-old Manny, who was probably the least dominant male in the colony. Manny was a chimp's chimp, one of the few who was raised by his own mother. As a result, his sexual instincts were very sound and unconfused. As soon as he saw Washoe presenting, Manny would become instantly erect and mount her through the wire mesh. Unfortunately for him, Washoe would usually lose interest and walk off before Manny was finished, which would send him into a screaming temper tantrum. Washoe seemed to take pity on poor Manny when she saw him writhing on the floor. She would sign COME HUG and then present her bottom again at the wire. He would mount her again, she'd pull away, he'd throw a fit, and she'd sign COME HUG again.

One day, Manny figured out that he didn't have to throw a fit to get Washoe's attention. All he had to do was sign COME HUG—which he learned from her—and she responded very dependably. From that moment on when either Manny or Washoe signaled COME HUG the other one knew it was time for a tryst.

This is interesting because in the wild, male chimpanzees almost always invite females to mate by making one or more gestures. They either clip a leaf, place a hand on a branch, or stretch one or both arms out toward the female. Manny was just adapting to his environment by learning a new mating signal.

When Washoe was ten she spurned Manny for Ally, a slightly more mature sexual companion. This didn't really surprise me. Of all the chimps, Washoe and Ally were the most proficient in American Sign Language, which they had both been using since age one. They almost always addressed one another in ASL. In addition, Washoe just seemed to like Ally, and she enjoyed his company. Ally was a low man in the chimp hierarchy, and had none of the macho bluster of Pan or the other males. He was all charm and comedy—a kind of sensitive guy. Washoe called him NUT, as we did, because of his antic personality.

Washoe's preference for someone gentle like Ally is not uncommon in the wild. In chimpanzee society, the alpha-male does not always get the girl. In fact a dominant male with many rivals may be so busy defending his power base that he has little time or energy left to sire many infants. Meanwhile, a less aggressive male may drop out of the power struggle and devote himself entirely to seducing females. That was Ally.

BY 1975 LEMMON'S GRAND EXPERIMENT in the maternal behavior of cross-fostered chimpanzees had pretty much fallen apart. The chimpanzee children he had farmed out to his patients had all died or were living back in the Institute. Except for Lucy. She was eleven years old and the only ongoing subject. She had been raised as a human child, away from her own species, for longer than any other chimpanzee.

But in 1975 there was an ugly falling out between William Lemmon and Maury Temerlin, Lucy's foster father and Lemmon's most loyal follower. Temerlin had just published his book, *Lucy: Growing Up Human*, which seemed as much a public purging of his therapist as it was an account of raising a chimpanzee. The portrait of Lemmon, who went unnamed in the book, was not only scathing but hard for anyone in Oklahoma to mistake. "I saw him as infallible," wrote Temerlin, "and I literally believed his most outlandish statements. I saw him as benevolent, and I ignored the most obvious evidence of human self-seeking and pettiness. I saw him as omnipotent, and I was blind to his dependence upon people who were dependent upon him."

Temerlin accused his ex-therapist of "psychological incest" for unethically entangling himself in his patients' lives and families outside of the therapeutic context. Lucy was clearly part of the umbilical cord that connected Lemmon to Temerlin, so I wasn't all that surprised when Maury told me one day that he and his wife, Jane, were looking for a new home for her. They wanted "to live normal lives" again, he said. Lucy was no longer a child needing constant affection, and her adult needs and demands were much harder to fulfill. In addition, Lucy had become overexcited one time recently and had bitten a guest on the arm, thinking mistakenly that the person was trying to attack Maury.

But where could they send Lucy? Temerlin wrote that Lucy was "biologically a chimpanzee to be sure, but psychologically able to live healthily and happily as a human being given no more supervision than a mentally retarded child would need." Of course, Lucy wasn't mentally retarded. She was a perfectly intelligent chimpanzee who was born in Florida instead of Africa, taken from her natural mother, and raised as a middle-class human. Now with a half-century left to live, she suddenly discovered that she wasn't human *enough* to live in society.

Sending Lucy to the Institute, the orphanage of last resort for Washoe, Ally, Bruno, and Booee, was out of the question. Temerlin burned that bridge by breaking with Lemmon. And one day he told me that Lemmon would kill Lucy if he ever got his hands on her. After more than two years of searching, the Temerlins finally decided in 1977 to send Lucy to Africa, the same chimpanzee "home" that I had considered but ruled out for Washoe. They wanted Lucy to enter a chimpanzee rehabil-itation project run by a woman named Stella Brewer in the tiny West African nation of The Gambia. In the 1960s Brewer started an orphanage for baby chimps who had been confiscated from the poaching trade before they could be shipped overseas and sold to the entertainment and biomedical research indus-tries. Once the youngsters grew too big for the orphanage, Brewer tried to return them to the wild in Niokolo Koba Na-tional Park, in Senegal. She had had some notable successes but mostly with chimps who had been born in the wild and held only briefly in captivity.

Lucy was an unlikely candidate, to put it mildly, for life in the African jungle. She had never even met another chimpan-zee. Not only was she fully grown and well past the optimal age to be reintroduced to the wild, but Lucy had grown accustomed to a lavish human lifestyle. She conversed in sign language, drank fine Chablis with dinner, loved television, and relied on *Playgirl* magazine to satisfy her monthly sexual cravings. Getting Lucy to build a nest in a tree, hunt her own food, and fend off spitting cobras went way beyond anyone's definition of rehabil-itation. "Rehabilitate" means "to restore to a former state or condition." Lucy couldn't be restored to something she'd never been. She was a middle-class Oklahoman. Not surprisingly, Stella Brewer said no to the Temerlins' request.

But the Temerlins weren't giving up. They took in a young female chimpanzee named Marianne so that Lucy would get to know one of her own kind, and they hired one of my graduate students, Janis Carter, to help take care of Lucy. After Janis and Lucy had established a good relationship, the Temerlins asked Janis if she would go with them to Africa and then stay on for a few weeks to help Lucy adjust to her new life. Janis said yes, and Stella Brewer agreed to take both Lucy and Marianne into her rehabilitation program.

I felt torn about Lucy's imminent departure. I sympathized with the Temerlins' dream of a simple solution to their predic-ament—"a happy ending," as Maury always called it. But by

1977 I was convinced that there was no such thing as a happy ending to a cross-fostering story. It was all well and good to want Lucy to have the life that nature intended, but she had never lived that life. For the Temerlins, though, Africa held more promise than handing their daughter over to Lemmon or a zoo.

The night before the Temerlins left in September 1977 I visited Lucy for the last time. As we sat on the couch grooming one another and signing, I was struck by how grown-up she seemed. The little girl who used to beg me for tickles was now a teenager, more a friend than a pupil. I couldn't imagine what lay ahead for her.

The next morning Lucy was tranquilized, put in a wooden crate, and loaded on a plane. She emerged several days later in Africa, a place she had never seen before. One week after that, the twelve-year-old girl signed good-bye to the only mother and father she'd ever known. The news from Africa over the next few months was not good. Lucy and Marianne were living in a cage in a small forest reserve, next to a city, until they could be moved to a wildlife park. Lucy was depressed, emaciated, and seriously ill. Night after night I was haunted by Jane Goodall's words of warning to me: "You just don't understand Africa."

I kept expecting to hear that Lucy had died, but that news didn't come. Somehow, Lucy was hanging on.

IN THE SPRING OF 1976 we noticed that Washoe's belly was swelling slightly. She was also regurgitating her breakfast, a sure sign of chimpanzee morning sickness. By June we were sure that ten-year-old Washoe was pregnant, perhaps in the final months of her eight-month gestation period.

Debbi and I were excited about Washoe giving birth. It seemed like a ray of hope in a very grim time. We felt that if Washoe was going to live in a prison, then at least she should get to enjoy the experience of motherhood. And scientifically I was curious to see if Washoe would sign to her child.

But our excitement was also mixed with concern. Mostly I was worried about Lemmon. Ally, the likely father, belonged to Lemmon, so Lemmon could claim ownership of the baby and make our lives miserable by demanding compensation. But that was a custody battle I was willing to take on.

Washoe gave birth on August 18, 1976. Her labor must have been very short. At 7:30 A.M. she was showing no signs of labor, but by 8 A.M. she had delivered the baby, with none of us there to witness the birth. When I arrived I could tell immediately that something was wrong with her infant. It was showing few signs of life. Washoe was cradling the baby and carefully grooming it, even sucking mucus from its nose and mouth to revive it. Every so often the baby moved, and Washoe would hold it close to her chest. Twice, when the baby stopped moving for a long time, Washoe laid it down next to her, signed BABY to it, and cried.

After failing to revive her baby, Washoe came over to us and stretched out her arms, as if she wanted to give us the newborn. She seemed to know that she could not handle this problem. It is extremely unusual for a chimpanzee mother to give up an infant, even a sick one, and Washoe changed her mind and moved back over to her bed. Finally we made the agonizing decision to anesthetize Washoe so we could take her baby to the university hospital. On the way the infant was given cardiopulmonary resuscitation, but it was already too late. The baby was dead. An autopsy showed a concussion to the back of the head. Washoe may have birthed over the edge of her bed, causing the baby to fall to the floor. But the actual cause of death was a congenital heart defect—a hole in one of the ventricles.

The next day Washoe began showing signs of depression. She hardly ate anything and was moping around her cage. I put Ally in with her to cheer her up. Within two weeks or so, she seemed to be on the mend.

.   .   .

BY FALL 1977 I felt as if I had hit bottom. Washoe's baby was dead. Ally, Booee, Bruno, and the others spent their days and nights in jail. The reports from Africa about Lucy were growing more and more alarming. And after Temerlin's rebellion, Lemmon had grown even more solitary and calculating.

Getting up in the morning, much less going to the Institute, seemed like a monumental task. I felt as if I had swallowed a poison that was slowly destroying my soul. Living with guilt is a fact of life for those who imprison primates. Some scientists, like Lemmon, deal with their guilt by degrading their prisoners and making them seem unworthy of compassion. Others put on white lab coats and pretend they are working with machines— treating chimpanzees with an icy detachment they would never show their own cat, dog, or hamster at home. Still others fall into the egocentric trap of self-pity and drown their sorrow in drink—as I'd been doing for almost four years.

Though my way of coping was different from William Lemmon's, there was no denying that I was on my way to becoming like him. I hated my work, and my relationship with the chimps was deteriorating. Washoe and the others didn't want to be around me anymore. I was just too depressing, even for them.

Then one evening, while sitting at a bar, it dawned on me that in a few years I wouldn't be *like* William Lemmon. *I would actually be William Lemmon.* I would be running the Institute, making the rules and commanding the jail guards, doing a little research on pigs and gibbons to pay the bills. The transformation of Dr. Roger Fouts would then be complete. There would be nothing left of the young idealist. Standing in his place would be just another faceless, alcoholic scientist who jailed chimps for a living.

And that was the image that finally scared me into taking action.

NINE

A DEATH IN THE FAMILY

I WENT TO MY OFFICE THE NEXT MORNING, sat down at my desk, and took out two sheets of paper. At the top of the first sheet, I wrote, in large letters, "Find a new home for Washoe." I wanted that home to be a sanctuary where chimpanzees could live relatively free with very little human intrusion. I had dreamed about such a place for years. Now I began laying plans. I would have to create the sanctuary myself, hopefully along with a university where I could teach. I made a list of schools that I was going to visit and speak at over the next year and highlighted the ones that might offer me a professorship. Then I studied a map to see which of those schools had the proper climate and suitable forest land nearby. Then I made another list, this one of provosts and chairmen to begin calling.

On the second sheet of paper, I wrote, "National Science Foundation." Then I began outlining a grant proposal to investigate the most intriguing unanswered question of ape language research, whether chimpanzees could transmit sign language across generations. Sooner or later Washoe would get pregnant again. When she did, I wanted to be prepared to observe and record all of the gestural communication between her and her child.

I'd contemplated this kind of observational study for years. I had grown tired of playing the master controller in experiments, running chimps through language tests. The traditional approaches left no room for the chimps to show me what *they* wanted to do with language—for Lucy to sign to her cat, for Booee and Bruno to argue over food, for Washoe and Ally to converse before mating. I wanted to study language in chimps as I'd studied it in autistic children. I wanted to report on it like an ethologist might, recording the natural and spontaneous behavior of day-to-day communication. I no longer saw chimpanzees as "research subjects" but instead as my research partners. I wanted their interests to come first, not last.

I planned to let Washoe and her child communicate without any human interference. This would address the skepticism of a handful of critics, mostly linguists, who said that Washoe and other chimps were highly trained animals who were mimicking their teachers or responding to their teachers' unconscious cueing. If Washoe transmitted sign language to her child—without any human involvement—it would prove that chimpanzees understood the correct use of signs and that they naturally used those signs to communicate.

I made one more resolution that morning, and it was so simple I didn't need to write it down: I was going to stop drinking. If I didn't pull my life together there was no way I could find a sanctuary for Washoe *and* carry out a major piece of research. Over the years I'd gone to enough Alcoholics Anonymous meetings to know that the key to my own recovery was the serenity prayer: "God grant me the serenity to accept the things I cannot change, courage to change the things I can, and the wisdom to know the difference."

Those words went against everything I had ever been taught in school. As an experimental psychologist, I had absorbed the scientific arrogance that presumes we can control animals, nature, and life itself. I believed I could do anything, therefore I was responsible for everything. But I had finally found something I couldn't control: William Lemmon's Institute. And if I didn't let go of the illusion that I could save the chimps, the drinking was going to kill me.

Letting go was easier said than done. I had no problem at

all accepting the strengths and weaknesses of others; I took every chimpanzee and every autistic child on his or her own terms. It was my own limitations I found so hard to accept. Somehow I would have to start treating myself with the same humility and compassion I showed my research subjects. I knew I was not in control of my drinking, and I knew there was no magic wand I could wave that was going to make things better for the chimps. The all-powerful, all-controlling scientist was a lie. *I was powerless*—and I felt a huge relief admitting it. I could still go to work every day and do my best to help the chimps, but there was no way in hell I was ever going to change William Lemmon.

Once I accepted this fact I didn't feel the same need to drink anymore. I immediately stopped going to bars after work. And pretty soon I stopped drinking completely.

LATE IN 1977 I submitted my grant proposal to the National Science Foundation (NSF) to study "the possibility that an infant chimpanzee will acquire sign language from its signing chimpanzee mother." At least one member of the NSF review committee, probably a linguist, said my proposal was preposterous, but recommended that NSF fund it anyway just to prove me wrong. And in early 1978, I was informed that my study had been approved, pending Washoe's becoming pregnant. This may have been the first time in history that an NSF grant hung on a pregnancy test.

In the meantime I had been searching for a chimpanzee sanctuary. I had made good contacts with several universities in Texas and the Southwest, and every time I got on a plane, I studied the landscape beneath me. I looked for rivers with oxbows—a natural refuge for chimps because the land is enclosed by water on all sides but one. When I spotted likely sites I'd look up the closest university to find out whether or not I could teach there.

As time went on, my search became more and more urgent.

I wasn't concerned just for Washoe but for all the chimpanzees at the Institute. After the collapse of his cross-fostering studies, Lemmon was increasingly embittered and less willing to put up with the logistical and financial headaches the chimpanzees were causing him. He became decidedly more businesslike, and began searching for a permanent solution to his "chimpanzee problem." Until then, Lemmon had always drawn a bold line between the "soft" behavioral research in his chimpanzee colony and the "hard" biological research being conducted on the siamangs and pigs in his pig barn. That was why Washoe and I had wound up at the Institute in the first place; it was the *only* research facility in America devoted exclusively to studying chimpanzee behavior, not biology or disease.

But one day in 1978, a group of men in business suits appeared at the Institute to tour the premises. Lemmon told me they were from Merck, Sharp, Dome, a large pharmaceutical company looking for a chimpanzee colony where they could test a hepatitis B vaccine. Lemmon was bidding on a contract that would turn the chimpanzees into subjects of disease research. He said that hepatitis B would not kill the chimps, which was true, strictly speaking. The research goal was to test the effectiveness of the vaccine, and for that the subjects didn't have to die and be autopsied. The hepatitis B vaccine is made with live virus, so it is considered too dangerous to test in humans. Researchers would inject the vaccine into chimpanzees and then "challenge" it with the hepatitis B virus to see if the chimps got the disease. Once a month or so, the chimps would be anesthetized so that blood could be taken and their livers could be biopsied. If there was no hepatitis B in their bodies, then the vaccine was effective. Most of the chimps would fall into this category.

The problem, as I saw it, was that some of the chimps *would* become carriers of hepatitis B, and many of them were likely to develop liver disease and, perhaps, liver cancer. Plus, those infected chimps would have to be isolated from the others and live in solitary confinement for the rest of their lives. Obviously,

Lemmon was willing to expose his chimpanzees to those risks. If his proposal was accepted, Merck, Sharp, Dome would build and maintain a state-of-the-art facility at the Institute that would end all of his financial problems.

Fortunately, Lemmon did not win the hepatitis B contract, which went instead to a biomedical laboratory in New York. But this had put an idea in his head and he began exploring other potential sources of revenue from hard science. The National Institutes of Health was looking for a chimpanzee breeding colony, and Lemmon put in a proposal that would have turned the Institute into a chimpanzee supply house that bred babies for use in disease research. He didn't win that grant either.

As the months went by Lemmon grew increasingly desperate about what to do with his chimps, and I knew that he would eventually find something.

IN JUNE 1978, Washoe's pregnancy test came up positive. I had the lab techs run it twice just to be sure. Then I sent the results off to the National Science Foundation. A month later, the first installment of my three-year, $187,000 grant was on its way. That was good news. The money would not only finance the study, but would also give me some independence from Lemmon and protect Washoe's baby from any plans Lemmon might have for him or her. I also wanted to protect Washoe's partner, Ally, in the same manner—whether or not I found a sanctuary. Ally would be a part of my study because of his excellent command of signs and because he was most likely the father of Washoe's baby. With Washoe and Ally housed together, the newborn would have a family of chimpanzee signers to learn from.

I began taking regular walks with Washoe in the woods. I'd fill my pockets with apples, dried fruit, and the other treats that she loved. Once in the woods, I would take off her lead so that she was free to climb trees. Later in her pregnancy, she preferred to just sit and relax *under* the trees. During these quiet times she would begin grooming me, picking through my hair and ears, and I would return the favor by grooming her arms, shoulders, and back.

There was no question that Washoe knew she was pregnant again.

WHAT IN YOUR STOMACH? I would ask.

BABY, BABY, she answered, cradling her arms in front of her.

Washoe's pregnancy gave me even more reason to speed up my search for a new home. That fall I was giving a talk on chimpanzee signing at the LSB Leakey Foundation, in Pasadena, California, when Joan Travis, the director of the foundation, told me that a screenwriter who was working on a movie about apes wanted to meet with me that night. I'd been contacted by screenwriters before, but all of them seemed to be writing movies about chimpanzees who were clownish sidekicks in human love stories. I just wasn't interested. But as a favor to Joan I agreed to meet the writer at her house after my speech. I expected to spend a polite fifteen minutes with him and that would be that.

The screenwriter was Robert Towne, the Oscar-winning writer of *Chinatown*, *The Last Detail*, and *Shampoo*, and the story he was working on was that of Tarzan, a character close to my heart. The Tarzan story is Washoe's story in reverse: a human foundling, raised by a community of apes in Africa, grows up to realize that he cannot go "home" to England again because the jungle is his home and apes are his family.

Towne had already spent years trying to get his Tarzan project off the ground. He was determined to make a movie that was more faithful to the original story by Edgar Rice Burroughs than its predecessors had been. I was impressed by his commitment to wildlife conservation and the homework he'd done on Africa and wild chimpanzees. Warner Brothers was investing millions in Towne's project, entitled *Greystoke: The Legend of Tarzan, Lord of the Apes*, and the studio had promised Towne that he would direct it.

Towne planned to shoot on location in Africa, with humans

dressed in chimp costumes. He wanted me to work with the actors to create the most realistic chimpanzee behavior possible. We talked about the Tarzan story until the sun came up the next morning, and at some point that night I was struck by an idea. Why go all the way to Africa just to film humans in costumes when there were real chimpanzees in Oklahoma? Instead of spending millions to film halfway around the world, the studio could build a replica of the African jungle on an island in Oklahoma. We could transfer Washoe and her friends to the "African" island, and the cameramen could film everything that Towne was looking for: grooming, displays, attacks, mothering, friendship, tool use. He could edit the chimpanzee behavior and then use humans in chimpanzee costumes for whatever we didn't capture.

The more we talked, the more Towne liked it. One obvious benefit had to do with an early and crucial scene in the movie in which Kala, Tarzan's chimpanzee mother, gives birth. After her newborn dies, Kala turns her maternal affections to the human baby, Tarzan, whom she finds in the jungle. If he filmed in Africa, Towne would have to shoot this birth scene with a human in a chimp costume and a doll as a baby. But in Oklahoma, Washoe would deliver her baby in January, just a few months away. I had already budgeted five thousand dollars from my NSF grant to build Washoe a special birthing cage that would let me film Washoe's very first interactions with her baby. But Towne, backed by Warner Brothers, could build an even better birthing environment and shoot feature-quality film. The birth of Washoe's child would be captured for the scientific record and for movie audiences around the world.

But the real long-term payoff of my plan was that once the filming was completed the "African island" in Oklahoma could become a permanent sanctuary for the Institute's chimpanzees. Maybe Washoe couldn't go back to Africa, but Hollywood could bring a piece of Africa to Middle America. Everyone would benefit. The chimps would be sprung from their lockup and saved from disease research. Lemmon's chimpanzee problem would go away once and for all. And I would have a safe haven for Washoe and the others, a goal that might otherwise take years of negotiations and fund-raising to realize.

Towne quickly sold the idea to Warner Brothers. He wanted to use all of the Institute's chimpanzees, so I had to get Lemmon's approval, which was no problem given his desire to dump his chimps. In fact, he was wildly enthusiastic, especially because there was money and media coverage involved. Predictably, Lemmon insisted that he take charge of the project. He thanked me and told me that he would now be dealing with Robert Towne and Warner Brothers. That was fine by me; I'd already gotten what Washoe and I needed.

Warner Brothers gave Lemmon twenty-five thousand dollars to build a large private birthing cage with blinds and camera ports for filming. Carlo Rimbaldi, who'd made the costumes for E.T. The Extra-Terrestrial, visited the Institute to see if the chimpanzee suits he was designing for Greystoke were realistic enough to be used alongside real chimpanzees. His costumes were so convincing that Ally charged one of the chimp-suited actors and swatted him right on his elongated mechanical arm. Terrified by this "chimp of steel," Ally ran off in the other direction.

Meanwhile, Robert Towne was scouting locations near the Institute where he could build his Africa set. Lemmon found a large island for sale in the middle of the Canadian River, which runs through Norman. It looked perfect. On one of Towne's visits, Lemmon prepared several horses to take him and Towne's crew down to the proposed site. Lemmon did not invite me along, but I hardly cared. I just wanted Towne to buy the land, which would seal the deal for the sanctuary. They headed off in the morning for their daylong scouting trip.

Later that afternoon I was working at home when an angry Robert Towne showed up without warning at my front door.

"I will have nothing to do with that man," Towne said, speaking of Lemmon. "He's insane. He beat his poor horse." Evidently, Lemmon's horse was too old for riding to begin with, but to make matters worse Lemmon had spent much of the day kicking the poor animal.

The horse-kicking episode was the last straw for Robert Towne. He already knew about Lemmon's insensitive and domineering personality, but he also suspected that Lemmon was trying to suck as much money as possible out of Warner Brothers. Unfortunately, Towne could not cut Lemmon out of the process. After all, Lemmon *owned* the actors—the chimps, that is. They were what brought Towne to Oklahoma in the first place. Lemmon was not about to hand over the chimps to Towne for a movie that didn't feature William Lemmon's participation. Towne was disappointed, but he felt he had to take his movie elsewhere.

I was crushed. My dream had been so close I could practically touch it. Now, the sanctuary was not only down the tubes but I had been so sure this would work out that I had broken off talks with a couple of universities that might have come through for the chimps. I was back to square one.

I wasn't the only one left disillusioned by *Greystoke*. After Towne had devoted nearly a decade of his life to the Tarzan story, the movie was taken away from him by the studio in 1982. The next time I saw Robert Towne, he said, "You know, Roger, seeing you is like seeing someone you haven't seen since the funeral of a good friend."

"I feel exactly the same way," I replied. *Greystoke* left us both heartbroken. Towne lost his movie. Washoe and I lost our sanctuary.

WARNER BROTHERS NEVER ASKED for the twenty-five thousand back that they had given Lemmon. I have no idea what Lemmon did with the money, but he never built the birthing cage. I kept pressing him about it, but he put me off. By December, one month before Washoe was due to deliver, I couldn't wait any longer. Without a cage Washoe would have no privacy and I'd have no film of the birth.

I wrote off the twenty-five thousand and decided to build the birthing cage myself with five thousand dollars from my NSF grant. I designed the cage and sent it out for bid. But again Lemmon outmaneuvered me. My NSF funds were held by the university, and Lemmon strong-armed the provost, a former student of his, into giving *him* the five thousand dollars. Lemmon threw out my design and started building the cage himself after New Year's. But it was too late.

At 7 A.M. on January 8, Debbi got a call from the main colony. A member of our team noticed blood mixed with water on the floor of Washoe's cage. Her water had broken, and she was in labor. I immediately rushed over to stay with her.

In the wild, a female chimpanzee wanders away from her group to bear her child in the privacy and seclusion of the jungle. This need for privacy is so intense that, even in captivity, chimpanzee birthing is shrouded in mystery. Lab workers often don't even notice that a chimp is pregnant. They walk in one morning and discover a new infant in the cage. Female chimps will wait for a lab technician to leave the room before they go into labor. Washoe had no chance of enjoying even this momentary privacy. She would be giving birth in a small five-foot-by-six-foot cage, next to a larger enclosure holding the main colony's twenty-five highly aroused and screaming chimps.

COME HUG, Washoe signed when I arrived at her cage.

Once her labor began, my concerns about her lack of privacy quickly vanished. She appeared to click into an altered state that was far removed from me and everything around her. She knew exactly what to do as she assumed various positions to ease the pain and advance the labor. One of her favorite positions was head down and rear up. Sometimes she was almost standing on her head while she held on to the cage with one hand for support. As the contractions became more intense, she grimaced and let out sharp "ah" sounds. Between contractions she lay on her side or on her back and asked me to get her things to eat and drink, like lollipops and ice for sucking. Having been with Debbi three times when she gave birth these were very familiar requests to me. During her contractions Washoe's communication became especially intense, and I was amazed that she could sign under such extreme physical and emotional stress.

In the wild, chimpanzee labor usually lasts for only one to two hours. But with the noise and cage banging next door, Washoe's labor dragged on. After four grueling hours, at 11:57 A.M., Washoe got into a tripedal stance, holding one hand behind and below her. Then she deftly delivered her infant into her waiting hand. Immediately, she brought the baby to her chest, where she greeted it chimpanzee-style by panting heavily with her mouth over its mouth. I couldn't tell whether the baby was male or female.

Washoe began grooming the baby's ear. Only then did I notice that the baby's umbilical cord was wrapped tightly around its neck. Washoe's baby did not appear to be alive. Washoe held the unmoving infant to her chest, and made a nest for the two of them to lie on, using an old tire in her cage. She began to kiss and suck mucus from the baby's mouth and nasal passages. Then, she breathed into its mouth several times. I held my breath, but despite Washoe's excellent maternal instincts, her baby lay still and lifeless.

Washoe soon began eating the umbilical cord from around the infant's neck, relieving any possible suffocation. Minutes later, she delivered the equivalent of a doctor's slap by gently squeezing one of the baby's tiny fingers with her teeth. Suddenly, the infant squeaked. I breathed a huge sigh of relief. A few minutes later, Washoe delivered and ate the highly nutritional placenta. Although somewhat startling to watch, this act is common among mammals—including some human cultures—and seemed to flow naturally from Washoe's state of heightened maternal instinct.

I could see that her infant was still not clinging the way it should. Washoe kept grooming and giving the baby mouth-to-mouth resuscitation. The infant grasped Washoe's hair briefly with one hand, but mostly it remained limp. Finally, after three hours, Washoe set the baby down, just as she had done with her first baby. This was not a good sign. In the wild a chimpanzee mother only puts down a dead infant. I made the same anguished decision and took Washoe's baby out of her cage. I could see that the baby was alive but clearly in distress and greatly weakened. As I held it close, I could also see that it was a boy.

When I got the baby to our house, I discovered he had a fever. Debbi and I stayed up most of the night hydrating him with fluids, and by the morning we had stabilized his temperature. We continued feeding him intravenously and through a bottle. We also named him Sequoyah, for the Oklahoma Indian chief who created the written language of the Cherokee people.

That afternoon we reintroduced Sequoyah to his mother. Washoe was very excited to see her infant, and she held him to her chest. But his nursing reflex was weak. Each time Washoe moved even slightly, he would dislodge from her nipple. This was not encouraging. If Sequoyah was too weak to nurse, then he would not survive. Once again I decided to take the infant from his mother so we could be sure he would be properly fed. But this time Washoe was not as accommodating. We had to anesthetize her to take her baby away.

I was now so determined that Sequoyah would be strong enough for nursing and clinging the next time we gave him back to Washoe that I decided to keep them apart for two weeks. We fed Sequoyah with human milk donated by local nursing mothers to help avoid any allergic reaction he might have to formula. We also used a different kind of nipple that made sucking more demanding in the hope that this would strengthen his nursing reflex.

We also had Washoe to think of. She was dispirited and I worried that her maternal instincts might lapse entirely during this separation. I gave her another infant chimp, Abendigo, on a foster basis. Abendigo was a two-year-old, no longer nursing, who had been living with his own mother until a week before.

Washoe took to Abendigo immediately and spent most of her time holding the new baby.

After two weeks, I came up to Washoe's cage to tell her in sign that her own baby was coming back. She was very excited and started signing BABY repeatedly. I returned Abendigo to the main colony, then brought Sequoyah in to Washoe. She immediately held him and groomed his ears and face. But when he began to nurse, Washoe grimaced and moved her body away, dislodging him.

It was time for a heart-to-heart talk with Washoe about her nursing. I got in the cage and signed that she must feed her baby. Washoe refused to do so. Pretty soon my "counseling" session turned into a face-to-face screaming match, in the middle of which I noticed that Sequoyah had begun to root again. I quickly positioned his head so that he was on Washoe's nipple and was sucking. Washoe looked down at him, then glared at me and let out a deafening scream. I grabbed a Tootsie Pop from my back pocket and slapped it on her exposed tongue. Startled, she took the pop out of her mouth and looked down at Sequoyah, who was now nursing quite nicely. She made a move to dislodge him, but I reprimanded her with a mild "ah ah" and, finally, she settled down and allowed him to nurse. After about seven minutes, Sequoyah dozed off to sleep. A few hours later he attempted to nurse again. This time, all I had to do was look at Washoe sternly to stop her from dislodging her son. After that, they nursed without any problems.

By the time Sequoyah was one month old things were beginning to look up. Lemmon finally had Washoe's new and more spacious cage ready in the pig barn. On a cold morning in mid-February we moved the entire family—Washoe, Ally, and Sequoyah—out of the main colony and into their new quarters.

It didn't take long for me to realize that Lemmon had built the new cage out of razor-sharp expanded metal instead of the safe chain link I had asked for. "Chimp-proof" is how he described this idiotic design—meaning that no chimp would dare try to unravel the sharp metal diamonds. My students and I immediately started hand-filing hundreds of deadly edges on the cage.

But there was no way we could get to all of them in time. The very first week Sequoyah cut his toe on one of the sharp edges. We applied topical treatments but his wound became infected, and he began growing even weaker. He could barely cling to his mother. As if that weren't enough, the propane heater in the pig barn ran out of fuel one night. Ordinarily, the lab techs told Lemmon when the fuel ran low, and he would order more. But this time, there was no new fuel forthcoming and the temperature in the barn dropped to 27 degrees Fahrenheit. The next morning, we found Washoe and Sequoyah huddled together in the cold.

Within days Sequoyah developed a serious respiratory illness, and Washoe stayed up until dawn for many nights, sucking mucus out of her baby's nose and mouth. Despite Washoe's diligent efforts at aspirating her son's nasal passages up to twenty times per hour, Sequoyah was deteriorating.

By March 8 Sequoyah's pneumonia was too severe for Washoe to cure, and once again we had to separate mother and child. When Washoe saw me coming with the needle and anesthetic, she started screaming at me and signing MY BABY, MY BABY. She knew right away that I was there "to knock her down." I rushed Sequoyah to the local community hospital in Norman, but the doctors refused to admit him. "No chimps allowed," they insisted. In desperation, Debbi and I created a makeshift infirmary in our dining room. Chimpanzees came to our house pretty regularly, but this visit was different and our kids, who were now three, eight, and eleven years old, knew right away that something was very wrong. Debbi and I couldn't hide our fear. All of us gathered around the dining room table where Washoe's son was resting, bundled in blankets. I held Sequoyah's tiny hand while I silently begged God to save his life.

Late that night I called our family pediatrician and friend,

Dr. Richard Carlson, and asked him to come over and tell us what to do to bolster Sequoyah's strength. After examining Sequoyah, Carlson determined that the pneumonia was bacterial, and had probably migrated from the staph infection in his toe and had settled in his lungs. The baby was so weak that he could no longer cling or grasp, and the prognosis was grim. We put Sequoyah in a mist tent with a vaporizer and began giving him ampicillin. Carlson inserted a tube in Sequoyah's nose to help aspirate fluid. He stayed with the baby until 11 P.M. Then Debbi and I took over.

Sequoyah died the next afternoon, March 9, at 4 P.M. Debbi, the kids, and I were numb with grief, hugging one another for comfort. It was hard to believe that this adorable infant, whose birth we had celebrated just two months earlier, was gone forever. Worst of all, it seemed so damned unnecessary. I stayed up all night agonizing over what I had done wrong. If only I'd fought harder for a better cage. If only I'd insisted on a better heating system. If only I'd realized sooner that the infected toe needed systemic antibiotics to prevent bacterial pneumonia. If only I'd left Sequoyah in his cage under Washoe's care. She had done a better job than we had of sucking the mucus out of her son's nose and mouth. Of course, the truth is that once Sequoyah developed severe pneumonia, nothing could have saved his life, not even the best mother in the world—which Washoe appeared to be.

More than anything, I dreaded telling Washoe what had happened. Early the next morning I went to see her. As soon as she saw me coming, she raised her eyebrows and signed BABY? She held her cradled arms in place to emphasize the question. Leaning in toward her, with all of the sympathy I could express in my face, I cradled my arms and put my two hands out in front of me, left palm down, right palm up. Then very slowly, I rolled both hands over in the sign for death: BABY DEAD, BABY GONE, BABY FINISHED.

Washoe dropped her cradled arms to her lap. She moved

over to a far corner and looked away, her eyes vacant. After sitting there for a while, I realized there was nothing more I could say or do.

I left Washoe and went over to Lemmon's office for our regular weekly meeting. As I sat there lamenting Sequoyah's death and Washoe's suffering, I noticed that Lemmon was uncharacteristically jolly. He actually seemed to be gloating at my misery. And when I began complaining about the dangerous cage he had built for Sequoyah, he interrupted me.

"You know, Roger," he said, "a long time ago I built a cattle carrier out of that expanded metal. By the time I got to market, the cattle were all cut to hell." He was laughing.

I had put up with a lot from William Lemmon, but now an innocent life had been taken. When I got up to leave, I knew it was for the last time.

SUDDENLY, I WAS STARING into yet another abyss. Only four months earlier, everything was going right. Washoe was pregnant, I had the biggest grant of my career, and best of all, I had found the chimps a sanctuary on the *Greystoke* island. My life had turned around so completely that I hadn't had a drink in months. Now I was teetering again on the brink. The sanctuary was gone. Washoe's baby was dead. Washoe herself appeared deeply, and understandably, depressed. And without Sequoyah my NSF grant was about to go up in smoke.

It took everything in me to keep from driving to the nearest bar. But I didn't. I was too angry at Lemmon to surrender and hand him a final victory. And I was too worried about Washoe. Finding her a new home suddenly seemed like a matter of life and death.

I headed straight to the provost of the university and demanded that the administration find Washoe and me a new home on university property. He hemmed and hawed. I told him that not only did I plan on returning all the grant money to NSF, but I was going to tell NSF exactly what had happened to Washoe's son. That got his attention.

A few days later the provost called to tell me we could move to some barracks on an abandoned airfield named South Base. It was not the African island I had dreamed of, but it was a step in the right direction, a step that would put Washoe beyond Lemmon's reach.

Having solved one problem, I now turned my attention to the bigger problem of getting Washoe through her grief. For the three days after her son died, Washoe greeted me each morning with the same question: BABY? And each time I would reply as I had before: BABY DEAD. Washoe now began sitting in the corner, refusing to interact with us at all. Her signing had dropped off to almost nothing. I tried putting Ally in Washoe's cage to help ease her melancholy. But even Ally, who was as energetic and playful as always, couldn't cheer up Washoe.

With every passing day we grew more concerned over Washoe's depression. There is no question that chimps recognize death, both in the wild and in captivity. This is part of the reason for their terrible fear of anesthesia and its deathlike state. I also sensed, from the way Washoe stopped signing and withdrew, that she understood the finality of Sequoyah's death. But chimpanzees, like people, do not seem to *accept* death easily. Washoe's daily question about Sequoyah was very much like human denial, our stubborn refusal to accept the sudden loss of a loved one. It was as if she were saying, "Are you absolutely sure he's dead? Isn't there some shred of hope I can hold on to?"

After three days, Washoe stopped asking me about Sequoyah. It seemed that she had at last accepted that he was gone. But that seemed to plunge her into even deeper anguish. She was no longer eating, and Debbi and I were panicked. We had never seen Washoe depressed, not even after the traumatic move to Oklahoma in 1970. Washoe was simply not a depressive type. She was a survivor. Seeing her so withdrawn was like seeing our most well-adjusted friend suddenly drop off into total darkness. But worse, we felt completely helpless to stop it. Washoe was telling us that she wanted her BABY, but there was nothing we could do to ease her pain.

I had already seen too many chimpanzees mourn themselves to death. If I didn't do something soon I knew that Washoe would perish from starvation and heartbreak. Letting her die was unthinkable. She was a member of our family.

There was only one hope left. Washoe had clung to Abendigo, her "foster child," just after Sequoyah's birth. Perhaps a new infant might rekindle her strong maternal instinct—and with it, a new will to live. Somehow, somewhere, I would have to find a baby for Washoe to adopt.

TEN

LIKE MOTHER, LIKE SON

I SPENT THE NEXT SEVERAL DAYS contacting primate facilities around the country, desperately trying to locate a chimpanzee infant to replace Sequoyah. After phone calls to dozens of places, the Yerkes Regional Primate Research Center, in Atlanta, Georgia, agreed to give us a ten-month-old male named Loulis. He was named for the two lab techs who cared for him: Louisa and Lisa. Loulis was already weaned, which was somewhat surprising because chimps normally nurse for at least four years. I assumed that they had separated him prematurely from his mother. His unusual eating habits made Loulis the perfect candidate for Washoe because we wouldn't have to reestablish a nursing bond.

The next morning, three of my grad students and I headed out in a van for the long road trip to Atlanta. I had always wanted to visit Yerkes, one of the country's largest facilities for the study of primates. The center's founder, Robert M. Yerkes, was a comparative psychologist who was both fascinated by great apes and respectful of them. Yerkes was one of the first researchers to raise chimpanzees in his home. In 1925, he speculated that the chimpanzee might be able to learn a gestural language. Forty years later, Washoe had confirmed his theory.

Yerkes retired in the early 1940s, and a decade later his successors moved the center from Orange Park, Florida, to a state-of-the-art research facility at Emory University. With its founder out of the picture, the focus of the Yerkes Primate Research Center shifted away from studying chimpanzee behavior and toward conducting biomedical experiments on the chimps. But I'd heard that the center was quite progressive, with outdoor runs and play areas where the chimps could be with one another.

We arrived at Yerkes early on March 22, 1979, after driving twenty hours straight. I stared in disbelief as we pulled up to the main building. It was a gray concrete fortress, surrounded by barbed wire fencing. It looked just like a maximum security prison I once visited in Oklahoma.

After meeting the director, Frederick King, we were taken to the chimpanzee nursery, where Loulis had spent his first few months. The NURSERY sign on the door made me think of a warm and friendly playroom for babies, the kind of environment Washoe enjoyed in the Gardners' backyard. But words had a more Orwellian meaning at Yerkes. The "nursery" was a barren room with two stainless steel cages on rollers. Seven baby chimpanzees were drinking bottles of milk, and my guide told me that researchers were using the milk in an attempt to infect the babies with live leukemia virus. I had to turn away.

We were then led down a long corridor that looked like a lockup for hardened criminals. We passed cage after cage of the same size and shape, all enclosed by giant steel bars, and all of them impregnable. Sitting or standing in each cell, the size of a small kitchen, were one or two chimpanzees. Some of them had a faraway vacant stare that reminded me of Washoe's recent depression. Others charged the steel doors and issued threatening pant-hoots, as though they wanted to kill us for what was being done to them. It was heartrending enough to see so many smart, sentient, emotional beings cut off from natural social contact. But to make matters worse, their cages were absolutely empty—not a toy, not a twig, not a blanket. I wondered what Robert Yerkes would have thought of this scene. Nearly a half century earlier he had summed up the chimpanzee's highly social nature when he said, "One chimpanzee is no chimpanzee."

I was completely numb as we walked the length of this tunnel. I almost slipped several times on the wet gray concrete that had been hosed down with water and chemicals. Technicians in uniform were patrolling the floor like prison guards. We passed what seemed like hundreds of cells, containing hundreds of chimpanzees—more than I'd ever seen in my life. At moments I tried to imagine Washoe in one of these cages, but it was too painful even to consider.

Finally, we got to Loulis's cage. He just sat there, staring out at us quizzically, with big saucer eyes and a cherubic face. He looked utterly incongruous—a helpless infant locked in a maximum security cell. His mother sat motionless in the far corner of the same cell. I took one look at her and didn't need to ask why Loulis was weaned so young. Protruding from the top of her head were four metal bolts, evidence of implantation research. Loulis's mother probably was the subject of a brain stimulation experiment, a popular focus of research in the 1970s. In these experiments, researchers tried to locate the brain's "pleasure center" and other control centers. Then they rewarded or punished the chimp's responses by administering electric zaps to those brain regions.

Whatever the experiment, it was clear that Loulis's mother could no longer care for or nurse her baby. I wondered if she even realized he was hers. While the vets separated Loulis from his mother, I went over the release papers. I thought it would be a simple matter of transferring custody, as one might do with a hospital patient. But I quickly learned that Loulis was the *property* of Yerkes. So I could either buy Loulis for ten thousand dollars or take him on loan. I didn't have ten thousand dollars, so I opted for the loan. They drew up the papers and I signed them.

Then a technician in a white lab coat appeared, holding a dog carrier. "That won't be necessary," I said. When I picked up Loulis in my arms, the young tech looked at me strangely, as if I were taking my life in my hands by uncaging a baby chimpanzee. Loulis clung to me tightly. As we walked out into the bright sunshine, I felt like I was jolted awake from a nightmare.

Yerkes had always been touted as one of the most humane research labs. If Yerkes was humane, I shuddered to imagine the inhumane. I would soon discover them for myself. Every time I visited a lab in the coming years it would rattle me, but no experience shook me to my core like that first tour of Yerkes. Until then I had lived in an academic ivory tower filled with captive but socially engaged chimpanzees. Apart from my one-day visit to Yale in 1973, I had been oblivious to the tragic fate of primates elsewhere. Yerkes was a wake-up call.

I SAT WITH LOULIS in the back of the van while my students shared the driving. Loulis pressed his small, warm frame against me and clung to my shirt. He seemed secure. But somewhere around Chattanooga, Tennessee, he tried to get down off my lap. I felt that he needed the physical contact, and I tried to hold him, but he let me know with his teeth that he had other plans.

He started checking all the corners and windows of the van for his mother. Occasionally, he would make a soft and mournful lost call: "hoo, hoo . . . hoo, hoo." When he got to the front of the van, a student took his hands and turned him around so he could start his search all over again. He did this many times until he was exhausted and fell asleep next to my students on the van's carpeted floor.

I should have been elated to have Loulis with me, but watching him search for his mother made me feel very ambivalent. In my determination to save Washoe's life, I separated Loulis from the only mother he ever knew. Loulis's mother may not have been up to parenting, but who was I to judge her maternal abilities, much less to take her child from her? How did I know that Loulis wouldn't just shrivel up and die? I was betting his life on a hunch.

We arrived home on March 24 before sunrise. It was my son Josh's twelfth birthday, and he stumbled into the living room to find George Kimball, my research assistant, stretched out on the sofa, with Loulis asleep on his chest. Josh was not happy about an infant chimp getting all the attention on his birthday, but before very long he was down on the floor wrestling with Loulis. They were soon joined by eight-year-old Rachel and three-year-old Hillary.

Dr. Carlson came over to examine Loulis, who was in very good shape. Loulis immediately took the bottle of formula Debbi prepared for him, and at 8 A.M., we were ready to introduce Loulis to Washoe.

We drove Loulis over to the pig barn, where he, Washoe, and Ally would be living until their new home at the air base was ready. I went inside alone to tell Washoe the news.

I HAVE BABY FOR YOU, I signed happily.

For the first time in two weeks, Washoe snapped out of her trance and became excited. BABY, MY BABY, BABY, BABY! she kept signing as she hooted for joy and swaggered on two legs. Her hair was standing on end.

I went out to the car and a minute later came back in carrying Loulis in my arms. But as soon as I went into Washoe's cage, she got one good look at Loulis and her excitement was gone. BABY, she signed calmly as she studied Loulis with mild interest. I had forgotten to make it clear to Washoe that it was A BABY, not YOUR BABY. Too late. I would have to hope for the best.

I thought that Washoe would want to hold Loulis, but instead she sat three feet away, watching him. I also expected Loulis to want Washoe to hold him, but instead he clung even tighter to me. I had to pry Loulis off me, turn him around, and hand him over to Washoe. The minute Washoe took him, I quickly left the cage and closed the door. Loulis struggled out of Washoe's arms and tried to race after me.

Washoe was already smitten with Loulis, though she knew better than to force herself on him. She went over to Loulis and touched him gently. Then she moved away, hoping to start a game of tickle and chase. But Loulis wasn't going for it. He sat alone on the floor, looking out at Debbi and me. So Washoe tried a new tactic: she got as close to Loulis as she could without making him run away, and there she sat, watching him, totally fascinated. That night, Washoe tried to get Loulis to sleep in her arms, as Sequoyah had done. But Washoe was obviously not his mother, and Loulis slept alone on the end of the metal bench.

At 4 A.M. the next morning, my students reported a dramatic turning point. Washoe woke up, stood up on two feet, and vigorously signed to Loulis with a loud slapping sound: COME BABY. Loulis was jolted awake and he jumped straight into Washoe's arms. Engulfed in this large and hairy pillow, Loulis fell back asleep.

From that night forward, mother and child slept together. Within days Loulis was dependent on Washoe for his comfort and security, especially around their neighbors in the pig barn, the siamang gibbons. Siamangs have inflatable throat pouches that they use to make ear-piercing shrieks from the canopy of the Sumatran rain forest. Inside a metal barn the effect is deafening. The siamangs were the bane of Washoe's existence. She would regularly fill up her mouth with water, run to the far corner of the cage, and shower them. But now with Loulis clinging to her, she couldn't leave her son alone on the bench because he would immediately start whimpering. She had to keep the very tips of her toes on the bench so Loulis wouldn't cry. Then she would stretch out as far as she could, holding on to a wall of the cage, and let the siamangs have it with a stream of water. Loulis loved this.

My own spirits were lifted for the first time in weeks. And with mother and adopted child now bonding, I would also get to keep my NSF grant and begin studying whether or not Loulis would learn ASL from Washoe.

.   .   .

I HOPED THAT BY STUDYING how sign language was passed from one chimpanzee generation to the next I could fill in a missing piece of the evolutionary language puzzle. Everything I'd learned to this point had led me to believe that language was *not* the product of some mutation that suddenly created a full-blown grammar in the brains of our hominid ancestors. Instead, language appeared to have been a system of communication that was passed down from generation to generation, evolving first in gestural form, then in spoken form. Along the way, our brain and voice box developed in ways that exquisitely prepared the modern human infant to learn and speak language. But language itself remains a rather tenuous cultural artifact. Its survival depends on each generation transmitting it to the next.

You can appreciate the cultural nature of language by conducting a simple mental exercise. Imagine what would happen if every member of our species over age one suddenly disappeared from the earth and the remaining infants somehow managed to survive. These infants would have all the highly evolved anatomical structures they would need to learn and produce language. But without adults there wouldn't be a language for them to learn. *They would have to reinvent language all over again*. And most likely they would start, like our ancestors, with a mixture of gesture and sounds that had some very simple rules. But it might take another hundred thousand years for them to build up the complex language systems we use today.

Culture—whether it's art, toolmaking, or language—is transmitted through learning. If language emerged culturally, as I believe it did, then our earliest hominid ancestors must have passed on their system of gestural communication to their young because gesture had social advantages. The best way to prove this would be to observe signing chimpanzees—whose capacity for cognition and gesture is probably similar to that of early hominids—and see if they gain enough social advantage from sign language to pass it on to the next generation. There is no better place to study this than in the relationship between mother and child.

In the 1960s, anthropologists would have thought this idea was absurd. My own college professors taught that only humans transmitted a communication system across generations. Language was also seen as unique because it was how tribes passed down the rest of their cultural heritage—their art, toolmaking, religious ritual, and so on. All other species were seen as slaves to instinct and incapable of transmitting language *or* culture. But by the mid-1970s Jane Goodall and other ethologists in Africa were reporting that young chimpanzees learned toolmaking skills from their mothers at a very young age. This was a clear indication that chimpanzee children acquired the knowledge and culture of their community in much the same way human children did. If language originated in the gestures and toolmaking of our common ape ancestor, then Loulis would be able to learn signs from Washoe.

From the day Loulis arrived we avoided signing in his presence so that we'd be sure he didn't learn ASL from us. We allowed ourselves to use only seven signs when we talked to Washoe in front of Loulis: WHICH, WHAT, WANT, WHERE, WHO, SIGN, and NAME. This would create a kind of control experiment to see if Loulis would learn these seven signs from us. In the meantime we answered any of Washoe's signed questions with either spoken English, which she understood fairly well by this point, or with chimpanzee vocalization. (If anyone accidentally signed around Loulis, we wrote it down; over the five years of the study, this happened fewer than forty times.) In this environment, Loulis could learn to sign only from Washoe or Ally.

On March 31, his eighth day with Washoe, Loulis learned his first sign: the name for George Kimball. His name was signed by moving the open hand down the back of the head, representing George's long hair. It wasn't surprising that this was Loulis's first sign, because George was the person who gave Washoe and Loulis their breakfast.

Loulis was soon using three more signs—TICKLE, DRINK, and HUG—that he learned by observing Washoe. But he was not just imitating her. At first, he was babbling the sign, exactly the way a deaf child plays with a sign or part of a sign when learning it. For example, the sign for TICKLE is made by drawing the right index finger across the top of the left hand. Loulis watched Washoe go over to a human and sign TICKLE either on her own hand or on the person's hand. Then Loulis babbled the sign to himself. Only after this babbling would he imitate Washoe's other behavior—by going to the person with a play face and signing TICKLE on himself or on them. He developed the sign for DRINK—resting the tip of the thumb on the lower lip—in exactly the same way: first watching, then babbling, and finally using the sign properly and at the right time.

All human infants babble vocally. This seems to be nature's way of getting them warmed up for language; this cooing develops into sounds and eventually into words. Of course, deaf children don't hear vocal feedback, so they soon stop babbling with their mouths altogether. Meanwhile, they begin babbling with their hands while playing with gestures and later with signs. Loulis was learning a gestural language from his mother as nature intended, so his babbling was emerging as naturally as any deaf human child's. Also he was confirming that language, like tool-making, can be transmitted by chimpanzees. Loulis learned to sign by watching his mother, interacting with other adults, and practicing by himself. Thanks to this flexible learning process, he was able to generalize his use of signs to new and different situations.

Loulis was a self-directed learner. Ninety percent of his signing was spontaneous and not prompted by Washoe. This led to creative breakthroughs. For example, one day after Loulis had learned the signs for HURRY and GIMME, I was giving him a drink when I accidentally took the cup from his mouth without warning. Loulis looked at me and signed HURRY GIMME—his first two-sign combination.

I knew that Washoe wouldn't need to do any active teaching because Loulis, like a human child, acquired language out of his deep-seated need to communicate socially. But from time to time she did tutor her son. Once, she placed a chair in front of

Loulis and showed him the CHAIR SIT sign five times. Another time, with Loulis watching, Washoe signed FOOD over and over when one of the volunteers brought her a bowl of oatmeal. Then Washoe molded Loulis's hand into the sign for FOOD and touched it to his mouth several times—just as I had done with her in Nevada and just as parents of deaf children often do. This maternal hands-on guidance seemed to work because Loulis promptly learned the FOOD sign. Again this was very similar to how chimpanzees transmit culture in the wild. (You will recall the example of the chimpanzee mother who gave her frustrated daughter some brief but explicit guidance in how to crack a nut with a hammer.)

After just eight weeks with Washoe, one-year-old Loulis was regularly signing to humans and chimps. Interestingly, Loulis did *not* pick up any of the seven signs that we used around him. He learned only from Washoe and Ally. Within eighteen months of his adoption, Loulis was using nearly two dozen signs spontaneously. He was the first nonhuman to learn a human language from another nonhuman. By doing so, he not only confirmed that language acquisition is based on learning skills we share with chimpanzees, but he showed that the transmission of language is a cultural phenomenon. Washoe transmitted a gestural system of communication to her son—and Loulis was motivated to acquire it—because it was socially beneficial to them. Language helped cement the bond between them. Probably one of the primary reasons language evolved in our hominid ancestors was the role it played in enhancing communication between mother and child. This communication is particularly important in any species with a long childhood that is focused on imitation and learning.

At the Psychonomic Society meetings in 1979, I presented the first results of the Loulis study. And beginning in 1982, Debbi and I published a series of scientific articles documenting Loulis's accomplishments. The news that a baby chimpanzee had learned American Sign Language from another chimpanzee was greeted with enthusiasm by biologists, ethologists, anthropologists, and sign language experts. These were the scientists who had already acknowledged the behavioral kinship between humans and chimpanzees. Many linguists, on the other hand, responded with silence. After all, what could they say? They had been claiming that Washoe and other chimps were either trained through reinforcement or were simply good mimics who could read human cues. But Loulis had not seen humans signing, so there had been no cues for him to read. And hundreds of hours of videotape showed how Loulis learned signs from Washoe. In my mind, the failure of ape language critics to rebut Project Loulis was the most eloquent testimony to its effectiveness.

IN JUNE 1979, about three months after Loulis arrived, we finally moved Washoe, Loulis, and Ally into their new quarters on the abandoned airfield named South Base, about five miles away. Lemmon and I communicated only through the provost, and I didn't ask his permission to take Ally. I knew this was a gamble. Even though Ally was part of my NSF-funded study, he legally belonged to Lemmon. I was hoping that Lemmon would just write Ally off.

South Base was a collection of pine barracks that had been thrown up quickly for naval air training during the Second World War, and it had some obvious problems. For one thing, it was a firetrap. Years before, one of the barracks had caught fire and burned down in three minutes. I installed a loft space so my students could stand watch around the clock in case of lightning or electrical fire. Also a visitor to the barrack could easily have thought we were studying cockroaches, not chimpanzees. You'd turn on the lights and thousands of roaches would scuttle up the walls.

Then there was the plumbing. We were housed in the infirmary barrack because its floor could accommodate pipes. I used to kid my brother, one of several plumbers in our family, that you only needed to know one thing to be a plumber: shit flows downhill. Apparently the plumbers at the University of Oklahoma were not familiar with this principle. They installed our small floor drain at the very highest point, in the center of the infirmary's floor. This meant that we had to squeegee everything uphill to the drain. In addition, the drain and pipe were too small to handle the monkey chow and other debris that wound up in them. As a result, Washoe spent a great deal of her time, mop and toilet plunger in hand, cleaning and unclogging her family's bathroom facilities.

Despite these drawbacks, South Base was a big improvement over the Institute because the university had installed air conditioning. Washoe, Loulis, and Ally had been roasting in the metal pig barn. Before moving to South Base, our most notable scientific finding in June 1979 was that when it's 120 degrees, chimpanzees don't do much signing or anything else, except lie around in front of fans.

Once they were sufficiently cooled off in their new home, Washoe and Loulis showed us some remarkable examples of chimpanzee learning, not all of it related to signing. Washoe taught Loulis a game I had played with her ten years before in Reno: blindman's buff—or PEEKABOO, as Washoe called it. Washoe would cover her eyes and then try to find Loulis. Of course, if she couldn't find him, she would cheat by peeking, just like she did when she was a toddler.

Loulis made up his own game that we called the "run-around" game. Loulis would start it by signing COME to Washoe or Ally. When they approached, Loulis would run in the other direction, just out of their reach. Then he would sign COME again—on and on until he was caught. Loulis was much smaller and quicker than Washoe so she never seemed to catch him. Finally, Washoe figured out her own devious way to win the run-around game. She would lie down on the bench and pretend to fall asleep. When Loulis came up to her she would suddenly grab him and the game would be over.

Like a young chimp in the wild, Loulis learned a lot of other important chimpanzee behavior, like nest building, by watching his adopted mother. In the African forest a chimpanzee takes five minutes or so to build her bed for the night, high up in a tree, by bending several branches into a comfortable pad. Washoe nested in her own special way by swirling her sleeping blanket around her on the floor. She would then place her toys in her nest before climbing in herself. At first, Loulis simply watched Washoe build her nest. From time to time he would help by giving her the toys. After a while Washoe began holding Loulis as she nested. Finally, after a year or so, Loulis was swirling his own blanket into a nest right next to Washoe's.

Like his mother, Loulis was a rambunctious child and loved to challenge authority. When he wanted attention he would spit water right at us. We didn't begin signing around him until he was six years old, so we had to lodge our complaints with Washoe in English. She would then tap Loulis on the head or grab him by the leg to divert his attention. After a while, if Washoe saw that Loulis was about to spit at someone, she would rush in and hold him tightly to stop him.

Washoe seemed greatly relieved to have Ally around to help with the parenting. Even with Sequoyah, Ally had been a very interested and gentle parent. Washoe had been quite reluctant to have Ally hold Sequoyah, but she had finally allowed Ally to groom him. By the time Loulis arrived Washoe seemed to trust Ally completely. When she got tired of Loulis's bottomless appetite for tickling, chasing, and "run-around," she would hand him off to Ally. Washoe would still sit nearby and watch the goings-on carefully, attentive to any sign that Loulis might be in distress. If Loulis did begin to cry, Washoe would rush over to retrieve her son, while Ally signed SORRY SORRY and HUG HUG.

IN THE END, my efforts to save Ally were in vain. That fall, the provost relayed a message to me from Lemmon demanding that Ally be returned. According to Lemmon's students, he was planning to sell his entire colony to a biomedical laboratory.

The thought of Ally winding up in a medical jail like Yerkes broke my heart.

It had been nearly a decade since I met one-year-old Ally, the antic young chimp who made the sign of the cross on his chest. Over those years, I had helped to raise him, taught him sign language, introduced him to his fellow chimpanzees, and perhaps saved his life while he mourned the loss of his mother. In turn, he had comforted, entertained, and taught me plenty. Ally was my friend. He was also Washoe's partner, Sequoyah's father, and Loulis's adopted father. But none of those bonds counted for anything now. According to the law, Ally was Lemmon's property, and I had stolen that property.

I was convinced that if I refused his request, Lemmon would call in the police, and the university would be forced to intervene on his behalf. Lemmon would never sell me Ally. I would have no course of action left but to kidnap Ally, which would be the end of my life as I knew it. I tried to imagine myself becoming a fugitive from the law, hiding out in my van with an adult chimpanzee. Sooner or later I'd be arrested.

I searched my soul for several days. In the end, I had to acknowledge that as much as I loved Ally I wasn't ready to sacrifice my own freedom for him.

One morning in October, I put a lead on Ally, loaded him into my van, and returned him to the Institute. After handing him over to Lemmon's lab techs, I stood there signing GOOD-BYE, NUT.

GOOD GO, he signed back.

That was the last time I ever saw Ally.

ALLY'S SUDDEN DEPARTURE was followed by an equally sudden arrival. In early December I got a phone call from Allen Gardner.

"Roger, we can't continue with Moja," he said. "We think the best solution is to send her to you."

Moja was the oldest chimp in the Gardners' second signing

study. Her name means "number one" in Swahili, and she had been with the Gardners since infancy and was now seven years old. Allen and Trixie had moved out of the small house where Washoe was raised and onto a seven-acre former "divorce ranch" where people used to stay when they came to Reno for its famous "quickie" divorces. The Gardners lived in the ranch's farmhouse while each of the three chimps had their own small cabin.

They had been calling us for nearly a year, asking for advice about Moja's increasingly strange behavior. She was biting people without warning and for no apparent reason. This didn't seem all that unusual to me. I've known plenty of *human* toddlers who think it's their job to bite anyone and everyone. Unfortunately, some of the Gardners' project assistants were refusing to work with Moja. The others started wearing whistles to let the students working with other chimps know that Moja was about to enter their territory. After a while, when no one would work with Moja, Allen brought back Greg Gaustad, my veteran coworker from Project Washoe. But now Greg had decided to move on again.

When Washoe was going through her bully phase, she would challenge and taunt me directly. But Moja was more emotionally manipulative. For starters, Moja would refuse to eat, which always got the Gardners' attention. Then there was the biting. Finally, she began mutilating herself. Moja would stay outside on freezing nights until her hands became frostbitten, and then she would chew her damaged fingers down to the bones. Apart from the danger to Moja's health, this practice raised the scientific question of how she could possibly sign with mangled fingers.

The Gardners were beside themselves. They no longer had the power to force Moja inside. And even if they could, no one wanted to work with her. Their only avenue of escape, apparently, was to send Moja to us. In a déjà vu of Washoe's transfer to Oklahoma a decade earlier, Allen told me that he had already planned Moja's move to our facility at South Base. Greg Gaustad would bring her by airplane, and he would stay for a few days to help in the transition.

Even after ten years, saying no to Allen Gardner was unthinkable to me. So Debbi and I agreed to take on Moja. Thanks to the NSF grant we could afford to care for her, and perhaps she could replace Ally as a friend for Washoe and a caretaker of Loulis. If we could get her to stop chewing her fingers, Moja would no doubt be a talkative companion; she had been signing since she was three months old and knew 150 signs. But we were also terrified. We didn't know Moja, nor she us. I had met her once for three days in 1977. How in the world would we handle a powerful chimpanzee who was, by most accounts, a neurotic tyrant?

As it turned out, we *didn't* handle Moja; Washoe did. Moja had never been around a chimpanzee who was older, bigger, and stronger than she was. She had gotten used to bullying her younger brother and sister, Dar and Tatu, and assumed that humans would respond to her every scream and fret over her every wound. Now, 70-pound Moja found herself dealing with a 150-pound female who had an 18-month-old child to take care of. Washoe had no time for Moja's self-pity, and she put her in her place in a hurry. The day Moja met Washoe was the day she began growing up.

It was a rocky transition. Moja's separation from Allen and Trixie tore her apart. Washoe had landed on her feet in Oklahoma and had just kept right on going, but Moja was an emotional basket case. She wouldn't eat. She had constant diarrhea. She screamed all the time. She groomed her wounds until they bled. And when all else failed Moja would simply sign to us, HOME? GO HOME?, holding the sign in place for several seconds to stress her urgency.

It was heartbreaking to see Moja suffer and it was all too easy to cater to her every demand. When she demanded SANDWICH—peanut butter on white bread was all she ate—Debbi and I would fall over ourselves to make it. When she screamed we would anguish over why she was upset. When she chewed her fingers we begged her to stop.

Washoe dealt with Moja more straightforwardly. If Moja wouldn't stop screaming, Washoe would swat her on the head, as if to say, "I'll give you something to cry about," and Moja would instantly snap out of her crying jag. If Moja refused to eat her dinner, then Washoe would eat it for her. If Moja was trying to manipulate us by grooming her wounds, Washoe would turn it into a group activity by joining in the grooming. And if Moja was listless, Washoe would put Loulis on Moja's back and let her play "aunt" for a while. Loulis loved Moja, and his energy and affection seemed to bring her out of herself. Meanwhile, Moja was spending so much time worrying about what Washoe might do to her that she had little time to be morose.

It took a year for Moja's wounds to heal. When she did emerge from her depression she was a transformed individual. The biting, the bullying, and the self-mutilation had stopped. Moja would always be rather neurotic and eccentric, but she looked up to Washoe and she was truly devoted to little Loulis. Washoe had done something the Gardners and we could never do: she had taught Moja to be a social chimpanzee.

A FEW MONTHS AFTER MOJA ARRIVED, the provost passed me another message from Lemmon, but this one caught me completely by surprise. Lemmon was now claiming that he owned Washoe, and he was already negotiating her sale to a biomedical research lab. The message said, "She's mine and she's going with the rest of the chimps." Lemmon insisted that the Gardners had legally transferred Washoe to the Institute back in 1970.

I was sure that the Gardners had given me custody of Washoe when we moved to Oklahoma, but I had no papers to back up my claim. And even if I did, they would be meaningless. As I had learned from the cases of Loulis and Ally, custody is irrelevant when it comes to chimpanzees. They may act and sign like children, but as far as the legal system is concerned, chimpanzees are nothing more than property—just like a car, a house, or a toaster. I either owned Washoe or I didn't. The university provost was asking for proof of my ownership of Washoe, and he wanted it fast.

I wasn't even sure that the *Gardners* owned Washoe. It seemed incredible that after thirteen years I didn't know who owned Washoe. Was it possible that Lemmon owned my chimpanzee sister?

Allen Gardner put that concern to rest without any hesitation. "Lemmon doesn't own her," he said. "She's an Air Force chimp." So I contacted the Air Force and a few weeks later I received a terse letter in reply, signed by an Air Force colonel.

> Dear Dr. Fouts,
>
> In response to your recent request concerning ownership of the chimpanzee #474 ("Washoe"), which previously had been in the colony at the 6571st Aeromedical Research Laboratory, we have reviewed available records. These indicate that ownership of the animal was transferred to Beatrix T. Gardner at the time of transfer in June 1966.

With this in hand, I was able to get Allen to acknowledge that he and Trixie had in fact owned Washoe. It was clear, however, that they had no desire to be considered her owners and no intention of bearing the responsibilities of ownership.

The Gardners could easily transfer ownership of Washoe to Debbi and me, but the idea of owning Washoe was as repugnant as the thought of buying or selling our own children. Unfortunately, society at large did not share our view that chimpanzees are more akin to people than to inanimate property. The legal system had its rules and we would have to play by them to protect Washoe.

Allen signed an agreement releasing Washoe to me indefinitely. I took the new agreement to the provost and showed him the paper trail of Washoe's ownership. That officially closed the whole matter.

THE FIGHT OVER WASHOE, plus the substandard conditions at South Base, convinced me that it was time to find a better home for the chimps. But where? In early 1980, I spoke at universities in California, Texas, Ohio, North Dakota, Oregon, Michigan, Tennessee, Colorado, and Manitoba. Everywhere I went I looked for primate facilities and possible sanctuaries. I began talks with one or two schools, but it was slow going.

Then a door opened in the least likely of places. In May 1980 I spoke at Central Washington University (CWU) in Ellensburg, Washington, but I gave no thought to any job prospects there. Central Washington awarded bachelor's and master's degrees only—no doctorates. They wouldn't have much interest in a professor who taught advanced courses in primate behavior and communication. But after giving my talk, I was approached by one of the school's administrators. He had previously worked at the University of Oklahoma, mostly in public relations.

"What would it take to bring you to CWU?" he asked me.

"You couldn't do it," I answered. "You don't have a primate facility."

But he became very persistent. He remembered that during some years in the mid-1970s Washoe, Lucy, and Ally had gotten more national media attention than the University of Oklahoma football team—and football in Oklahoma is something akin to a state religion. The next day he called with the surprising news that CWU did have a primate facility, and it was almost new. The third floor of the recently built psychology building had a wing with four rooms designed to house monkeys, but no one was working with monkeys. I went to have a look at the rooms.

They had excellent trench drains, individual heat and humidity controls, floor-to-ceiling windows (in one room), and, luxury of all luxuries, a kitchen! (Cooking for three chimpanzees is a full-time job.) Compared with our cockroach-infested, windowless firetrap at South Base, the monkey wing looked palatial.

I hadn't formally been offered a job yet, but when I got back to Oklahoma I found myself making the case to Debbi for moving to Central Washington. Ellensburg is a charming small town nestled in a beautiful valley on the eastern slopes of the Cascade Mountains, about two hours east of Seattle. We agreed that it would be an improvement for the chimps and a good place for our kids to grow up.

The drawbacks were just as clear. Compared with Oklahoma, whose psychology department was ranked a respectable thirteenth in the nation, CWU was an academic backwater. Unlike the University of Washington at Seattle, CWU was not a major magnet for federal funding. I would not have Ph.D. students to assist and advance my work. And Debbi would have to give up her lifelong dream of getting her own doctorate.

These were all significant sacrifices. But deep in our hearts, Debbi and I both felt that CWU offered us the freedom to create the best possible environment for Washoe's family. In fact, CWU being a backwater was one of its greatest attributes. It had no established primate protocols, no technocrats, guards, guns, or rules. It was a blank slate. We would finally be free to create a chimpanzee research setting based on mutual respect and compassion, instead of fear and domination. Obviously I would have preferred a large sanctuary where the chimps could spend most of their time outdoors, but the president of CWU was encouraging about the possibility of building an outdoor facility in the near future. In the meantime, my priority had to be getting Washoe, Loulis, and Moja into a safer and more comfortable home.

In June 1980 I accepted a tenured position as Professor of Psychology at Central Washington University. The dean of Oklahoma's graduate school was aghast when I told him the news. "Where?" he kept saying, as if he hadn't heard me the first time. After urging me to stay, he promised that Oklahoma would build me a proper facility in a year or two. If they didn't, he added, I would certainly get offered a position at an Ivy League school. He still didn't seem to get it. In the previous year I had visited a dozen university-affiliated primate labs. I'd seen firsthand how they treated nonhuman primates, and I wanted no part of it.

I WAS ALMOST EXPECTING LEMMON to spring some last-minute surprise on us, so I kept our moving plans under close wraps. I quietly made arrangements to borrow a double-long horse trailer from a student who worked for an animal training facility near Los Angeles. Other than six graduate students, no one affiliated with the university or the Institute knew how or when we were leaving. I recruited two of my students to help me transport the chimps to Ellensburg.

One morning before dawn in late August, I backed up the trailer to our South Base barracks. The trailer was certainly big enough to hold two chimpanzee cages, with Washoe and Loulis sharing one cage and Moja in the other. I would ride alongside them in the trailer while one of my students drove the truck. But persuading Washoe and Moja to get into the cages was even tougher than I had expected. Both of them loved car rides, but they could tell by the strange trailer and my shameful begging that this was more than a ride around the block. The more I offered them—soda, candy, yogurt—the more suspicious they grew. I was stuck. I couldn't muscle them in, because they were stronger than me. I could anesthetize them, but that would require a "capture gun" that shoots a thick dart, tipped by a needle, that actually opens a wound and can be quite painful.

Finally I just flat-out lied to Moja. YOU ME GO HOME, I signed to her, meaning that I was taking her to Reno. Her eyes lit up and she bounded into the trailer. Washoe was tougher. I told her we were moving to a new home and a better place, but she wouldn't go farther than the back ledge of the trailer. She reached into the cage and cleaned out the goodies inside, holding the trailer door open so I couldn't close it behind her. Next I tried yelling and intimidating. But in the end I had to threaten her by taking out the capture gun. Washoe screamed and climbed into the cage, with Loulis right behind her.

She sat in the back corner of her cage, mad as hell that I had forced her in. Two hours later, when we stopped at the first gas station, she still wouldn't look at me through the side door of the trailer. But when I came back from the convenience store with ice cream, her whole attitude toward our road trip changed. HURRY, HURRY, GO, GO, she signed, pointing down the road. Suddenly she was an enthusiastic traveler, and every two hours we had to stop for gas and ice cream.

The trip gave me plenty of time to reflect on the life we were leaving. The moment we pulled out of South Base I breathed a long sigh of relief. Ever since Sequoyah's death I had become so fearful for the chimps that all I could think about was getting Washoe, Loulis, and Moja out of Oklahoma alive. Now they were beyond Lemmon's reach, and I was almost gleeful.

My elation was quickly tempered by thoughts of those we had left behind. I had a terrible feeling about what would happen next. Lemmon might finally lure some biomedical researchers to the Institute, or maybe he would auction off the chimps to a lab. Either way, I had trouble imagining a happy ending for Ally, Booee, Bruno, Cindy, Thelma, Manny, and all the others.

As for Lucy, who had left three years before for Africa, there was still a glimmer of hope. She was no longer living in her cage in the forest reserve. Janis Carter, who originally went with Lucy to Africa "for three weeks," had recently moved Lucy, along with Marianne and seven wild-born chimps, to one of the five Baboon Islands on the River Gambia. Janis was trying to teach Lucy wild chimp behaviors like nest building and food gathering, but so far she hadn't had much success. Poor Lucy was still emaciated and pleading in sign language for Janis to find food for her—MORE FOOD, JAN GO. Janis refused, and Lucy finally began to collect her own fruit out of a baobab tree, but only after Janis set up a ladder for Lucy to climb. It sounded like both Janis and Lucy were near the breaking point. But I told myself that as long as there is life, there is hope.

I turned my attention back to the road, to the three chimpanzees in the trailer with me, and to the van ahead of me carrying Debbi and our three children. It seemed hard to believe that just a decade earlier, when I got off that plane in Oklahoma, Washoe was the only chimpanzee I knew. In the years since then Washoe's chimpanzee family and my human family had become kin. But now as we headed toward an unknown future and a hoped for sanctuary, I felt a terrible sense of loss. I knew in my heart that no matter how wonderful a home we would build in the years to come, it would always feel incomplete. As long as Ally and the rest of my extended family were locked away, I could never be at peace.

PART THREE

THE SEARCH FOR SANCTUARY

ELLENSBURG, WASHINGTON: 1980-1997

Though the difference between man and the other animals is enormous, yet one might say reasonably that it is little less than the difference among men themselves.
—Galileo, 1630

How smart does a chimpanzee have to be before killing him constitutes murder?
—Carl Sagan, 1977

ELEVEN

AND TWO MORE MAKES FIVE

EBBI ARRIVED IN ELLENSBURG ahead of me, along with our three kids, two dogs, and one cat. When I pulled up to our new house with the horse trailer in tow, she greeted me with the news that the chimps' home in the psychology building was still not ready. The university had kicked in twenty thousand dollars to build room-sized wire enclosures inside the four primate rooms, and the chimps would be able to climb and swing and move between rooms, thanks to overhead tunnels. Everything was being built to our specifications for safety wire, secure doors, feeding windows, and the like. But we soon discovered that there was some "assembly required."

We arrived at the psych building and found mountains of disassembled caging, tunnels, beams, poles, and hardware. And like a parent's worst Christmas Eve nightmare, the instructions made no sense at all. While we were trying to make holes line up so that we could bolt the walls together, there were three chimps locked in a horse trailer with no place to go. Right away, we began drilling new holes and piecing the whole thing together. After hours of labor, we knew that it was going to take several days.

Washoe, Moja, and Loulis wound up spending their first night in Ellensburg parked in front of our house in the horse trailer. Some local residents were already nervous about the chimps coming to Ellensburg. A front-page story in the *Daily*

Record had announced, "The chimps are coming! The chimps are coming!" The strange hooting and other jungle noises coming from the horse trailer in the dead of night probably didn't ease our new neighbors' sleep.

I didn't sleep much that first night either because the horse trailer would be returned to Los Angeles the very next day and my classes were about to start. We needed an angel, and we found one in Pautzke Bait, a local supplier of salmon eggs, who let us use an empty warehouse. I parked our Honda in the warehouse and attached Washoe to it by a twenty-foot lead. (Loulis didn't need a lead and Moja was on a long lead attached to a nearby wall.) Everything was fine until I left Washoe in my students' care when I went to class. Washoe immediately began blackmailing them by threatening to smash the car windows and tear off the windshield wipers unless they gave her soda. By the time I came back that afternoon, she and Loulis were surrounded by a mountain of empty soda cans.

About a week later we had two of the enclosures assembled, and we moved Washoe, Loulis, and Moja into their new home on the third floor. These two rooms were the smallest of the four so Moja had to stay in her own room for the time being. She was extremely lonely and began mutilating herself by grooming a boil on her leg down to the bone. Pretty soon I had a contingent of students from my Psych 101 class keeping Moja company from dawn until dusk.

Moja was easy to engage if one brought along the right fashion accessories. She was extremely conscious of her appearance, and there was nothing Moja loved better than putting on an old dress, shoes, and makeup and studying herself in the mirror. She insisted on red dresses, but she wasn't choosy about her footwear. The irrigation boots we wore to clean the cages made her just as happy as party shoes. After she was dressed she would ask us to brush her long hair, which could keep her entertained for hours.

Washoe and Loulis were more challenging. The problem wasn't getting them to play but getting Loulis to stop. Loulis was in that difficult phase, familiar to any human parent, when a child insists on being the center of attention at all times. He could play TICKLE CHASE for hours, but when playtime was over and we needed to clean the cages, collect the signing data, and prepare the meals, Loulis would become a little terror. He'd throw a tantrum and spit streams of water at us. At times like these, Washoe would stop calling Loulis BABY and would address him as DOG, as in, COME DOG!

Finally, after months of separation, Moja was reunited with Washoe and Loulis in one large room that had seven-foot-high windows that looked out over the college football field. Moja was now back to socializing with her family, especially with little Loulis. Washoe immediately tended to Moja's latest self-inflicted wound, and Moja was soon on her way to a full physical and psychological recovery.

But in February 1981, just one month after Washoe's family settled into its new home, I received another life-changing phone call from Allen Gardner. He and Trixie wanted to send us Moja's two foster siblings, Dar and Tatu.

The "problem child" this time was four-year-old Dar. The Gardners had acquired Dar when he was an infant from Holloman Air Force Base in 1976 and named him for Dar es Salaam, the capital of Tanzania. There was nothing wrong with Dar. He was just entering those "fearsome fours" when chimpanzee boys, like most human boys, begin showing their aggression and testing their physical power. Dar's was somewhat more intimidating because he was already a lanky 60 pounds and growing fast. His chimpanzee father, Paleface, was the largest chimp ever held at Holloman Air Force Base. Paleface was five feet four inches tall and weighed 235 pounds. Dar's mother, Kitty, gave birth to a series of immense chimpanzee infants, including one who was fondly known as The Hulk. Dar also inherited his father's distinctive white skin and his mother's huge floppy ears.

The Gardners had never dealt with a male chimpanzee past infancy, and they didn't know what to do with Dar. He was

becoming known around his Reno neighborhood as The Aya-
tollah. (This was during the time that Iranian revolutionaries
were holding scores of Americans hostage.) Dar would often
escape from his student companions and park himself by the
side of the road, a picture of innocence, with his freckled face,
enormous ears, sweet expression, and child's T-shirt. As joggers
came by, Dar extended his hand and beckoned them over.
When the person shook hands, Dar would not let go. When the
Gardners' students arrived, Dar opened his mouth and threatened to bite his prisoner, as if to say, "Don't make me hurt
him!" One time this hostage drama dragged on for hours.

Dar also took great pleasure in his newfound powers of de-
molition. One day some grad students were driving the chimps
to get hamburgers. Dar was so excited he began thumping the
windshield until it popped right out of the frame. The windows
at home were not faring any better. Dar regularly escaped his
cabin at the crack of dawn and got into the Gardners' house by
breaking a window. He would then crawl into bed with Allen
and Trixie. The Gardners had decided not to find out what
havoc Dar would wreak when he turned six or seven.

Five-year-old Tatu, on the other hand, was the family's precious angel, the good girl. The Gardners acquired her from William Lemmon in 1975, and named her for the Swahili word
meaning "three." (Between Moja and Tatu, the Gardners had
acquired a "number two" chimp named Pili, but he died of leukemia in 1975.) There was no need for padlocks and keys
around Tatu. Unlike Washoe, she never ransacked the cupboards or raided the fridge. Lubriderm oil could be left out in
clear sight and Tatu would never empty it. Her room was neat
and tidy with the toys lined up in a perfect row. After she played
with one toy she would put it back in its place before playing
with another. The Gardners often said that Tatu was the kind
of no-fuss chimpanzee they could live with forever.

But Dar and Tatu had grown up together, and the Gardners
didn't want to separate them. With their federal funding running out, Allen and Trixie wanted to unload Dar and Tatu right

away. Allen always said that a scientist should never do research without funding. I tried telling Allen on the phone that Debbi and I had our own money problems. My three-year NSF grant was almost used up, and it hadn't been renewed yet. If I lost my funding, we would have trouble just feeding and caring for Washoe, Loulis, and Moja.

"If you don't take them, we're going to send them to a zoo," he warned.

"No you won't," I protested.

"I'm going to call Washington Park," he said, referring to the zoo in Portland. "If they wind up there it'll be your fault, because you didn't take them."

I knew that a zoo wouldn't take Dar and Tatu. Zoo directors think that home-reared chimps act too human. It's bad business for zoos if their chimpanzees dress up in clothes, talk in sign, and leaf through magazines. It tends to unnerve the visitors, who expect dumb beasts. But if the Washington Park Zoo wouldn't take Dar and Tatu, I knew who would: a biomedical laboratory. Allen and Trixie had gotten all their chimps, except Tatu, from such labs, and unlike me, they were comfortable in both the behavioral and the medical research worlds. They spoke highly of certain biomedical researchers, like the ones conducting hepatitis research, who claimed to have the chimpanzees' best interests at heart. Besides, Allen was always saying that the chimps belonged to science.

I found myself shouting at Allen, "They'll end up in biomedical!" And he was screaming back, "You don't know what you're talking about! I have lots of zoo contacts. I won't have any problem getting rid of them!"

Debbi was standing next to me and she could tell exactly what Allen was proposing. When I got off the phone, I looked at her and she looked at me. Take on Dar and Tatu? I had met them only once, in 1977, when they were infants. We would have to comfort them through the trauma of separation from their home and parents and slowly integrate them into Washoe's family. We would have to steer them through childhood and adolescence, which was already a full-time job with Loulis—not to mention with our own three children. But unlike our kids, Dar and Tatu would be completely dependent on us for the rest of their lives, and that could be another forty or fifty years (if we lived that long). When we took on Moja we at least had some grant money, but now we were just getting by.

But still we couldn't say no. Dar and Tatu were *babies*. Neither of us could live with the possibility that they might wind up in the kind of place we had rescued Loulis from. Our fear was heightened by the anguish we'd felt at leaving Ally, Booee, and the others behind in Oklahoma. We had been legally helpless to take them or protect them, and it was only a matter of time before Lemmon sold them. We knew that once Dar and Tatu left the Gardners', they would also be beyond our reach.

I called Allen back and told him we would take Dar and Tatu.

It was May and school was still in session so Debbi and I had only two days to drive the seven hundred miles from Ellensburg to Reno, pick up Dar and Tatu, and return home. When we got to the Gardners' ranch late Saturday evening we were exhausted. Dar and Tatu were already asleep. Two of the chimps' human friends, good old Greg Gaustad and another student, Pat Drumm, were going with us back to Ellensburg. At 4 A.M. the next morning, Greg and Pat got Dar and Tatu out of bed, onto the potty, and into their clothes. We met the two sleepy youngsters while they were waiting next to their suitcases. Greg told Dar and Tatu that they were going for a RIDE, and while it was still dark we drove away.

Dar and Tatu sat in the backseat in their friends' laps. Everything was fine for the first two hours or so. The chimps played games, shared some treats, and got on their travel potty when they needed to. But as the sun began rising I could see their curiosity turning to alarm. They could tell that something was not quite right about this car trip.

GO OUT, GO OUT, Dar began demanding.

NOT NOW, WAIT, Greg kept stalling him.

Dar must have decided that he and his sister were being kidnapped by total strangers. Sure, his human friends were in the backseat with him, but they were probably being kidnapped, too. Dar got up on two legs and began banging on the doors and throwing things at me in the driver's seat. Meanwhile, Tatu was huddled quietly in the corner, looking completely petrified. This melodrama went on all day and all night until we finally arrived back in Ellensburg with the "stolen" children.

WE KEPT DAR AND TATU APART from Washoe, Loulis, and Moja for the next six months. There were just too many strong personalities involved and no way of predicting the family dynamics. Even though they were separated, the two families could see and touch each other through a gated tunnel that connected their rooms. They spent a lot of time in the tunnel sizing one another up. Dar and Loulis frequently kissed through the gate and it was obvious they would be friends. Tatu seemed happy to see her foster sister, Moja, after two years apart, but Dar was terrified of Moja. He may have remembered that his older sister had a habit of biting. Washoe stayed close behind Loulis, occasionally displaying to let the newcomers know who was boss.

In December we opened the tunnel gate and there was instant pandemonium—a lot of running around and screaming on all sides. With a mixture of excitement and fear, Dar and Tatu crept slowly into Washoe's room. Their own room had no windows, so Tatu sat down and stared out Washoe's windows at the snow. Then she pointed to the asphalt parking lot and signed BLACK.

With a play face, Loulis signed HURRY COME to Dar, who signed the same thing back. The two boys began tickling, laughing, and chasing around the room. At one point Dar was dragging Loulis by the leg, and Washoe grabbed her son back. Dar was swaggering, and he was clearly too macho for Washoe's taste. Even though Dar weighed eighty pounds he was no match for Washoe. After reprimanding him a few times, she held out her arm and Dar kissed it in submission. A little while later he turned his back to her and she patted him and tickled his neck.

Every so often Loulis ran back to Moja and touched her repeatedly, as if to let her know that everything was OK. Tatu stayed back in the tunnel, probably scared of Washoe. When Loulis and Dar rolled by she grabbed at Loulis somewhat aggressively. After Debbi told Tatu that Loulis was a BABY, her touching became gentler. An hour later, Dar and Tatu were playing with Loulis on the bench when Washoe rushed into the room, startling all three of them. Loulis signed HURRY HUG to Tatu and she gave him a big hug. When I closed the tunnel gate before dinner, separating the two families again, Loulis began crying. He didn't want the day to end.

As time went on, each chimp found his or her own way of adjusting to the new and bigger family. At sixteen years old, Washoe was clearly the matriarch. Dar found a great friend in Loulis and ample maternal love from Washoe, but Tatu was depressed, withdrawn and struggling to fit in among so many chimps. To top it all off, in early 1982 Tatu reached puberty and her first estrus. The sudden physical and hormonal changes propelled her into wild and unpredictable mood swings. She'd be crying one minute, taunting Washoe the next, and grabbing at the male students in between.

After her first menstrual cycle, Tatu became a rather intense adolescent. She would go back and forth between rough play with the boys and quiet time alone or with Moja. Moja was already the dreamy teenager, happiest when brushing her hair, looking through magazines, and vying for the attention of young human males in the lab. Moja and Tatu spent a lot of time lying on the floor, holding magazines with their feet, leaving their hands free for conversation and comments. Tatu, especially, loved to find photos of men's faces and sign to them, THAT FRIEND TATU, which would be followed by many variations on this romantic theme.

Like any sisters, Tatu and Moja had their share of screaming fights. In the summer of 1982, they both went into estrus during the same week, and there was hell to pay. They could barely stand the sight of one another, and for days they shrieked, pinched, poked, and pulled each other's hair. Washoe, Dar, and Loulis gave them a wide berth. But as soon as their cycles were finished the sisters were back to brushing one another's hair and sharing their magazines.

Moja and Tatu loved playing with Loulis, whom they called PRETTY BOY. But sometimes PRETTY BOY would flaunt his special status as the favorite son by setting up his new sisters. He would start play-crying to Washoe and then point his finger at the girls. Whenever Loulis pushed Moja and Tatu too far, you could just see them barely holding back the impulse to smack him. They didn't want to provoke the matriarch. Instead, the girls would look pleadingly at Washoe and scream for mercy.

By the spring of 1982, Loulis had turned four and was becoming less Washoe's baby and more a juvenile chimp. His play was like Dar's, with lots of aggressive displays, foot stomping, and charging. For the first time, Washoe actually began gently but firmly disciplining her son by holding him back, shooting him dirty looks, and even lightly thumping him on the back when he annoyed someone. But Loulis's tough-guy act fell apart the second he was frightened or hurt. He would run to Washoe, signing HURRY HUG, and she would sit next to him, her arm around his shoulder, and groom him until he calmed down.

AROUND THIS TIME DEBBI AND I DECIDED to nurture the bonds between Washoe's family and our own children. In Oklahoma our kids came to the Institute occasionally and, in the early years, Lucy, Booee, and Ally sometimes visited our house. But we tended to discourage these encounters for a couple of reasons. Children carry a lot of the colds, flus, and other respiratory infections that chimps are especially vulnerable to.

And chimps who are not used to children can become quite aroused and frightening in their displays. As a result, Washoe and our kids were not well acquainted.

That suddenly seemed unfortunate. Washoe was now the head of a household that seemed very similar to our own. Her days, like ours, were also filled with teenage growing pains, sisterly spats, and childhood traumas. Clearly, both the chimps and the children were missing out on a great opportunity to watch one another grow up, and we parents might learn a few things ourselves.

When Joshua, Rachel, and Hillary began spending their afternoons with the chimps, we discovered that Washoe understood our human family configuration quite well. Debbi and I had never hugged one another or been demonstrative in Washoe's presence. This precaution went all the way back to the Reno days when Washoe would sometimes misinterpret physical affection and attack the "offender." Washoe had rarely been to our house since Reno. As far as we knew, Washoe thought Debbi and I were friends or coworkers. Out of habit, we kept up this act in Ellensburg for the first year, but on one of six-year-old Hillary's first visits to our lab, Washoe asked to hug her good-bye before she left. After they hugged I asked Washoe, WHO THAT?, pointing to Hillary. Without hesitating, Washoe signed ROGER DEBBI BABY. Nobody reads nonverbal behavior like a chimpanzee. And all those years we thought we had Washoe fooled!

Washoe called Rachel, our middle daughter, FLOWER GIRL, the same name she called Debbi. We don't know *why* Washoe associated Debbi with FLOWER, but some coworkers said it was because Washoe smelled Debbi's fragrant lip balm when she greeted her with a kiss every morning. Both Rachel and Hillary spent many long afternoons in our lab that year, and the chimps always looked forward to their visits. Rachel, who was then in fifth grade, already knew some ASL because she had taken ASL classes to communicate with a girl in her school who had cerebral palsy and couldn't speak. Hillary began learning ASL as well, and the two girls practiced together at home.

Washoe loved to teach Rachel and Hillary games. In one of them, Washoe would say GIVE ME SHOE, and the girls would line up their feet in front of her. Then Washoe tickled the toe of one shoe until that person laughed like a chimp (top teeth covered, lower teeth exposed, and breathy sounds). Then she'd move to the next foot until she was playing their feet like xylophones and the girls were hysterical. Washoe loved this game so much that she wouldn't let the other chimps play it. They were only allowed to watch. One time, Washoe let Loulis tickle the girls, but all he wanted to do was untie their shoelaces.

The girls really bonded with Loulis, and he became very possessive of them. When the girls tried to include Dar, Loulis did his best to drive Dar away and win back all the attention. If he couldn't, he just sat and cried until the girls comforted him. Hillary and Rachel also brought their friends to play with Washoe and Loulis. One day I overheard Hillary reassuring another girl who didn't even want to *look* at the chimpanzees. "Washoe's not pretty to look at," my six-year-old daughter explained, "but she's pretty on the inside, and that's why she's my friend."

Moja and Tatu didn't take to our daughters as easily as Washoe and Loulis did, but that was understandable because they hadn't known me nearly as long as Washoe had. Moja and Tatu were always testing the girls, as they would any strangers, chimpanzee or human. Once, when Rachel was handing Tatu an apple, Tatu grabbed her hand just long enough to show dominance and scare her. I rushed over to scold Tatu, signing, MY BABY CRYING. Tatu was so taken aback that she looked genuinely remorseful as she signed SORRY SORRY to Rachel. After a few incidents like this, Moja and Tatu figured out that Rachel and Hillary were fun to play with, but they were the wrong kids to pick on.

My son, Josh, who was fourteen years old, began volunteering in the lab on a daily basis in the summer of 1981. Josh's involvement took a great deal of courage. As a teenager he was having a hard enough time finding his place in a new community. He also had to deal with some bullies in his new school who tagged him as the "chimp kid." This led to a number of fights and destroyed bicycles, but Josh persevered with the chimps anyway. In addition to becoming good friends with Dar and Loulis, Josh helped to get the lab up and running smoothly during a difficult year of transition.

Family relations got a little too close for comfort the next year, however, when Washoe developed a head-over-heels crush on Josh. It seems that my son's looks and sexuality had matured just enough that Washoe's own teenage hormones now began raging at the mere sight of him. Whenever Josh entered the lab, Washoe literally threw herself at his feet and began shrieking like a desperate, lovelorn suitor. It was bad enough, Josh said, that he couldn't get the girls at school to pay attention to him. To have a female chimpanzee throwing herself at him every day really added insult to injury. After a few months of Washoe's entreaties, Josh decided to avoid the lab for a while.

AFTER OUR MOVE TO ELLENSBURG I continued to focus my signing research on Loulis. Every day we recorded a sampling of his utterances. We were still not signing in front of him and, by the end of 1981, he had acquired thirty-two reliable signs from Washoe and Moja. His two-sign combinations had become three- and four-sign combinations like HAT HURRY COME PLAY (a visitor to the lab was wearing a hat) and COME GIMME DRINK HURRY. Now with Dar and Tatu in Washoe's family, we suddenly had a chance to study the social interaction of no less than five signing chimpanzees. Dar and Tatu had both been signing since infancy and each had a reliable vocabulary of more than 120 signs.

Our new study of social signing was fortuitous because a famous controversy had just engulfed the entire field of ape language research, and the media attention it was getting was causing people to think that chimpanzees were incapable of using sign language socially as human children do. The whole brouhaha erupted over Ally's younger brother, who was born at Lemmon's Institute in 1973. Lemmon loaned the infant chimpanzee to Herbert Terrace, the same psychologist who had taken Bruno to New York in 1968. Terrace named the infant Nim Chimsky (a joking allusion to Noam Chomsky) and studied his language development for four years before returning him to Oklahoma in 1977. Terrace published his findings in 1979, and his book, entitled *Nim*, provoked a scientific backlash against chimpanzee signers.

The purpose of Project Nim was to teach an infant chimpanzee to use ASL and to prove more conclusively that chimpanzees could create sentences. Terrace was hardly alone in doing this. In the 1970s numerous ape language experimenters, including myself, followed in the Gardners' footsteps. There was Duane Rumbaugh (who communicated with a chimpanzee at Yerkes named Lana by using a keyboard and a computerized language he called "Yerkish"), Sue Savage-Rumbaugh (who also used Yerkish with two chimps named Sherman and Austin and, later, with a bonobo named Kanzi), Penny Patterson (who taught a gorilla named Koko to use ASL), and Lynn Miles (who taught an orangutan named Chantek to use ASL). By 1979 there was no longer a question of whether great apes could use language but only to what extent.

Project Nim departed from the Gardners' proven method in two very crucial respects. Nim was not raised like a human child, nor was he immersed in a natural ASL environment. The central premise of Project Washoe was cross-fostering; a chimpanzee would learn a human system of signing if she were reared as a child and allowed to learn spontaneously like a child.

Herbert Terrace, a student of B. F. Skinner, had a very different approach. He explicitly instructed his assistants *not* to treat Nim like a child. From the age of nine months, Nim was driven every weekday to Columbia University, where he was subjected to a pair of three-hour training sessions in a window-less eight-by-eight-foot cell. "This was by design," Terrace wrote in his book. "I felt that Nim would not romp around too much in a small area. . . . I also felt that a bare room would minimize distractions." Terrace later described Nim's environment: "I had paid little attention to the cold cinder block walls of the complex. . . . I wondered how I and other teachers could have spent so much time in these oppressive rooms."

Nim, in short, was treated a lot like a rat in one of Skinner's operant conditioning chambers. Despite Terrace's later claims, Project Nim was nothing like either Project Washoe or my own studies. Project Nim was an experiment in social deprivation. Nim's learning environment was so devoid of natural human interaction that the linguist Philip Lieberman described Nim as "the Wolf-Ape," referring to so-called "wolf-children" who were raised in such aberrant environments that they failed to develop normal linguistic ability.

In addition, Nim was taught by no fewer than sixty different teachers who, according to Terrace, ". . . cycled through Project Nim in a revolving door manner." In describing their training procedure, Terrace wrote, "Typically, Nim reached for something he might want to play with, eat, or inspect. The teacher withheld the item, molded the object's name sign and then asked Nim to sign for the object." In other words, Nim was conditioned to beg for food, toys, and other goodies. You will recall that the Gardners abandoned Skinner's principles of conditioning in 1967 because they undermined Washoe's natural inclination to learn through observation.

After three years of training Nim in this rigid manner, Terrace came to believe that Nim had learned more than one hundred signs and was creating primitive sentences. But after returning Nim to Oklahoma and reviewing videotapes of him, Terrace discovered that Nim's signing was not spontaneous like that of a child. Viewed in slow motion, most of Nim's signs were clearly prompted by his teachers and were imitations of what had just been signed to him. This fact surprised no one but Terrace. After all, Nim had been *rewarded* for imitating his teachers.

The one thing Terrace actually proved was that a socially deprived chimpanzee locked in a prisonlike environment will *not* learn American Sign Language. But in 1980 Herb Terrace found a way to turn his flawed experiment into a dazzling media success. He claimed that all chimpanzee signing was an optical illusion. "Nim had fooled me," Terrace later wrote. Terrace said his videotape analysis would show that Washoe's signs weren't any more spontaneous than Nim's.

Many of us in ape language research could see why Terrace had been fooled. Project Nim had absolutely no testing procedures to guard against the Clever Hans effect (named for the famous horse who could "count" by watching his trainer's unconscious cues). The Gardners and I eliminated any possibility of human cueing by using elaborate double-blind testing. Herb Terrace was the only ape language researcher who hadn't taken at least some precautions. The data and conclusions of Project Nim were suspect on the basis of this egregious lapse alone.

But that didn't stop Terrace from claiming that he was right and everyone else wrong. He said that when films of Washoe's signing were viewed in still frames (without motion) they looked less like human verbal behavior. This charge made a big impression on linguists who were unfamiliar with sign language. But signing, like speech, is a time-coded signal. When you play back a film of a human signer slowly enough, the signing becomes totally unintelligible, in the same way that vocal speech is unintelligible on a slowed-down audiotape. Frame-by-frame film analysis also loses all the inflections of ASL that are contained, after all, in the *movements* of a signer's eyes, hands, and body.

Terrace also charged that Washoe frequently interrupted her human companions and cited this as proof that Washoe didn't know how to take turns in a conversation. By comparison, Terrace said, "children show a good sense of when to listen and when to talk." In a freeze-frame video Washoe did occasionally begin to sign while her human companion was still signing, but this is a normal turn-taking device of ASL. Signers, unlike speakers, overlap in their conversation about 30 percent of the time. The reason is obvious: you can read signs while you're signing, but it's hard to listen while you're speaking. ASL experts who viewed and "scored" the Washoe films at the customary normal speed confirmed that her turn taking was in line with that of deaf humans.

I have no doubt that Terrace's charges would have quickly been dismissed if they'd been subject to normal scientific review and debate. But that's not what happened. Terrace trumpeted his claims in the popular media and he quickly became a cause célèbre among the followers of Noam Chomsky. For linguists of the "human uniqueness" school, Herb Terrace was a dream come true. Here was an ape language researcher confessing that he'd been duped by his own chimp!

In May 1980, the critics of ape language held a conference under the auspices of the New York Academy of Sciences entitled, "The Clever Hans Phenomenon: Communication with Horses, Whales, Apes, and People." A parade of scientists and nonscientists charged that ape language was little more than an exercise in deception or self-deception. There was an Alice-in-Wonderland quality to this whole conference. It was supposedly devoted to exposing the Clever Hans effect, but those present were rushing to lionize Herb Terrace, the only researcher who *failed* to take precautions against the Clever Hans effect. The conference ended with a proposal to halt all funding to ape language experiments, a move that would have prevented any further evidence of ape language from undermining the Chomskyan party line that human language is unrelated to animal communication.

The media had a field day with Terrace's sensational accusations, and ape language became a victim of that peculiar American syndrome in which the media gleefully attacks and destroys a phenomenon it once celebrated. Article after article appeared in prominent newspapers and newsweeklies touting new "evidence" from Project Nim that ape language was a hoax. These articles almost never pointed out that comparing Nim to Washoe was like comparing a wolf-child to a normal child. In the end, Terrace's energetic crusade made sure that a good deal of the American public came to believe that ape language was just another passing intellectual fad.

Trying to rebut Terrace in the media was a losing proposition. I wish that more people outside academia had read the less sensational but more important discussion in the professional literature. For example, the linguist Philip Lieberman concluded that Terrace was guilty of "the systematic misrepresentation of other investigators' work, particularly that of the Gardners." Two comparative psychologists, Thomas Van Cantfort and James Rimpau, published a fifty-page article in the journal *Sign Language Studies* that detailed Terrace's distortions of the scientific record.

But the most compelling rebuttal of Terrace came from Nim himself. After he returned to Oklahoma in 1977, a new study showed that his spontaneous signing increased dramatically when he was allowed to socialize naturally under relaxed conditions. Nim's "language deficit" had nothing to do with his intelligence and everything to do with Terrace's rigid training procedures. Terrace had deprived Nim of social conversation, then accused him of not having spontaneity and other elements of social linguistic behavior.

To this day, certain linguists of the Chomsky school still cling to Project Nim as if it had never been discredited. They claim that signing chimps are highly trained animal acts, that they don't sign spontaneously, have to be drilled and coerced, don't take turns in conversation, and only sign to get things they want. As evidence, they point to Project Nim.

These ongoing, specious attacks on Washoe and other signing chimps were effectively refuted back in 1983 by the pioneering linguist and ASL authority William Stokoe in an article aptly entitled, "Apes Who Sign and Critics Who Don't." Stokoe is the chief author of the *Dictionary of American Sign Language*, whose study of visual syntax led to the recognition of ASL as a natural human language in the 1960s. Stokoe has observed Washoe, Moja, and Loulis firsthand, has viewed films of Washoe signing as many as a dozen times, and has carefully reviewed my studies, as well as the Gardners', from the past two decades.

"There can be little doubt," Stokoe wrote in his most recent book, "that chimpanzees have well-developed abilities to communicate using signs." In Stokoe's opinion, the reason Washoe, Moja, Tatu, and Dar have evolved linguistically like human children is, that they were *not* conditioned, trained, or even overtly taught. They learned to sign, Stokoe says, in the same way that the deaf children of deaf parents do, through spontaneous interaction with signing human adults.

Of course, Herb Terrace or any other critic of chimpanzee signing could have discovered this simply by visiting Washoe's family in Ellensburg. There, he would have witnessed a scene altogether different from Project Nim: five chimpanzees signing to one another about paint colors, dress-up clothes, and photos in magazines.

BY AUGUST 1981, a year after the Clever Hans conference, my National Science Foundation grant had run out, and my application for more funding was rejected. Federal funding for all ape language studies was drying up completely. Some researchers have placed the blame for this squarely on Herb Terrace's shoulders, but I disagree. He certainly didn't help matters, but there were larger political forces at work.

Ronald Reagan had recently been elected president and had made the "cure of cancer" a top scientific priority. His administration was busy diverting all available grant monies to biomedical laboratories. When I asked about my own application with the federal government, the grant officer told me I'd have no problem getting funding if I was willing to conduct biomedical studies on Washoe and her family.

The death of behavioral funding had been coming since that

moment in the 1960s when molecular biologists discovered that chimpanzees were our genetic next of kin. Medical researchers viewed chimps as the next best thing to human subjects and began injecting them with every imaginable disease. At the same time, lab psychologists and field ethologists were discovering that chimpanzee intelligence and family behavior were very much like our own. For a long time, scientists on either side of this divide stayed out of each other's way, as if to say, "You're doing your science, I'm doing mine." But these two very different approaches to the chimpanzee were destined for a head-on collision.

As recently as the early 1970s, a federally funded researcher could, and did, experiment with unanesthetized chimpanzees by smashing their skulls with steel pistons, and there was little or no public outcry. Yet barely a decade later—thanks largely to Jane Goodall's field studies, *National Geographic* television specials, and Washoe's use of sign language—the public seemed to grasp the psychological and emotional kinship between chimpanzees and humans. This put biomedical labs in a rather uncomfortable position. Their chimpanzees were looking less and less like hairy test tubes and more and more like thinking, feeling people. What if the public discovered that nearly two thousand chimpanzees were being used in painful and deadly experiments? The NSF and other government agencies responded by cutting off funding to studies like mine that reminded taxpayers that chimps are just like us. Then they mounted crusades against various deadly diseases, and claimed that the cures could only be found if we sacrificed chimpanzees and other nonhuman primates.

The fallout for my own research came down to dollars and cents. Somehow, somewhere we had to come up with forty thousand dollars every year just to feed and care for Washoe's family. That meant we would have to become full-time fund-raisers. Debbi and I immediately founded a nonprofit organization, called Friends of Washoe, that could accept donations from individuals. When our plight was reported on television and in newspapers,

280 / THE SEARCH FOR SANCTUARY

we began receiving gifts from around the nation. But it was the people of our new hometown, Ellensburg, who really came to our rescue. One businessman donated four hundred T-shirts with Washoe's picture printed on the front. We sold them on the streets of downtown Ellensburg, and the proceeds kept Washoe's family alive during the month of September 1981.

Keeping the chimps fed became a community-wide effort. A professor and his family let us dig one hundred pounds of carrots and potatoes out of their garden. College students began a weekly fruit-collecting program in their dorms. Dairy Queen gave the chimps coupons for chocolate-dipped cones—their favorite. One generous family let us pick apples, pears, and nectarines from their orchards. We spent four days there and wound up with seven hundred pounds of fruit. For the next two months, I spent all my time between classes helping volunteers use loaned dehydrators to turn fruit into fruit leather.

The chimps also did their part. Washoe, Dar, Tatu, and Moja all contributed paintings to the first-ever show of their artwork, held at an Ellensburg café in October 1981. It was called "Chimpressionistic Works by Washoe and Friends." Although painting and drawing are not a part of natural chimpanzee culture (as far as we know), cross-fostered chimps love to make art. Each of the chimps has a very distinct style. Washoe's paintings are bright and energetic, with titles like ELECTRIC HOT RED. (The chimps title their own paintings.) Moja was the first nonhuman to paint representationally, birds being her favorite subject. (It is very possible that other chimps before Moja painted representationally, but it could never be proven because they didn't use signs to title their work.) Tatu is extremely serious about her art and will not put aside an unfinished painting, even for dinner. Her creations have a great sense of color and composition. Dar is a more temperamental artist. His paintings are tight and energetic, but when he loses interest he begins eating the paint, an artistic faux pas that infuriates his sisters.

Several months after we started fund-raising, I realized just how dangerous it had been to rely solely on federal support.

When federal funding disappears, all work grinds to a halt. But now we enjoyed a broad base of support from hundreds of volunteers and a caring community, and in return we deepened that community's knowledge about chimpanzees. Even so, raising money was exhausting, and there were many long nights when Debbi and I wondered how long we could keep up such a frenetic pace.

The answer to this question came to me one night a few weeks later. I was looking through the mail before going home to sleep, when a newsletter caught my eye. It contained a report about ongoing biomedical research at Holloman Air Force Base, where Dar was born. It described a study of adrenarche, the onset of puberty, in which a researcher had recently castrated six chimpanzee adults and six five-year-olds and then removed their pituitary glands. In his postsurgical report, the doctor referred to the chimpanzees as "monkeys."

Dar was now five years old. If he were still at Holloman, he would have been the perfect candidate for the castration study. Six chimpanzee boys had had their testicles sacrificed "for the good of science" by a researcher who didn't even know the difference between a chimpanzee and a monkey. Before I went home I looked in on Dar. He was already asleep, curled up next to Tatu. Even through the thick hair on his inner thigh I could make out the number 445, tattooed on his skin in large blue numerals. All chimpanzees in biomedical research are tattooed the day they are born. Dar's number was supposed to stay with him through a lifetime of experimentation. Instead, his tattoo is a constant reminder to me of the painful life and lonely death he managed to escape.

Dar's seven-by-ten-foot enclosure in our own lab, which had looked so small to me just that morning, suddenly seemed like a very sturdy lifeboat. Keeping that lifeboat afloat was no longer a matter of science or even a matter of choice. It was a matter of one chimpanzee family's survival.

Within a few more months I had to accept that our very best efforts were barely keeping Washoe's family alive and properly cared for. Things got so bad that Debbi, the kids, and I started going to the local Albertsons every evening to look through leftover bins for fruits and vegetables to feed the chimps the next day. In desperation we even allowed Rainier Beer to use Tatu in a television commercial in exchange for a five-hundred-dollar donation to Friends of Washoe. Tatu played a bartender pouring a beer for Tarzan; she had a great time but kept demanding MILK! MILK! between takes.

Then, just when we seemed to hit bottom, Hollywood bailed us out. After years of delay Warner Brothers was finally getting Greystoke into production, though not with Robert Towne directing. Hugh Hudson, the British director of Chariots of Fire, called to tell me that he was making Greystoke and wondered if I would be involved. He flew up to meet with me and to observe the chimps. Hudson was committed to shooting on location in Africa and he wanted to use humans in ape costumes. "If you can teach chimps to act like humans," he said, "you can certainly teach humans to act like chimps."

I wasn't enthused about leaving my family and Washoe for what would probably be a six-month-long project, but with my family scrounging through leftover bins, I was in no position to refuse. Warner Brothers was offering a mind-boggling amount of money—one hundred thousand dollars—for my services. That was the size of a major NSF grant!

I asked the studio to cut two checks: one made out to me for forty thousand, which would cover my teaching salary for the academic year I would have to take off, and another to the Friends of Washoe organization for sixty thousand. Sixty thousand dollars, tax-exempt, would support the chimps for a year, with some left over for a down payment on an outdoor play area.

After years of waiting for him, Tarzan finally came to our rescue.

TWELVE

SOMETHING TO TALK ABOUT

I WAS PREPARING TO LEAVE FOR AFRICA when I got a phone call from Chris O'Sullivan, one of my graduate students in Oklahoma. Chris was one of the researchers who had shown that Nim was capable of signing socially and spontaneously after he returned to the Institute.

Chris was very upset. She had just learned that Lemmon was selling his entire chimpanzee colony, including Nim, to the Laboratory for Experimental Medicine and Surgery in Primates (LEMSIP) in New York. LEMSIP, owned by New York University, had won the Merck, Sharp, Dome contract for hepatitis B testing that Lemmon had tried to bring to Oklahoma in 1978. The chimps were winding up in hepatitis research after all, and it was going to happen soon. Chris knew someone at CBS News, and I urged her to tip off the network about the impending sale.

CBS ran a news story about what was about to happen to the famous "signing chimpanzees," but it didn't stop Lemmon from going ahead with his plan. At the end of May and the beginning of June 1982, a specially outfitted tractor trailer made several round-trips between Oklahoma and New York, carrying more than two dozen chimpanzees to LEMSIP. Moja was born at LEMSIP in 1972, and I knew enough about the place to fear for the chimps being sent there.

After being unloaded at the laboratory each chimp was locked in a solitary cage that was five by five by six feet—the size of a coat closet. The steel-bar-bottom box hung from the ceiling, like a birdcage, dangling above the floor so that the chimp's feces could drop through the cage onto plastic sheets below. There were two rows of these hanging cages, facing each other across a walkway. The chimps could see one another and call—or sign—to their friends, but there was no group contact or access to the outdoors. There wasn't even daylight because the metal buildings didn't have any windows.

The entire facility was designed to make it easier for the workers to have access to the chimps' blood. Sterile-gowned technicians came around on schedule to inject the chimps with hepatitis B vaccine, or to challenge the vaccine with live hepatitis virus, or to draw their blood to see whether the vaccine was effective. The chimps did not stop signing, though the technicians didn't understand them. We heard from visitors that Booee, Bruno, Nim, Ally, and others kept asking the techs in ASL for food, drinks, cigarettes, and the keys to their cages.

Although the CBS story didn't stop the chimps' transfer, it did set off an avalanche of bad press for William Lemmon, New York University, and LEMSIP. Chimpanzees were transferred between biomedical labs all the time, of course, but these were "talking" chimpanzees. Two of them were celebrities. Ally had appeared in *People* magazine, and his brother Nim had been the focus of the widely reported Clever Hans controversy just two years earlier. Caring citizens and animal welfare groups were spurred into action, and a campaign of letter writing, phone calls, and protest marches was launched.

I was pleased by the public's outrage, but I could see that neither camp in this dispute really understood the underlying tragedy. The researchers at LEMSIP didn't care that their newest research subjects could sign GO OUT, SMOKE, and HUG. All they wanted was the chimps' blood. On the other hand, the people protesting the chimps' harsh treatment seemed to care *only* that the chimps could sign, as if that made them somehow more special and worthy of compassion.

In my eyes, there was no difference between Pan, who

couldn't sign at all, and Booee, who knew thirty signs. Or between Manny, who had two signs—COME HUG—and Ally, who had 130 signs. All of the chimps felt the same pain of loneliness and a terrible fear about their strange new surroundings. Each of them had the same deep need that you or I would for the comfort of physical contact and affection. *That* was the tragedy of putting these social creatures in solitary cages that dangled above the floor. Ally and Nim weren't suffering because they knew sign language; they suffered because they were chimpanzees.

I tried to get this point across whenever reporters and talk-show hosts called to interview me that spring, but the battle lines were already drawn around the *signing* chimps. This gave New York University and LEMSIP a public relations opportunity that it quickly seized. They sent Ally and Nim, their two best-known chimps, back to Oklahoma, and the uproar immediately died down. The public didn't seem to be as excited about the ongoing solitary confinement of less famous chimps like Booee, Bruno, Thelma, and Cindy.

Ally's and Nim's paths diverged again almost as soon as they got back to Oklahoma. Lemmon was determined to unload them without any more negative publicity. He sold Nim, the more famous of the two, to the Fund for Animals, a group run by the popular writer and animal advocate Cleveland Amory. Nim went to live at Amory's Black Beauty Ranch, in Texas, where he was housed in a large enclosure with access to the outdoors. As the only chimpanzee among horses he must have been lonely, but he was joined a few years later by an older female chimpanzee named Sally. Sally died recently, and Amory is now looking for another female companion for Nim.

Ally was not as fortunate. One of Lemmon's students told me later that Lemmon wanted to send Ally to a place where no one would ever find him. He succeeded. On November 15, 1982, Ally was shipped to the White Sands Research Center in New Mexico, a private laboratory that tests drugs, cosmetics, and insecticides on animals. Officials at White Sands have never acknowledged receiving a chimpanzee named Ally. The most they'll say is that they received two nameless chimpanzees on November 19, 1982. Presumably, one of them was Ally.

What happened to Ally inside White Sands, if he is indeed there? Nobody will say, and nothing is known. One source told me later that Ally died after being injected with insecticides as part of a toxicity study. I don't know if it's true, and I probably never will. But Ally is not forgotten. In October 1983, four years after Washoe last saw Ally, she and the other chimps were watching a slide show of themselves. They were laughing at the funny pictures and signing about their friends. A picture of Ally came up on the screen. Washoe looked at it intently and moved closer to the screen. WHO THAT? one of the students asked her. HUG HUG NUT, Washoe replied, staring at the screen.

As I looked at Ally's picture I traveled back thirteen years to the first day I met the antic one-year-old who made the sign of the cross on his chest. The words his foster mother had used on the happy occasion of Ally's baptism came back to me now but with a darker meaning: "Why hasn't my baby a right to be saved, like anybody else?"

GREYSTOKE WAS THE FIRST MOVIE that ever showed how apes actually move, communicate, and socialize. Movies like King Kong and Planet of the Apes portrayed apes as grotesque caricatures. I wanted the chimpanzees in Greystoke to resemble the highly intelligent and emotional personalities that Jane Goodall was documenting in the field and that I knew in captivity.

Christopher Lambert, the actor playing Tarzan, came to Ellensburg to study how young male chimps, namely, Dar and Loulis, walked, groomed, played, and fought. He also got to watch Washoe, the model for Tarzan's strong and compassionate chimpanzee mother, as she mediated fights between the two boys. After a couple of weeks with Christopher, I flew to London, where I trained about a dozen actors who would portray Tarzan's chimpanzee friends and family.

The first thing I had to do was get rid of their stereotypes of ape behavior. Many of them tended to cock their heads, as if they were King Kong giving Fay Wray the once-over. "You're not a rooster," I'd say again and again. "You're a chimpanzee with two eyes in front, just like a human." Getting them to walk like chimps was the toughest challenge. When humans bend at the knees and walk in a crouch they look more like Groucho Marx than Dar or Loulis. After watching a lot of videotape of chimpanzees, the actors mastered the smooth, rolling action of a chimp's bipedal walking. One of the actors, an athletic Tae Kwon Do champion, even managed the strenuous feat of running on all fours.

Next, we focused on creating individual characters. Each chimpanzee was given a name, and we carefully put together a family history for every member of the community. In addition to Kala, Tarzan's adoptive mother, and Silverbeard, the dominant male and Tarzan's surrogate father, there was White Eyes (Tarzan's rival), Blush (a young female), Balino (an ambitious and sometimes foolish male), and a number of others. Each chimp knew who his mother was, who his brother was, who his allies were.

I encouraged the actors to inhabit a chimpanzee world where one is always running a kind of social cost-benefit analysis: If I sit next to this person, am I likely to get hurt or feel safe? What can I get away with? Who should I challenge and who should I build alliances with? Many of the actors told me later that they found themselves feeling and acting "chimpanzee" long after the filming was completed.

The chimpanzee society portrayed in *Greystoke* had an authenticity that has rarely been duplicated. Audiences could hardly believe that there weren't actual chimpanzees in the film. A few years later, *Gorillas in the Mist* picked up with gorillas where *Greystoke* left off with chimps, but many recent films about apes have gone back to pre-Greystoke caricature. I have worked on several films where the most authentic chimpanzee behavior winds up on the cutting room floor because the producer wants the scenes to be "freakin' hilarious," meaning that the humans playing chimps should act like clowns.

I personally will not work on movies that use real chimpanzees. I have spent enough time in Hollywood—and quit enough films when the "cattle prods" came out—to know what it takes to get a chimpanzee or a gorilla or an orangutan to perform. The ape we find so funny was most likely beaten, electrically shocked, deprived—or frequently all three. Exploiting great apes for the sake of human entertainment is a tragic endeavor.

The filming of Greystoke took place in a dense West African rain forest on the slopes of the Mount Cameroon volcano. This is where many of the chimpanzees used in America's space program were captured, and it is possible that Washoe was born in the same rain forest where we were working. There are only about eight thousand chimpanzees left in Cameroon, and I saw only one of them, from a distance, deep in the rain forest.

While I was in West Africa, I had hoped to see my old friend Lucy. She was in The Gambia, about two thousand miles from where we were filming. I had recently learned that Lucy's life had taken a turn for the better. Though still dependent on Janis Carter, Lucy was finally gathering her own food and her health was improving. She was now the dominant female in her group of rehabilitated chimpanzees, and, like Washoe, she had even adopted an orphaned baby chimp boy. Unfortunately, my film schedule didn't allow time for anything but work, so I wasn't able to visit her.

Though I didn't encounter any chimpanzees in Africa I did meet plenty of well-adapted human primates—Pygmies. Hugh Hudson brought in an entire tribe of them from Nigeria for several scenes in the film. They were the genuine article, right down to their filed teeth, loincloths, and quivers full of arrows. During breaks in filming the Pygmies would climb into the trees and build sleeping nests, vanishing into the dense foliage like a band of chimpanzees. I stood there marveling at their ability to become invisible. It was remarkable evidence of the kinship between Washoe's species and our own.

I couldn't help but wonder, during the filming of *Greystoke*, how Tarzan, the English foster child of chimpanzees, would have viewed the fate of my friend Ally. At the end of the movie Tarzan wanders into the anatomical research wing of the British Museum, which was the crowning glory of nineteenth-century Western science. To his horror, he discovers several chimpanzees pinned to the dissection table. He hears a grunt, and sees another chimpanzee, still alive, inside a small cage. This animal is just one more research subject to the scientists, but Tarzan knows this caged animal. It is his chimpanzee father, Silverbeard. Tarzan opens the cage, frees his father, and they embrace.

This scene was my personal tribute to Ally. The original script for *Greystoke* called for Tarzan to discover his father in a zoo, but I persuaded Hugh Hudson to change his script. Turn-of-the-century British anatomists, I argued, were the real forerunners of today's biomedical researchers, the loudest proponents of the Cartesian view that chimpanzees are unthinking, unfeeling machines. In my mind when Tarzan opened that cage he was freeing not only his father but Ally, Booee, Bruno, and every other chimpanzee prisoner of science as well.

AFTER I RETURNED FROM AFRICA Debbi and I worked intensively toward our goal of creating a research environment based on mutual respect between chimpanzees and humans. Our first priority was training human volunteers who would have positive relationships with the chimps. Washoe's family was always given a chance to size up the candidates, and Washoe was especially good at fingering anyone who acted arrogant or bossy. We gave her the right to fire anyone she didn't like, which she usually did by spitting on them.

Every trainee underwent a year's instruction in ASL, chimpanzee behavior, and analyzing chimpanzee conversations on videotape before they could collect scientific data on chimpanzee signing. But our guiding principle was summed up by a sign that hung outside our laboratory during the 1980s: LEAVE YOUR EGO AT THE DOOR. Many people had a hard time accepting that Washoe's family was not there for human amusement or gratification. Some of our own volunteers had always wanted to touch a chimp, or laugh at a chimp, or control a chimp. Some thought that chimps, like horses, needed to be trained or "broken." Others turned the water hose on Loulis when he spit at them. Debbi and I found ourselves reciting this speech, or parts of it, almost daily:

> This laboratory is Washoe's home. You are guests in her house. Conduct yourself as you would expect your own guests to behave. You do not have the right to discipline her child, threaten any member of her family, or even to touch them if they have not asked to be touched. In this house the chimpanzees' welfare comes first, the research comes second, and your needs come last. You are free to walk out of here anytime you like, but Washoe's family can never leave. They are prisoners in their own home. Your job is simple: to make their lives as pleasant, as social, and as interesting as possible.

Volunteers quickly discovered, usually on their first day, that they were *not* in control of the chimps. They found themselves being easily manipulated by the chimps, who understood English, were more fluent in sign language, were experts at reading nonverbal behavior, and knew all the workings of the lab. One of Washoe's favorite scams was to ask a new volunteer for ICE CREAM before lunch, as if that were what she had every morning. It was like watching some kids tell the new baby-sitter that they *always* jump on the furniture.

The volunteers who survived were the ones who accepted that they were visitors to the home of some very complex and hairy people who happened to be chimpanzees. These volunteers left their "chimp myths" behind and built respectful relationships with the individual members of Washoe's family. They were rewarded with one thing only: chimpanzee friendship.

One of our longtime volunteers, Kat Beach, once told me that when she first met Washoe she was amazed that a chimpanzee could use human language. But after getting to know the chimps, she was instead amazed by *what* Washoe communicated. In the summer of 1982 Kat was newly pregnant, and Washoe doted over her belly, asking about her BABY. Unfortunately, Kat had a miscarriage, and she didn't come in to the lab for several days. When she finally came back Washoe greeted her warmly but then moved away and let Kat know she was upset that she'd been gone. Knowing that Washoe had lost two of her own children, Kat decided to tell her the truth.

MY BABY DIED, Kat signed to her. Washoe looked down to the ground. Then she looked into Kat's eyes and signed CRY, touching her cheek just below her eye. That single word, CRY, Kat later said, told her more about Washoe than all of her longer, more grammatically perfect sentences. When Kat had to leave that day, Washoe wouldn't let her go. PLEASE PERSON HUG, she signed.

CHIMPANZEES ARE FIRST AND FOREMOST social creatures. That means loneliness and boredom are the two biggest enemies of chimpanzees in captivity. Chimpanzees in the wild thrive on change. They forage in different fruit trees every day, build different sleeping nests every night, and socialize with different members of their community in the course of traveling through the jungle. It is hard to think of a worse candidate than the chimpanzee for confinement in a concrete cell and adherence to a monotonous, institutional routine. In the case of Washoe's family, we could not alter their laboratory surroundings. They had to live in the same four rooms until we could raise a lot more money to build them a large outdoor area. That made it even more important to enrich every aspect of their daily life.

A typical day at our laboratory began at 8 A.M. when we opened the lab and woke Washoe, Moja, Tatu, Dar, and Loulis with signed greetings and chimpanzee head-bobs and pant-hoots. Before breakfast we asked the family to clean out the previous night's dinner bowls, spoons, and sleeping blankets. Tatu, who was the most reliable housekeeper, would gather up all the items and push them under the cage for us to wash.

For breakfast each chimp got a smoothie, made out of seasonal and frozen fruits as well as vitamins and minerals. Chimpanzees in the wild eat more than 140 different kinds of plants and fruits, in addition to medicinal plants, so we tried to vary their breakfast fruits as much as possible. We always served Washoe first, out of respect for her status as the oldest and most dominant. Notice that I use the word "serve" and not "feed." We learned that the words we chose had a big impact on our own attitude and behavior. We "served" meals to Washoe's family just as we "served" our family and guests in our own home. It is quite different from "feeding the dogs" or "slopping the hogs." Mealtime is much more peaceful when the diners are treated with respect.

After breakfast, we cleaned the family's compound. Instead of being confined during cleaning, the chimps were invited to help. Washoe and Tatu loved cleaning. They'd take a pail of soapy water and a brush and scrub down the inside of their home while we cleaned the outside. We made a point of turning cleaning into a social event with plenty of games and lots of signing, such as asking to DRINK from the hose, CHASE with the mop, and trade FOOD for work. As always, a trained observer would record the chimps' behavior and conversations.

After cleaning we announced a SURPRISE, which was always greeted with joyful and excited hooting. This could be anything from a snack, such as fruit leather, frozen fruit, herbal tea, gum, or cornstalks, to something more exotic like snow or water balloons filled with fruit juice. The chimps loved a raisin-board device that was suggested to us by Jane Goodall. This six-inch-long board had about twenty holes drilled through it that could be stuffed with raisins or marshmallows. We'd give the chimps willow or apple branches that they'd use to get the raisins out, much as chimps "fish" for termites in the wild.

During the summer or fall, we might declare a TREE DAY, when we dragged in apple branches with green apples, willow branches, giant sunflowers, or other vegetation. It was amazing to watch the chimps come alive when we introduced this small piece of nature into their concrete world. Washoe asserted her dominance by picking through all the branches first, searching for the best ones. Then she'd drag them over to the bench and sit cross-legged, chewing on the leaves. Moja picked out pairs of leaves, made them into whistles by blowing into them, and then ate them. Tatu loved to sift through the leaves and nibble on galls, the bumps that contain insect eggs, all the while keeping up a stream of signing: THAT TREE, THAT FLOWER, MINE TATU.

Dar and Loulis staged mock fights that were right out of the African rain forest. Both of them would pick up branches, stand bipedally, and stomp their feet loudly. They'd shake the branches above their heads threateningly and then charge one another. When they collided they'd fall on the floor, wrestling and tickling until they were both laughing hysterically. Loulis had experienced the outdoors only a few times in his life, but he could spend entire afternoons occupying himself with just a few small branches.

After SURPRISE time or TREE time we brought out toys and other objects for the chimps to play with before lunch. Each chimp had his or her favorites. Loulis loved Halloween masks, especially ones of monsters and the Lone Ranger. After putting on a mask he'd run around chasing the others. Then he would order a lab worker to put on the mask so he could laugh at them. Seven-year-old Dar was partial to toy dinosaurs, which he would tickle and sign to. Sometimes he and Loulis, still masked as the Lone Ranger, would play hide-and-seek in the overhead tunnels with their dinosaurs.

Washoe, Moja, and Tatu all loved to play dress-up. Moja, the most elaborate dresser, would tie a scarf around her head, a belt around her waist, and then, using a small mirrored compact, apply bright pink lipstick. She was also intensely fascinated by Velcro. She could lie on the floor opening and closing it for hours. After a while, the lab workers began calling her VELCRO WOMAN.

Tatu was partial to black, as in BLACK PURSE, BLACK LIP-STICK, or BLACK SHOES. Her makeup of choice was nontoxic black oil pastels. Tatu's infatuation with anything black went back to her childhood, when she started signing BLACK to refer to anything "cool," desirable, or beautiful, as in THIS FOOD BLACK or SHE BLACK.

Washoe preferred red, especially in shoes. She loved nothing better than to share a good cup of coffee with a lab worker while looking for her favorite red shoes in what she called the SHOE BOOK, the shoe section from a fashion magazine. Washoe had very eclectic taste in magazines, as the following conversation with Debbi revealed:

Washoe: MORE BOOK!
    (Debbi retrieved two catalogs: men's clothing and household furniture.)
Debbi *(in English)*: Which one do you want?
Washoe: BOY BOOK.
    (Debbi gave her the men's catalog. Washoe leafed through it for a while, then put it down.)
Washoe: GIRL BOOK GO!
    (Debbi looked but couldn't find any women's fashion catalogs.)
Debbi *(signing)*: I CAN'T FIND GIRL BOOK.
Washoe: MEAT BOOK!
    (Debbi returned with a Williams-Sonoma catalog featuring tasty delights that Washoe studied very intently.)

Lunch each day was a choice of cooked soups such as split pea, bean, or lentil, all flavored with different vegetables and spices. Sometimes we offered other proteins, like tuna or chicken, though only Tatu was fond of meat. We served the chimps their food in bowls with spoons, but we also let them forage by putting lettuce leaves and other vegetables on top of the chain-link ceiling.

After lunch there were quiet activities—grooming, looking at picture books or photo albums, painting—followed in the late afternoon by more cleaning. As usual, Tatu would clear out the magazines, toys, and stripped branches. At 4 P.M. we served dinner, usually rice, cooked cereal, steamed vegetables, or, occasionally, sandwiches, tortillas and beans, or pizza and popcorn feasts. After dinner we gave each chimpanzee a blanket for sleeping, as well as straw, willow branches, cornstalks, or other vegetation for bedding. The chimps had their own preferences when it came to sleeping arrangements. Washoe and Loulis shared a sleeping nest until Loulis was about ten years old. Washoe would make a circle on the floor with her two blankets and then invite Loulis inside. Chimpanzee children, like human children, are not always thrilled about going to sleep, and Loulis would often climb the walls and hang from the ceiling rather than get into his mother's nest. Washoe would patiently sign COME COME, or else physically retrieve him and groom him until he calmed down and fell asleep.

Moja, meanwhile, prepared the bench bed by laying her blanket out flat and then crawling under it. Dar and Tatu, like wild chimps, liked to climb *up*—into the overhead tunnel— before going to sleep. Washoe and Loulis also liked the tunnel, but it was too narrow for their circular nest so they usually settled for the floor.

We have always tried to make one or more days of each month extra special for Washoe's family by celebrating every possible holiday we could think of: Valentine's Day (red and white streamers, balloons, and heart cupcakes), Easter (colored hard-boiled eggs and treasure hunts), Mother's Day for Washoe (we completely fill a room with lilacs, and the chimps smell them, sign about them, eat them, and nest in them), Independence Day (we tried fireworks and sparklers, but they scared the chimps, so we use Kool-Aid balloons), Halloween (the chimps help in carving the pumpkins, eat the meat and seeds, put on costumes, and hunt for trick-or-treat purses). We have even celebrated Saint Patrick's Day, Rosh Hashanah (the Jewish New Year), Hanukkah (Jewish festival of lights), and any other occasions we think Washoe's family will enjoy.

Thanksgiving and Christmas are big holidays in our lab. Thanksgiving dinner features BIRD MEAT, POTATO SWEET, BERRY SAUCE, and PUMPKIN SWEET, with the chimps each requesting their favorite part of the turkey. The weekend after Thanksgiving we set up a Christmas tree, which the chimps call CANDY TREE, in a room where they can easily see it from their tunnel. The CANDY TREE becomes the most popular topic of chimpanzee conversation for the next month as it grows bigger every day with garlands of nuts, popcorn, cranberries, fruit, gum, and raisins. On Christmas Day, we give them the edible tree ornaments, stockings filled with treats, and special gifts, such as dress-up clothes from SANTA.

We also celebrate the birthdays of all five members of the family. Tatu's birthday, because it falls only five days after Christmas, has always seemed a little bit excessive, but we try not to let that dampen our enthusiasm. In 1983, for example, we had large gingerbread cookies with eight candles on top as well as yogurt, Tatu's favorite food. When we sang "Happy Birthday" (signing it rhythmically) Tatu began screaming in such delight that she needed help from Washoe to blow out her candles. (All the home-reared chimps blow out their birthday candles; Loulis eats his.) After the treats it was time to open gifts. The first was from the Gardners: a fuzzy—what else?—BLACK blanket. Tatu signed THAT BLACK over and over as she wrapped herself in it. Then there was a filmy yellow negligee that Moja put on, and five Velcro wrist wallets, one for each chimp, with dried fruit hidden inside. Loulis ate his fruit, then put the wallet on his head and signed HAT HAT to Washoe. Once the Velcro wallets were emptied Moja wrapped them around her ankles and wrists, then took them over to the corner, where she lay on the floor opening and closing them.

The birthday girl disappeared to another room, where she applied a thick coat of lipstick to her lips and cheeks. When she finished she sat there quietly rolling the stick up and down carefully. Washoe saw Tatu through the doorway and rushed over, but before she could get to the lipstick, Tatu rolled it down and popped it in her mouth. Washoe looked all over but couldn't find the lipstick so she gave up and left. Tatu then pulled the tube out of her mouth and went back to applying the lipstick. All in all it was a very BLACK day for Tatu.

ALL OF THIS SOCIALIZING AND CELEBRATING created a family environment where language flourished quite naturally. Getting a group of signing chimpanzees together and letting them converse may seem like an obvious idea, but it had never been done before. Partly it was a matter of numbers. Most researchers worked with a single chimpanzee. (The Gardners were the notable exception, and they had reported spontaneous signing among Moja, Dar, and Tatu in the 1970s.) But it was also a matter of approach. Some ape language researchers were teaching chimpanzees computerized languages that they used by selecting symbols on a keyboard. This doesn't lend itself particularly well to normal, social conversation, and the consequences were sometimes amusing. One computer-using chimp at Yerkes named Lana typed out sentences like "Please machine tickle Lana" when her human companions went home at night. It was rather absurd for critics to charge that chimpanzees like Nim and Lana signed only for human rewards when these chimps were never given the chance to converse with other chimpanzees.

Loulis had learned to sign from Washoe, and this transmission of signs across generations was evidence that language probably evolved in the family. But human children eventually move outside of their families and speak or sign to their peers to express thoughts and feelings, to form friendships, to make plans, to resolve disputes. Though language emerges in the family it finds its fullest expression in the wider community, where conversation animates every human encounter and facilitates culture, commerce, and education.

A community of chimpanzees, our evolutionary siblings, would very likely integrate language into its own daily life for these same social reasons. If they did so, we would have more evidence that language evolved in our hominid ancestors as a way of building social relations. Equally important, if a group of chimpanzees conversed among themselves, it would provide the ultimate proof that chimpanzees used sign language because they naturally needed to communicate, and not because they were prompted or rewarded by humans.

Washoe, Loulis, Moja, Tatu, and Dar were like an extended family, or a small community, and I wanted them to have the same conversational freedom that humans do. From the first day we reunited Moja, Washoe, and Loulis in one large enclosure in January 1981, my research focused on their spontaneous conversations. During random forty-minute sessions every weekday, we recorded the number of signed interactions within Washoe's family. By the last months of 1981, Washoe, Moja, and Loulis were having signed conversations about once per hour. Loulis initiated most of the signing, usually by trying to get his mother or Moja to play with him: COME TICKLE, PEEKABOO, HURRY HAT (one of his favorites toys), and COME GIMME SHOE (another favorite toy). Moja was then nine years old, an age when young female chimps are absolutely fascinated by infants. Like Washoe, she signed most often while playing with Loulis or reassuring him.

We were well on our way to finding out how a group of chimpanzees would use language socially when a breakthrough came in January 1982. That was the month Dar and Tatu moved in. Suddenly there was a lot more to talk about, and Washoe's family began conversing five times as frequently. And by February, they were up to nearly ten conversations per hour, a rate they pretty much maintained for the remainder of 1982. Loulis was the main catalyst. At four years old he was moving away from his mother and shifting his focus to the social group around him. This is true of human children as well; four-year-olds don't speak to their mothers nearly as much as they speak to their friends. Bringing in Dar and Tatu was like introducing two new kids into a household with a single mom and her one child. The family's rate of signing took off.

Loulis initiated nearly thirteen hundred signed interactions in the first three months of the new year. (His total amount of signing was much greater; we were only observing and recording for three and a half hours each weekday.) He signed much of the time to Dar, who responded in kind. All their rough play led to more fights, which meant that both boys were asking for more reassurance and grooming from Washoe. Dar had replaced Moja as Loulis's main playmate, but Moja now had more time to interact with her sister, Tatu. But when Moja was in estrus, most of her signing was directed toward human males and Dar.

These chimpanzee conversations were very similar to those of two-year-old children who combine gestures with words—or in the case of deaf children, gestures with signs. When they're about ten months old children begin using "deictic gestures"—showing, giving, pointing—to request things and to express themselves. But once words or signs begin emerging, children don't stop gesturing. They string together what the linguist Virginia Volterra calls "sequences of communicative signals." Instead of using two words in a row, the hearing child may combine two gestures to communicate "me that," or one gesture plus one word for "you eat." Instead of using two signs, the deaf child may combine one sign and one gesture to communicate "MOMMY there." In the same way that spoken and signed languages evolved from gestures in our hominid ancestors, there is an unbroken continuum between a child's first gestures and his or her later language.

These face-to-face conversations between young children are not the idealized linguistic exchanges that academics diagram on blackboards. Deaf children seamlessly weave together signs and body language. So do chimpanzees. For example, Tatu once came over to Moja, who was about to receive a fruit drink. Moja screamed at Tatu's intrusion. Tatu then signed SMILE to Moja, prompting Moja to turn away from Tatu.

They used only one sign in this exchange but there were four turns that combined linguistic and nonlinguistic elements: approach, scream, SMILE, turn away. We might translate their conversation as follows:

Tatu: GIMME DRINK.
Moja: NO.
Tatu: SMILE.
Moja: GO.

The chimps not only communicated like young deaf children, they also displayed some of the same strategies that are common in human families. For instance when the two boys were fighting, Loulis always blamed Dar for the commotion. When Washoe rushed in to break it up, Loulis would sign to her, GOOD GOOD ME and then point at Dar. Washoe would then discipline Dar. After several months, Dar caught on and would throw himself on the floor when he saw Washoe coming. Then he would begin crying and frantically signing COME HUG to her. Washoe would then scold Loulis by swaggering toward him and signing GO THERE, pointing to the overhead exit tunnel.

The most significant finding of this study of chimpanzee conversations was that chimpanzees do not use language to get rewards as Terrace and others had charged. When given the freedom to talk in a loving and supportive environment, Washoe's family used language the way a human family does, to build and maintain personal relationships in the course of daily life. The vast majority of their signing was related to play, discipline, housecleaning, and reassurance. A lot of signing occurred when the chimps talked to themselves, while they looked at photos, painted pictures, or gazed out the window at people or things outside. Only 5 percent of their conversations had anything to do with eating. Interestingly, even their conversations about food were more like human talk around the dinner table than anything one would call "begging."

Some chimpanzees, like some humans, love to talk about food, especially when they're hungry. Tatu was fixated on getting regular meals. Every day about one hour before lunch she began telling everyone in the family that it was TIME EAT. Then she demanded to know what was on the menu. If she didn't smell anything cooking, she'd remind any humans nearby that it was TIME EAT. If she still didn't see any food coming, then she'd act like an irate customer at a restaurant and demand to see the manager: ROGER ROGER ROGER! One time we told her that her room had to be cleaned before she could have a banana, and she began telling all the chimps, HURRY CLEAN! BANANA! BANANA!

Tatu was also the family's timekeeper. On the morning after Thanksgiving dinner in 1986 Debbi and I entered the lab and began cleaning. The students had gone home, so it was especially quiet. Snow was falling outside. After staring at the snow, Tatu began following us around signing CANDY TREE? CANDY TREE? as if to remind us to set up the Christmas tree. Debbi responded, NO, NOT YET. Tatu persisted: CANDY TREE, CANDY TREE! When Debbi told her once more that she would have to wait a few days, Tatu plopped down on the bench, stuck her thumb in her mouth, and dejectedly signed BANANA? Now almost every year right after Thanksgiving Tatu starts asking for the CANDY TREE.

Since the time of Plato it has been a truism that only humans can remember the past and plan for the future. But when it comes to holidays and food, Tatu not only remembers them, she seems to know when they are supposed to occur. When our Halloween party was over Tatu would ask for BIRD MEAT, suggesting that Thanksgiving must be right around the corner. One time, after celebrating Debbi's birthday, Tatu kept asking us, ICE CREAM DAR? ICE CREAM DAR? Dar's birthday was the next day.

ALL OF THIS CHIMPANZEE SIGNING was recorded in the same way that Jane Goodall records wild chimpanzee behavior, by writing down observations in a logbook. Then, in 1983, an interesting chain of events led to a breakthrough in how we documented sign language in Washoe's family. Debbi was then working toward her master's degree in experimental psychology, and for her thesis she wanted to focus on Loulis's growing use of language. Her thesis committee was made up mostly of Skinnerian psychologists who didn't like this idea at all. They were skeptical that Washoe's family was signing at all, and they recited the charges of Herb Terrace. Her advisers suggested instead that Debbi mount some cameras on the wall and videotape the chimps when no humans were around at all. They were certain that this would prove once and for all that chimpanzees only signed because they were cued by humans.

Debbi was thrilled at this idea. If the remote videotape could disprove chimp signing, it could also *prove* it once and for all. The only question was why we hadn't thought of it ourselves. We had been using handheld cameras to document how Loulis learned signs from Washoe, but Skinnerians were still charging that our mere *presence* somehow cued the chimps to sign to one another. Remote videotaping was the next logical step.

Debbi mounted four cameras in the chimpanzees' rooms and connected them by cables to monitors in a separate room. Then she began recording twenty-minute segments three times a day at random hours. In just the first fifteen hours of tape, there were more than two hundred instances of chimpanzee-to-chimpanzee signing. It quickly became clear that our old method of live observation had underestimated the amount of signing in Washoe's family. Human observers could look only in one direction at a time, and even using a kind of shorthand notation they could only write so fast. We had been missing a lot, especially when three or four chimps were signing at the same time. But videotape misses nothing. (Jane Goodall's research station at Gombe Stream has also added videotape recording and analysis.) The chimps' conversations were so clear that independent ASL observers agreed nine out of ten times about the meanings of these videotaped exchanges.

From 1983 through 1985, Debbi videotaped forty-five random hours of chimpanzee conversations. The tapes showed Washoe's family signing while they shared blankets, played games, ate breakfast, and got ready for bed. The chimps were signing to one another even in the middle of screaming family fights, which was the best indication that sign language had become an integral part of their mental and emotional lives. When they discussed their favorite food, it wasn't to get the food (there were no humans present, after all) but just to comment on it. Several times, Dar looked out the window and signed COFFEE to himself. Each time we went to our own window to see what Dar might be referring to, a person was walking by the building holding a coffee cup.

But Debbi's videotape of Loulis was the best evidence yet that chimpanzees, like humans, learn language in the immediate family and then use it to build other social relations. As Loulis grew up he still turned to Washoe when he was scared, hurt, or angry, but, overall, he talked to her much less. This was especially true by the last year of the study, 1985. By then Loulis had learned fifty-five reliable signs inside his family, and he used them in one out of every eight interactions with the other chimps. But when he was with Dar, his best friend, he signed more frequently, in one out of every five interactions. Overall, Loulis now talked to Dar three times as much as he did to Washoe. Like a human six-year-old Loulis was using language to cement his friendships.

Interestingly, Loulis did not want to converse with humans in ASL. For five years, since he went to Oklahoma, Loulis had almost never seen humans signing. And when we did begin signing to him, on June 24, 1984, he promptly ignored us. His look seemed to say, "That's my language, not yours." He responded only to our vocal speech. It took four more months for Loulis to acknowledge our signs and to respond in kind.

Debbi's thesis advisers were not pleased with her findings. But they couldn't really dispute the videotaped evidence, and so they awarded her the master's degree. She and I screened the videotape at the International Congress of Psychology in 1984 and at a meeting of the American Association for the Advancement of Science in 1985. For all the controversy over Nim and Washoe a few years earlier, very few scientists had ever actually *seen* a chimpanzee signing. The last footage people had seen was the Gardners' old 16-millimeter films of Washoe from the late 1960s. Seeing, as they say, is believing. The sight of Washoe, Loulis, Moja, Dar, and Tatu signing to one another on videotape had these scientific audiences enthralled. Here, at last, was stunning proof that chimpanzees could use language like humans after all.

AFTER LUNCH ONE DAY IN MARCH 1985, Washoe, Moja, and Dar were lying on the bench and Tatu was sitting in the tunnel. Loulis was hanging from the room's wire ceiling, when suddenly he began clicking his teeth and shaking his head back and forth. He jumped down and started bouncing off the walls like a rubber ball. As Moja screamed and bounded away, Washoe went over to Loulis with open arms and sat him down. She opened his mouth and peered in, like a lion tamer, to see what was bothering him. But Loulis closed his mouth and moved away from her. Washoe grabbed a hose, swirled it into a nest, and made Loulis sit down next to her. As the other chimps gathered around to watch, Washoe looked in his mouth and began poking around. She must have been too rough because Loulis jumped back up and climbed to the ceiling, where he hung, clicking his teeth. A few minutes later he came down again, and this time Moja took a stab at investigating the problem. But Loulis finally pulled away and lay down on the bench, shaking his head from side to side.

Suddenly we heard the loud CLINK! of something falling off the bench. All the chimps ran to look, and Washoe was the one who held up Loulis's baby tooth. For the rest of the day, Loulis ran around with his mouth open, showing everyone, chimp and human, where his tooth had been. The next morning the tooth fairy brought him some newly dried pears.

At six years old Loulis was beginning the long transition from chimpanzee childhood to adolescence. He no longer had the little white tuft of baby hair on his bottom. His face was turning darker and his thinning hair was exposing the impressive musculature of his growing body. His aggressive displays, which had once seemed so cute and comical, were now truly intimidating, like that of the fully grown adult male. He was taking his arguments with big Dar much more seriously, and he was no longer running to Mom every time he had a fight.

At this age in the wild, the young male usually starts challenging his mother and her friends until he wears out his welcome with the females. Though he will return home to his mother throughout his life, he now begins the awkward and aggressive process of joining the adult male hierarchy. Loulis was now rebelling by taking Washoe's favorite spot on the bench during dinnertime, provoking her into physically removing him. He was also blocking the tunnel, and she would have to push and shove until he whimpered and got out of the way.

In spite of these new outbursts Loulis was still the only child of the dominant female, and he remained a kind of perpetual child. Even at eight years old he was still a holy terror, banging and spitting at any chimps or humans who didn't give him their undivided attention. But the next moment he could be the sweetest boy, holding hands and asking for kisses. And he was always a sucker for a sad face. If you pretended to cry, Loulis would open his eyes wide and offer a kiss to make it better.

Loulis wasn't the only one going through major developmental changes. After a rambunctious childhood Dar had matured into a very relaxed and easygoing ten-year-old. Despite his large physique Dar did everything with a minimum of physical effort. Even when he was performing the typical male display—crashing around the cage, stomping his feet and swaggering—he did it with an I'm-just-doing-my-job attitude. Dar was also turning into quite a tinkerer. He loved to play with tools and take things apart to see how they worked. When the plumber or electrician visited, Dar followed him around inspecting his work. One day we heard a plink-plink-plink sound coming from the enclosure and found Dar taking a music box apart and picking at the metal tines like a toddler at a piano.

But Dar was also finding out that adolescence brings its own set of problems. He had never particularly liked females, chimp or human, except for Washoe and Debbi. But when he reached puberty, he seemed to change his mind and decided that Moja and Tatu were interesting after all. We now saw Dar engaged in long grooming sessions with the two girls, and he even succeeded in getting Moja's beloved hairbrush and mirror away from her. He was every bit as awkward as a human teenager. He would pester Moja, slapping and poking her until she would go to the water fountain, fill up her mouth, and let Dar have it in the face. Stunned, Dar would sputter and sneeze and finally give up his obnoxious pestering.

Moja, who was fourteen years old, acted like an unabashed groupie whenever a human male was around. She'd try panting loudly to get his attention or else she'd brush her long hair very self-consciously. Though Moja was the first to befriend any new lab volunteer, she was still very insecure and somewhat neurotic. She screamed if someone was looking at her the wrong way, if two family members were fighting, or if she got too much attention at her own birthday party. She even screamed if she thought something terrible was about to happen. For example, if we gave Moja a tube of lipstick her first concern seemed to be that Washoe would take it away. She would run to Washoe, hold up the lipstick, and scream, as if to say: "I just know you're going to take this away from me." And guess what? Washoe would take it away from her.

Tatu, at eleven years old, was the opposite of Moja when it came to humans. She was a very reserved young lady who gave to any relationship exactly what she decided and no more. Tatu may have cleaned out the family's enclosure every day, but she always named her price in food, toys, or magazines. She was practical, determined, and extremely stubborn. Tatu's personality reminded me of her mother, Thelma. Like Thelma, Tatu rarely showed her emotions unless she was extremely frustrated. And when Tatu did let loose, the tantrums sounded exactly like her mother's. I could swear Thelma was in the building when I heard Tatu screaming. This was all the more amazing because mother and daughter had spent only one day together in Oklahoma before the Gardners adopted Tatu.

As she moved through adolescence, Tatu became more contrary. During estrus, especially, she was always looking for a fight. The girl who rarely signed CHASE was suddenly bounding into rooms and starting aggressive games. She would slyly set up anyone, chimp or human, by offering them something and then grabbing it back. At her most willful, Tatu once went on strike for weeks and refused to clean up the enclosures. The other chimps picked up a spoon here and a bowl there, but they were totally dependent on Tatu. After every meal, they kept looking at her hopefully. Finally, when it seemed she couldn't stand the mess anymore, she again took charge of the household chores.

WASHOE, LOULIS, MOJA, DAR, AND TATU had been together for five years, and in that time they had formed a complex and cohesive family unit. Their culture was distinctly human (American Sign Language, Christmas trees, and fashion magazines) but very chimpanzee (gestural communication, tool use, and creative play). It was a far cry from the repressive conditions in Oklahoma.

Still, I knew there was one important element missing from

their life: the daily experience of being outdoors. The chimps were safe, healthy, and often happy on the third floor of the psychology building, but they never breathed fresh air, felt the sun on their faces, or climbed a tall tree. Unlike humans, chimps are tree climbers by nature, not by choice. More than anything, we wanted Washoe's family to have the freedom to climb outside in the sunshine.

In 1985 Debbi and I began putting our dream of an outdoor sanctuary on paper. We commissioned plans for a new, freestanding facility that would let Washoe's family live together more naturally, and we began our efforts to raise the estimated half million dollars Washoe's new home would cost. By the end of 1986 we had the first seventy-five thousand. But then events outside our own walls intervened on December 29, 1986.

That afternoon Debbi and I were going over the latest architectural plans when an unassuming videotape arrived in the mail from a group called True Friends. The tape, which documented the living conditions of chimpanzees in a biomedical lab in Maryland, was only sixteen minutes long. But those sixteen minutes changed my life forever.

THIRTEEN

MONKEY BUSINESS

DEBBI AND I WENT INTO OUR DATA ROOM and slipped the videotape from True Friends into the VCR. It begins with the camera following a woman down a hallway. She is wearing a mask and is obviously the member of an animal rights group that has raided this facility to document the conditions. First stop is a room full of small cages, stacked from floor to near ceiling, each one containing a single monkey. Many of the monkeys are running in tight circles, a sign of severe stress. One squirrel monkey is lying dead in his cage. Another bangs his head against the cage, while others are ill and vomiting. Some have clearly mutilated themselves.

The next room is lined with what appear to be stainless steel refrigerators. The front of each refrigerator has a thick Plexiglas window. As the camera zooms in on one window I can make out the shape of a full-grown chimpanzee inside. The sign outside the steel box indicates that chimpanzee number 1164 was infected in February 1986. He has been hermetically sealed inside of this stainless steel "isolette," which is designed to filter out viruses. We can hear the loud hum of a fan, circulating air into and out of the sealed chamber.

The woman in the video unlatches the steel door of the isolette, revealing an internal cage made of iron bars. The chimp does not respond at all to the door opening. He is crouched inside the sterile chamber, rocking back and forth, back and forth, moving his mouth as if muttering to himself. He appears to have gone completely insane.

Yet another room is lined with smaller isolettes, about two feet wide and three and a half feet high. There are two rows of these steel boxes, one row stacked on top of the other, on either side of the room. With their small windows, they look like over-sized microwave ovens. I am already dreading what I know I am about to see.

The woman unlatches one of the isolettes and opens it. The chimpanzee inside cannot be more than four years old. He, too, is rocking from side to side. He doesn't stop his repetitive motion, even for a moment, when the visitor peers in. His expression is completely blank, his eyes dead. There is barely room for him to turn around inside the box. He suffers complete sensory deprivation—no sight, no smell, no touch. The only sound is the incessant hum of the fan.

Next, the woman introduces us to two baby chimps, named Kyle and Eric, who have been selected for injection with HIV or hepatitis and isolation in the steel boxes. The two boys have been crammed together for three months in a cage the size of a large cat carrier. They thrust their tiny hands through the mesh, obviously starved for physical contact and affection.

When the camera gives a close-up view of their faces, Debbi and I stare in disbelief. One of them looks uncannily like Dar. These two were born in an Air Force breeding facility, presumably Holloman, in New Mexico. Dar's father is still alive and breeding at Holloman so it's very possible that this is Dar's brother. As the woman grooms the two babies, the chimps in neighboring cages bang on their own bars. They want affection, too.

In the final scene we meet a young chimp named Barbie, who is locked in one of the small isolettes. Barbie cannot have been sealed in her box for long, because when the door opens she looks directly into the woman's eyes. She and the woman touch, then hold hands. The woman gently kisses Barbie, who responds with a kiss of her own. When the woman finally goes to close the steel door on the isolette, Barbie hugs herself with both arms and begins to shriek uncontrollably. The look on Barbie's face as the door closes—her sheer terror at being sealed into a box alone—will haunt me for the rest of my life.

The tape ended. Debbi and I were speechless. I was trying to make sense of the ghastly images we'd just watched. I had seen chimpanzees in prisonlike conditions before, even solitary confinement, but sealing great apes in steel boxes? The scientists who operated these horrid devices had clearly crossed the line between the pursuit of medical knowledge and cruel and unusual punishment. How could they ignore the torment and anguish on the faces of those chimps?

HERE I NEED TO GIVE YOU SOME BACKGROUND. Two years earlier, in 1984, researchers first injected chimpanzees with HIV, the virus that causes AIDS. With the AIDS pandemic ravaging human populations, there was tremendous pressure to find an "animal model" for the virus—a nonhuman species in which to track the disease, test treatments, and hopefully develop a vaccine. As our closest biological relative, the chimpanzee seemed to be the obvious candidate. But chimpanzees do not get AIDS in nature so researchers had to induce the disease in the laboratory. After being injected with HIV, the chimpanzees showed an antibody response, indicating that they were infected with this human virus. Researchers began demanding an ample supply of chimpanzees from the National Institutes of Health (NIH), the federal agency that funds most medical research, to carry out their studies.

But *that* presented a problem. Until the mid-1970s biomedical researchers bought their chimpanzees from international animal dealers, used them up, then ordered some more. Chimps were viewed as renewable resources, like trees. Meanwhile, wild chimpanzee populations were plummeting as humans destroyed

African tropical forests, hunted the chimpanzee for food, and captured thousands of chimps for sale to American and European laboratories, circuses, and zoos.

In 1975, a new worldwide treaty, the Convention on International Trade in Endangered Species, classified the chimpanzee as endangered and clamped down on their export from Africa. This made it illegal to bring wild chimpanzees into the United States and it caused a chimpanzee supply problem for labs. NIH responded to the shortage by getting into the chimpanzee breeding business. In 1986 NIH proposed an ambitious Chimpanzee Breeding and Research Program that would house 327 "dedicated breeders" and produce a "yield" of thirty-five chimpanzee infants per year, enough to supply future AIDS studies with baby chimps.

It was at this point, just three months before the videotape arrived, that I found myself embroiled in a political controversy. I decided to challenge the new chimpanzee breeding program, and I made my views known in a scientific evaluation that I sent to the congressional committee that appropriated the NIH budget. I had two main concerns about the breeding program, one moral and one scientific. Morally, I questioned whether it was right to breed chimpanzee babies in order to take them from their mothers, infect them, and kill them. Scientifically, I questioned whether experimenting on chimpanzees was the best possible use of scarce AIDS research dollars.

I felt it was crucially important that our society begin to reckon with the ethical implications of experimenting on our closest living relatives. After all, the previous twenty-five years of research, including my own, had revolutionized our understanding of chimpanzees. We now knew that the animals we had viewed in the 1950s as "mindless beasts" were highly intelligent, emotional, and social beings we could communicate with in our own language. Was it morally acceptable to inflict pain, suffering, and death on our evolutionary siblings in the name of human health?

To pose this question was not meant in any way to diminish

the terrible suffering of those with AIDS or their urgent need for a cure. Human disease has always forced us to face tough moral questions. The frightening reality of AIDS does not excuse us from the ethical dilemmas raised by the search for a cure. I have always thought that public debate is one of democracy's crucial checks on the potential abuses of unregulated science—whether the issue is doctor-assisted suicide, genetic engineering, cloning, or animal experimentation. That's why I asked Congress to debate whether this use of chimpanzees was morally acceptable to the American people.

Personally, I was opposed—and still am opposed—to the breeding and use of chimpanzees in any harmful experimentation, including AIDS research. For me it would be unthinkable to send someone from my extended chimpanzee family, or any other chimpanzees, into painful, often terminal, research. And I would feel that way no matter how much it benefited humans, even humans in my own family. My conviction was rooted in the twenty years I had spent with individuals like Washoe, Loulis, Moja, Tatu, and Dar. My own guess is that many people, had they known chimpanzees as I had, might feel similarly.

I had no illusions about the outcome of the debate. I knew that Congress would most likely determine that experiments on chimpanzees were acceptable and that they should proceed. But the researchers still had to prove that chimpanzees were, in fact, the correct animal model. After three years of studies not a single chimpanzee out of one hundred infected with HIV had shown any symptoms of AIDS-related illness. Increasingly, scientists were speculating that chimpanzees might simply carry the virus but not get sick from it. If so, then the chimpanzee was not a viable model for testing drugs and other treatments for AIDS-related illness.

There was also the chimpanzee's apparent resistance to HIV itself. The virus simply did not replicate as quickly in chimps as it does in humans, and their immune system displayed a much more limited antibody response than that of the human immune system. This raised the disturbing question of whether data from chimpanzees were applicable to humans at all, and, specifically, whether chimpanzees were of any use in developing an HIV vaccine. Researchers were planning to inject chimpanzees with experimental vaccines and then "challenge" them with a dose of HIV to see whether the vaccine worked. But the vaccine's effectiveness in chimpanzees might simply be the function of their natural resistance to AIDS, and the vaccine would *still* have to be tested in humans.

By 1986 we knew that the chimpanzee was an imperfect and perhaps useless model for AIDS research. Meanwhile, AIDS researchers were hardly rolling in cash. The disease had been ravaging the homosexual and IV-drug-using populations for five years without so much as a mention by President Reagan. The National Academy of Sciences was calling for the government to spend two billion dollars annually on fighting AIDS, but the actual budget was barely one hundred million dollars.

Everything of value that we'd learned about AIDS had come from studying humans, not chimps. And those human studies— clinical trials of experimental drugs, in vitro cell research, epidemiological tracking, prevention programs—were starved for funding. Was it rational to spend ten to twenty million dollars over four years just to breed chimpanzees? Didn't it make more sense to spend those scarce dollars on human research that had a proven track record and thousands of sick and willing volunteers? I also warned Congress that we might wind up with a generation of HIV-infected chimps who required housing for another fifty years. Tens of millions of dollars for their care would *also* have to be diverted from AIDS research and prevention programs.

But in the end, there was no debate about the moral or scientific issues involved. The biomedical community was screaming "animal model" like a person screaming "fire" in a crowded theater, and the ensuing panic won them their millions of dollars in government funding.

. . . .

AFTER DEBBI AND I WATCHED THE VIDEO made by True Friends, we read a report that came with it, prepared by another group, People for the Ethical Treatment of Animals. The report said the laboratory was Sema, Inc., located in Rockville, Maryland. Sema conducted AIDS and hepatitis research, all of it paid for with $1.5 million taxpayer dollars disbursed through NIH. The report documented an appalling record of negligence and deaths at the lab. Many primates had perished over the past five years, not from actual disease studies but as a result of poor veterinary care and accidents. A separate report by the United States Department of Agriculture (USDA), the agency that inspects animal laboratories, stated that Sema violated animal care regulations governing cage size, feeding, veterinary care, and so forth. But despite these violations, Sema was still very much in business with taxpayer dollars.

There was also a letter from Roger Galvin, an attorney and vice president of the Animal Legal Defense Fund, another animal rights organization. He was asking me to file an affidavit giving my professional opinion of the conditions at Sema, especially the isolation chambers. He believed there was sufficient evidence of "unnecessary suffering" to demand that the Maryland State Attorney's Office prosecute Sema under that state's anticruelty statute. He was careful to point out that this action was not meant as an attack on research in general, but just to stop Sema from using isolation chambers and "to promote a sincere regard for animal well-being."

I didn't need much convincing. Within two weeks I had written and filed an affidavit arguing that the Sema chimpanzees were very likely suffering psychological and neurophysiological damage from exposure to life-threatening conditions. These facts could not have been news to NIH, the agency paying for Sema's isolettes. It was NIH that funded Dr. Harry Harlow's notorious primate isolation experiments in the 1950s. Harlow began by separating infant monkeys from their mothers and ended years later by raising monkeys totally alone in the bottom of a V-shaped metal chamber that he called the "pit of despair" or the "hell of loneliness." The monkeys raised in Harlow's "pit of despair" developed the most extreme symptoms of human depression and schizophrenia. Meanwhile, a psychologist at Yerkes named Richard Davenport was rearing baby chimpanzees alone in small boxes for two years at a time. The isolated chimps soon developed stereotypies like rocking, head banging, and on.

In February 1987, one month after I filed my affidavit against Sema, Jane Goodall called me. She had also gotten the videotape and had watched it with her family during their Christmas holiday in England. They were shattered by the experience. Jane had always stayed away from biomedical labs, focusing her considerable energies on defending chimpanzees in the wild. But I could tell that this videotape had changed everything. She was distressed and very angry, and she had filed her own affidavit, calling the conditions at Sema "psychologically damaging" and "totally unacceptable." More significantly, Jane went public with her opinion, breaking a code of silence among scientists about inhumane conditions in labs.

The president of Sema, Dr. John Landon, denied that there were any problems at his lab and chided Jane for lending credence to a videotape from an animal rights group without ever visiting the lab herself. Jane then asked to see the lab and the chimpanzees firsthand. To her amazement the lab had agreed, and now she was calling to ask me to accompany her on a tour of Sema. Jane knew wild chimpanzees, but she wanted someone along who had expertise with chimps in captivity.

In late March Jane and I were picked up in Washington, D.C., by a top-level NIH official and driven to the Sema facility in nearby Rockville, Maryland. I don't know what I was expecting, but when we pulled up to the building I was dumbstruck by how ordinary it looked on the outside. It was a one-story suburban office building, curtains drawn, next door to a bank. Hundreds of people must walk by this place every day, I thought. Did anyone suspect that inside there were five hundred monkeys and apes locked up in steel boxes?

We were greeted by Dr. Landon, then led on our tour by the chief veterinarian. He was extremely businesslike. I suspected that Sema officials had prepared for our visit by cleaning the place up and bringing in larger cages. But as we moved down the corridor, stopping briefly in each room, I could see that they hadn't. The conditions were exactly as portrayed on the videotape. There were the squirrel monkeys running in circles. There were the other monkeys banging their heads against the bars. There were the pairs of infant chimpanzees, crammed into cages, twenty-two inches by twenty-two inches, waiting to be separated, infected with HIV, and boxed in isolettes. There were the juvenile chimps, thirty-two of them, each one sealed in a stainless steel isolette that was twenty-six inches wide by thirty-one inches deep by forty inches high. I peered into an isolette but it was so dimly lit that I could barely make out a face inside.

Jane and I asked to see Barbie, the young chimp who had screamed in such terror. The vet assured us that Barbie didn't always scream like that; she was just upset at the intrusion. He led us to Barbie's isolette and a lab tech opened the door. Barbie was not screaming. She wasn't even rocking back and forth like the others. She was clinging to the bottom of the cage, facedown, lost in her despair. She turned her head slowly and looked up at us. Her eyes were vacant. It was the same "two-thousand-mile stare," Jane later told me, that she saw in the eyes of starving African children who had seen their parents slaughtered and their homes burned down. Barbie was gone.

"Take her out," the vet told the lab tech. "Take her out."

The tech grabbed Barbie out of the isolette like she was a sack of potatoes. He didn't speak to her or try to comfort her. She just lay there in his arms, not even clinging.

"Give her the apple." The vet pointed to an apple that had been placed on the lab table, presumably for our benefit. Barbie ate the apple robotically, showing no interest or pleasure.

"See, she's fine," the vet said. "She didn't scream at all."

. . . .

WHAT I FOUND MOST ASTOUNDING about our tour of Sema was how nonchalantly my colleagues went about their business. There was no doubt in my mind that if I gathered up a few customers from the bank next door and gave them a quick tour of Sema, most of them would grow physically ill and beg to be let out of the building. But the scientists and lab staff, who were presumably decent family people who loved their cats and dogs, showed little feeling at all about the cruelty they were inflicting. It didn't even occur to them to hide anything from us.

"How do they stand it in here?" I asked Jane as we were walking out.

" 'All pity choked by custom of fell deed,' " she answered, quoting Shakespeare. Whatever compassion they once felt had long since vanished.

After the tour, we met with a group of laboratory and government officials. Jane talked knowledgeably and movingly about the psychological makeup of the chimpanzee—the prolonged mother-infant bond, the sheltered and playful childhood, the complex social relations, the capacity for laughter, mourning, and despair. These are facts that can be found in at least a hundred scientific articles every year and are well known to any watcher of public television shows. But as I looked around the table I saw blank faces, as if the year were 1959 and nothing at all were known about chimpanzees. Once scientists have convinced themselves that their chimpanzee research subjects have no capacity for suffering, it's a lot easier to dismiss suggestions about how to alleviate that suffering.

For example, the Sema officials said that the isolettes were necessary for "biocontainment" and to protect lab workers against HIV or hepatitis infection. Jane and I pointed out that there were probably HIV-infected humans working at Sema, but surely NIH wasn't proposing that *they* be quarantined in isolettes. Based on our knowledge of how HIV and hepatitis B are spread, there was no medical justification for subjecting chimpanzees to such cruel and unusual measures. This view was supported by many leading researchers, including Dr. Alfred Prince of the New York Blood Center, who discovered the hepatitis non-A non-B virus.

We asked the Sema officials why they couldn't house the chimps infected with HIV together in groups or in pairs so that they would have one another for company.

That's inefficient, they responded. *With two infected chimps in a cage we'll have only one "data point" to study. It's a waste of a chimp.*

Jane then suggested that they house one monkey with one chimp. (Most monkey species are not susceptible to HIV, although they can carry a related virus called SIV.) Young monkeys and chimps often get along quite well.

That's too much trouble.

Well, then at least enlarge the cages so they can move around.

That makes it too hard for techs to inject them and draw their blood.

At least give them some toys so they have something to play with.

Toys can carry disease.

You have autoclaves (a steam-heated machine for sterilizing lab equipment). You can autoclave the toys.

The cages will be harder to clean.

And so it went. The mental health of a chimpanzee carries no weight in a system that is designed to be as cheap, efficient, and space-saving as possible. The ultimate irony is that by ignoring the psychological needs of their subjects, researchers produce scientific data that is extremely suspect, if not useless. It is well known that psychological stress can profoundly depress the immune system in all animals, making them more prone to a variety of physiological disorders. Yet laboratories routinely house chimpanzees under conditions stressful enough to cause aberrant behavior. These abnormal chimpanzees do not represent a normal human population. No researcher would test a vaccine, intended for use in healthy individuals, on human subjects who are living in solitary confinement, deprived of all physical affection, social contact, and mental diversion.

If researchers want optimal test data they will have to address the chimpanzee's need for social contact and mental stimulation. And there's the rub. Every biomedical researcher operates within a contradiction: "We need to experiment on chimpanzees because, *physiologically*, they are just like us." Why, then, is it acceptable to isolate, torture, and even destroy animals that are just like us? "Because, *psychologically*, they're *not* like us."

The scientist who begins to acknowledge that his chimpanzee subjects have emotional needs will quickly find himself sinking in moral quicksand. It is much harder to inflict pain on a chimpanzee once you've looked into his eyes and acknowledged the person inside.

As Jane and I left the Sema lab, I wondered why NIH had let us take the tour at all. We sat in the backseat of the car without speaking as the NIH official drove us back to Washington.

"Jane," the official said over his shoulder, "at least you will agree that these chimpanzees are well cared for. I'm sure you'll have no problem writing a letter stating that this lab is up to USDA regs and that there are no violations."

I guess he couldn't see Jane in his rearview mirror, or he never would have made that comment. Tears were pouring down her face.

She composed herself, then spoke slowly and deliberately: "By no means will I write you any such letter."

NOT ONLY DID JANE NOT WRITE NIH its hoped-for letter of endorsement, she published a scathing account of our visit to Sema in *The New York Times Magazine*. Her article, "Prisoners of Science," described the life sentence served by chimpanzees under conditions "worse than those we accord to even the most evil human criminals." She called for newer and higher standards of humane care in labs. Her unprecedented broadside was the "shot heard round the world." She revealed the well-guarded government secret that nearly two thousand chimpanzees were being subjected to the cruelest imaginable treatment. What's more, NIH could no longer dismiss these accusations as the unsubstantiated charges of a few animal rights "fanatics."

The time was right for us to push for humane laboratory care and, coincidentally, an opportunity presented itself the next month, April 1987. Two years earlier, Congress had strengthened the Animal Welfare Act and ordered laboratories to provide "a physical environment adequate to promote the psychological well-being of primates." Now, the USDA was convening a panel of primate experts—including Jane Goodall and myself—to make recommendations for the new rules that labs would have to follow in caring for their chimps.

Jane couldn't attend the panel meeting, but she and I decided in advance that I would focus my efforts on getting the panel to recommend larger cages. That was the simplest, most tangible way to improve the life of captive chimps. The current regulations said you could legally put a 230-pound chimpanzee like Dar's father—or even a 600-pound gorilla—into a five-by-five-foot cage and feed and water him once a day. That was unconscionable, and I assumed everyone would agree.

I was wrong. The panel of nine experts was stacked with six representatives from large NIH-funded biomedical labs who were clearly hostile to the three of us who specialized in chimpanzee behavior. A coterie of NIH officials, who of course funded the work of virtually all the panel members, took note of everything that transpired in the meeting. When I suggested that larger cages had to be a top priority, I was met with a hail of criticism about how expensive it would be. The head of one NIH-funded lab summed up this position, saying: "If they increase cage size, I'll turn half my monkeys into dog food." I reminded them that the Animal Welfare Act did not order us to promote psychological well-being for primates only if it was cheap and convenient for labs. We were there to propose more humane regulations *no matter what the cost.*

The next day I introduced a specific motion to increase cage size to twenty feet by twenty feet, enough to give a chimpanzee physical exercise and room to play. No one seconded my motion—not even the two other members of the panel who were experts in chimpanzee behavior. Dr. Michael Keeling, the panel chairman and head of an NIH-backed chimpanzee research center, asked, sarcastically, if I cared to make another motion that might be more amenable to those present. I repeated the same motion, and again, no one seconded it.

Finally, I said, "Well, let's at least give them fifty square feet." Again, there was silence.

"Well, Roger," Chairman Keeling said with a smile, "it looks like no one is interested in seconding your motion."

I felt betrayed by the two other behavioral scientists on the panel, and I confronted one of them after the meeting. "Roger," he said, "you've got to realize that these people support my research." He was clearly speaking for everyone there. I was beginning to appreciate the long arm of NIH. It was one thing for me to file an affidavit against a lab like Sema or to challenge an NIH policy in a report to Congress. But it was another thing entirely to stand up in front of my colleagues and tell them to change the way they conducted their research. I had broken ranks with the good ol' boys' club of primate experts and their federal funders. My next grant proposal to NIH would not be looked on kindly.

Why hadn't I kept my mouth shut like the other behavioral scientists on that panel? I had nothing to gain and everything to lose by speaking up. Why was I endangering my career in this way?

I thought about the chimpanzees suffering in solitary confinement at that very moment. Ally was in White Sands Research Lab, if he wasn't already dead. Booee was in LEMSIP, and so was Tatu's mother, Thelma, and Moja's mother, too. Loulis's mother was in Yerkes. Dar's father was in Holloman, and his brother was in Sema.

Who was going to speak up for them if I didn't? I had no

desire to be a martyr, but I couldn't just go back to Ellensburg, hole up with Washoe's family, and pretend that biomedical laboratories didn't exist. I was in a real fight now. The government had forced me to choose between my scientific peers and the chimpanzees they were experimenting on.

I DIDN'T THINK my position was unreasonable. If scientists were going to force chimpanzees to spend their entire lives in service to human welfare, I felt they should at least acknowledge the contribution of those chimps by giving them bigger cages, exercise, and companionship. That much seemed like simple decency and gratitude.

The law of the land was clear. Congress passed the Animal Welfare Act because it wanted to improve conditions for captive primates, and I was prepared to work through the system one more time to uphold that law. Fortunately, I had an ally in Jane Goodall. Jane adamantly believed that biomedical scientists were simply victims of ignorance. They worked within systems that had been set up long before we understood the nature of a chimpanzee's intelligence and emotional needs. She was convinced that if we could only open enough laboratory doors and educate enough scientists, then they would see the light and agree to improve conditions.

That led us to a new approach. Jane and I proposed to NIH that we put together an international scientific summit made up of biomedical researchers, behavioral experts, and ethologists from Africa and try to find common ground on new and more humane standards for chimpanzee living conditions. NIH officials were in damage-control mode after the Sema fiasco, so they were enthusiastic about our suggestion for building bridges among scientists. They asked us for a grant proposal right away so that they could fund and sponsor the high-profile conference. But our idea was causing growing alarm among biomedical researchers, and after months of encouragement and assurances, the deputy director of NIH summoned Jane and me to his office in Washington, D.C. He told us the conference was off. I could see from the look on Jane's face that she was getting the picture. Scientists like us were persona non grata at NIH.

When I walked out of NIH headquarters that day, I knew that I had been banished from the scientific establishment for good. The officials who ran that system were determined to go right on conducting business as usual. Jane Goodall and I were soon branded as heretics and animal rights extremists by NIH spokespersons. According to them, anyone who wanted to improve the welfare of animals was out to abolish biomedical research and ought to be excommunicated.

Unlike Jane and other ethologists, I was a laboratory scientist. If I wanted federal funding I had to apply to NIH or its sister agency, NSF. Though I had stopped relying on federal funding in 1980—when the government suggested I do biomedical research on Washoe—I did submit applications for small grants every year, mostly for equipment.

Both NIH and NSF began rejecting all my requests for funds. They denied that their rejections had anything to do with the position I had taken, but the comments on one pink rejection slip from NSF explained it all: "The investigator [Fouts] is a member of a number of organizations which oppose animal research of all types." At the time I belonged to only one organization, Psychologists for the Ethical Treatment of Animals, which advocated only the humane treatment of animals in labs.

The truth is that my own evolving viewpoint about animal research did not fit neatly into the biomedical or "abolitionist" camp. I was troubled by the use of apes in all forms of research, not only biomedical. From my experiences in Oklahoma, I was all too familiar with the suffering of chimpanzees in behavioral studies. As far back as 1974 I sought, in vain, for a way out of ape language studies for both Washoe and me. Finding none, I resolved to make life better for the chimpanzees in my care and, later, for all captive chimps. By the time I visited Sema I knew that trying to draw a clear moral distinction between a biomedical experiment and a language experiment was a losing proposition. As horrific as many biomedical labs were, the underlying problem was captivity; captive environments differed only by degree of their cruelty. The humane solution, in my mind, was to work toward the gradual termination of *all* research on captive apes. And if I were to have any moral authority on the subject, that would have to include my own research as well.

Just as I found it difficult to draw moral boundaries between different kinds of research, I was finding it harder and harder to draw moral boundaries between species. Surely a baboon or a dog in a biomedical experiment suffers no less than a chimpanzee or a human would. And the same troubling questions apply: Is it right to kill other animals to prolong human life? Must we inflict suffering on one species to relieve suffering in another species?

It was Washoe who taught me that "human" is only an adjective that describes "being," and that the essence of who I am is not my humanness but my beingness. There are human beings, chimpanzee beings, and cat beings. The distinctions I had once drawn between such beings—distinctions that permitted one species to imprison and experiment on another species— were no longer morally defensible for me.

NIH would have been correct to say that my goal was the *ultimate* removal of all animals from research. But they were wrong to say that I "opposed animal research of all types." After all, I was applying for funding to do animal research. It would have been the height of hypocrisy for me to take a public stand against all research, and I never did. Here is where I disagreed with those who wanted to abolish all animal research immediately. I knew that it might take decades to achieve my goal of phasing out research on animals and to care humanely for the subjects, including my own. That's why I advocated a pragmatic approach: to reduce the pain and suffering of lab animals whenever possible and to find alternatives to animal subjects whenever feasible, which is quite often. By following this humane path, I hope that one day we finally empty all the cages in all the labs, including my own in Ellensburg.

But my humane, gradualist approach was lost on NIH officials. They were especially fond of invoking the specter of AIDS to ridicule the intentions of anyone who spoke out on behalf of chimpanzees. Any scientist who dared to criticize laboratory conditions was branded "irrational" or "antihuman," or both. All this mudslinging obscured the plain fact that I was asking only that the government treat its chimpanzees with the decency and compassion required by law.

In December 1987 Jane Goodall and I held our conference on chimpanzee enrichment, without the support or participation of NIH. The workshop was hosted instead by the Humane Society of the United States, and it was attended by experts from major laboratories, zoos, and chimpanzee colonies. We came up with a list of reasonable recommendations: that chimps always be housed in groups, that their cages be large enough (four hundred square feet) to allow exercise and socializing, that infants remain with their mothers as long as possible, and that chimps' lives be enriched with toys and activities. We submitted our recommendations to the USDA, which was just then writing its regulations to promote the psychological well-being of primates, as Congress had ordered.

In 1991, the government finally issued its long overdue regulations for the care of chimpanzees. The new rules said nothing about larger cages or any of the other measures we had recommended. The biomedical industry had been lobbying behind the scenes to kill any improved standards, but I had assumed that the will of Congress—the Animal Welfare Act—would prevail in the end. I knew many scientists, including quite a few biomedical researchers, who thought the new regulations were a complete outrage. This was not only bad for chimps, but it would diminish the public's faith in science. The scientific community should have been up in arms, and I waited for someone to sound the alarm. But no one said a word.

I knew it was time to stop trying to change the research establishment from the inside. Someone would have to go to federal court and sue the government for its failure to uphold the Animal Welfare Act. I soon discovered that a few others were thinking along the same lines. Christine Stevens from the Animal Welfare Institute asked me to join with her organization and the Animal Legal Defense Fund to file just such a lawsuit. They were looking for a research scientist to be a coplaintiff.

I knew how it would look for me to join with two animal rights groups in a legal action against the scientific establishment. I was already a whistle-blower, but now I would be portrayed as consorting with "the enemy." My twenty-five-year journey from scientist to activist would be complete, and I would have to live with the consequences.

I already knew the Animal Legal Defense Fund attorneys from the case against Sema, and I was very impressed by them. They were not wild-eyed extremists. They were successful professionals who had wrestled with their consciences and decided to devote themselves to protecting animals in laboratories. I felt comfortable having them as allies.

On July 15, 1991, I joined in a lawsuit against the USDA. Almost two years later, in February 1993, U.S. District Court Judge Charles Richey not only found in our favor, he delivered a stunning rebuke to the government and the biomedical industry. He said that the government's refusal to enlarge cages and set standards for the psychological well-being of primates was "arbitrary and capricious and contrary to law." In Judge Richey's eyes, Congress meant what it said, and it said what it meant. Laboratories would have to take real, measurable steps to improve the lives of captive chimpanzees. If those measures cost the biomedical industry a lot of money, *too bad*—Congress had been aware of that when it wrote the law. My faith in government was restored, for the moment.

Unfortunately, Judge Richey's ruling did not stand. The National Association of Biomedical Research joined the government in appealing Judge Richey's decision. In July 1994, the court of appeals overturned Judge Richey's order for tougher laboratory standards. The court did not challenge Judge Richey's logic but ruled that humans hadn't suffered any injury and therefore had no standing to sue. According to the court, it was the *chimpanzees* who had suffered from the government's failure to uphold the Animal Welfare Act.

But chimpanzees are caught in a Catch-22 under our current legal system. Chimpanzees are property, and, legally speaking, a piece of property cannot suffer injury or file suit. If I damage someone's car, it is the car's owner that suffers legal injury, not the car itself. It is the same with chimpanzees. Researchers can inflict physical and psychological damage on a chimpanzee, but as far as the law is concerned, a chimp cannot suffer an injury and so there is no injury to be remedied.

This will all change one day when our legal system finally recognizes that chimpanzees are *not* inanimate objects but thinking, feeling individuals who do suffer and therefore need legal protection. Our lawsuit against the government was just the beginning, not the end, of the legal fight to protect chimpanzees. I believe that over the next decade the federal courts will hand down precedent-setting decisions that will forever change how our society treats our fellow apes.

I remain hopeful, especially because there have been some significant improvements in individual laboratories. For example, Sema, which is still run by Dr. John Landon and is now known as Diagnon, has thrown out its isolettes and today houses its chimpanzees in Plexiglas cubicles that are larger than current regulations call for. Although the chimps still lead mostly solitary lives, they regularly interact with humans and sometimes get to play with other chimpanzees. It's far from ideal, but it's a great improvement over the nightmare that Jane Goodall and I witnessed back in 1987.

Still, I don't expect to see industry-wide improvements unless researchers are forced to make them. So I will continue working with animal rights attorneys until all biomedical laboratories are promoting the psychological well-being of chimpanzees as Congress ordered them to do way back in 1985.

· · · ·

AT THE SAME TIME WE WERE BATTLING over laboratory conditions in America, the Jane Goodall Institute produced a landmark scientific report on wild chimpanzee populations in Africa. The numbers were staggering. At the turn of the century there had been five million chimpanzees on the African continent, but now the number was down to about 175,000. Once abundant in twenty-five countries, chimpanzees had already disappeared from four nations, were about to vanish in five other nations, and were likely to be exterminated in another five in the near future. In West Africa, the chimpanzee population had crashed from one million to an estimated seventeen thousand. Much of the population decline in nations such as Guinea, Sierra Leone, and Liberia had been caused by demand for biomedical research.

After reviewing the Goodall Institute's report, the U.S. Fish and Wildlife Service announced that it wanted to upgrade the chimpanzee's status from *threatened* to *endangered*. I strongly endorsed this move because it would require laboratories to report to the public on how they were using and treating chimpanzees. But this sensible proposal was strenuously opposed by another government agency—NIH. NIH officials began claiming that a critical shortage of research animals was creating an urgent need to acquire more wild chimpanzees from Africa, something that hadn't been done since the early 1970s. They questioned the "alleged decline" in chimpanzee populations and challenged the Goodall Institute's report.

The U.S. Fish and Wildlife Service received 54,212 letters in support of the *endangered* classification from private citizens, wildlife organizations, scientific experts, and African governments. The only people who opposed the new classification were eight biomedical researchers and one person from a circus.

Even so, the biomedical lobby got its way. NIH negotiated a deal with the U.S. Fish and Wildlife Service that split the chimpanzee's classification. Captive chimps would still be classified as *threatened,* which meant that researchers could continue experimenting on them in secrecy and without accountability. Only wild chimpanzees would be upgraded to *endangered* status.

By failing to give full protection to *all* chimpanzees, the United States was sending a message to the rest of the world that the survival of chimpanzees should take a backseat to human need whenever it was convenient. After all, if Americans can trade, use, and kill captive chimpanzees without answering to any international authority, then why should an African poacher refrain from hunting wild chimpanzees? NIH made "extinction" a meaningless abstraction and gave moral support to those who would kill the last 175,000 wild chimpanzees on the African continent.

These chimpanzee victims may have been faceless animals to NIH officials, but in 1988, as the battle raged over *endangered* status, I learned that my old friend Lucy had been found dead on Baboon Island in The Gambia. Janis Carter found Lucy's skeleton by their old campsite. It appeared that Lucy had been shot and skinned by human poachers. Lucy was always the first to approach human interlopers on her island. She was, after all, unafraid of humans. Whoever had killed her had cut off her hands and feet. They were probably sold as trophies in one of the many African markets that also offer gorilla skulls and elephant feet.

It seemed like a lifetime since Lucy and I were featured in that 1972 issue of *Life* magazine. Lucy did have two lifetimes. Born in a Florida carnival, she spent her first thirteen years as the pampered daughter of human parents—a chimpanzee who'd never met another chimpanzee. She then lived her last ten years in the African jungle, where she succeeded, against all odds, at integrating into a community of her own species.

Janis had recently visited Lucy on Baboon Island after a six-month separation. She gave Lucy a mirror, a doll, a hat, books, and other tokens from her life in Oklahoma, hoping the memories would no longer upset her. After they embraced and groomed, Lucy turned and walked away into the jungle, leaving the mementos behind on the ground.

Like Washoe, Lucy had been a survivor. She evaded the clutches of William Lemmon, as her parents had hoped. She avoided the solitary confinement and biomedical experimentation that Ally and Booee endured. But she could not escape human beings altogether. Humans raised Lucy, taught her language, sent her to Africa, and rehabilitated her. And, in the end, they killed her.

Lucy's extraordinary life was reduced to an endangered species statistic. She had been one of those remaining 175,000 African chimpanzees, the embattled remnant of humankind's closest living relatives. With the savage killing of my friend, the chimpanzee was pushed one step closer to extinction.

FOURTEEN

HOME AT LAST

WHEN WASHOE WAS A LITTLE GIRL, and we were living in Reno, she would run to the trailer door every morning after breakfast and sign GO OUT, GO OUT. Rain or shine, she would wait there, until I opened the door. Then she would charge into the backyard and head straight for her sandbox or favorite willow tree. Once she was outside it was almost impossible to get her back inside. For Washoe, to be OUT was to be happy.

By the early 1990s, it had been an awfully long time since Washoe or the other chimps had been OUT. The last time Washoe had felt the sunshine and heard the birds sing was when she took her newly adopted son, Loulis, for a walk in Oklahoma in 1979. Moja, Tatu, and Dar had last felt the grass under their feet in 1982.

A trip we took with Tatu to Dairy Queen illustrates why we couldn't take the chimps OUT anymore. We were sitting in our car, waiting our turn at the drive-up window. Suddenly, a woman in the car ahead of us saw Tatu in her rearview mirror. Her car door flew open, and before I knew it she was running toward us, screaming, "It's a monkey, it's a monkey."

For a chimpanzee, nothing in the world is more threatening than a primate charging on two legs—waving her hands and screaming, no less. It would have been understandable if Tatu bit her in self-defense. I quickly started rolling up the car window, but the woman got to our car in time to thrust her arm inside. I kept rolling the window, which pinned the woman's arm against the door frame. She looked at me like I was insane. But if she had grabbed Tatu and provoked an attack, I knew perfectly well that Tatu was the one who would have been blamed.

That was our last outing with the chimps. They had to settle for watching nature from the third floor of the psychology building. Sitting by their window they conversed with each other about FLOWERS, CARS, GRASS, TREES, and DOGS. After a few years, when the chimps signed OUT they no longer meant *outside*. GO OUT now meant *I want to go across the hall to the playroom*. Being in nature wasn't even a possibility to them, and they didn't even have a sign for it.

But worst of all, this life indoors was somehow taking a terrible toll on Tatu and Moja. A mysterious illness was progressively crippling both of them. In late 1991 we first noticed that Tatu seemed stiff and lethargic, a year later she had constant diarrhea, and her weight had dropped from ninety to sixty pounds. She was so skeletal her menstrual cycle stopped completely. Moja was also growing stiff in her joints. She couldn't even grip the fence with her hands.

We suspected that the girls had some kind of environmental poisoning, maybe from the zinc on the steel fencing in the psychology building. But a series of tests turned up nothing. We enriched their diet and gave them plenty of supplements, in case they had a vitamin deficiency.

But by 1993 Tatu could hardly crawl. At seventeen years old she looked like an eighty-year-old hunchbacked woman. Our doctor, Rick Johnson, snuck her into a local hospital in the middle of the night, and a sonogram showed a blocked and enlarged colon, but that did not explain the other crippling symptoms.

Finally, I sent a videotape of Tatu and her test results to a respected physician that we know. His reply was devastating: "You should prepare for Tatu to die soon."

• • •

ON MAY 7, 1993, WASHOE WOKE UP, looked around, and rubbed her eyes. She was not in the room where she'd gone to sleep the night before. And when she looked out the glass door, she could see that she was no longer on the third floor of the psychology building. Instead, she was facing out on a vast grassy area—more than five thousand square feet in size—that had giant climbing poles and structures, earthen terraces, and hanging fire hoses, all enclosed in a wire-fence dome that vaulted thirty-two feet above the ground. Sunshine was pouring through the dome, flooding the grass below.

Washoe was in her new home. After fifteen years of planning, including ten years of fund-raising, eight years of designing, and two years of building, our dream of a Chimpanzee and Human Communication Institute had finally come true. At 2 A.M. that morning we had tranquilized the chimps and given them physical exams. Then we bundled them in blankets and took them one by one, in a van, from the psychology building to the new institute. It was all of four hundred yards, but on the last trip, while ferrying Moja, we ran out of gas. No one had thought to buy any. We wound up carrying Moja, who was by now awake and sitting up, on a makeshift sedan chair across the campus. She looked like the Queen of Sheba and loved every minute of it.

The few weeks leading up to the big move had been difficult. Washoe, in particular, had grown reserved and withdrawn when she saw us packing boxes, putting toys away, and disrupting her family's routine. She may have been remembering the terrible upheavals when she left Reno at age five and Oklahoma at age fifteen. Perhaps she was wondering who of her family and friends was going to be left behind this time.

To ease the chimps into their new home, some of our volunteers produced a videotape preview, with Debbi and me showing them the new sleeping quarters, kitchen, indoor exercise areas, and outdoor play area. Two days before the move,

Washoe, Loulis, Moja, Dar, and Tatu gathered around a television set to watch the videotape.

LOOK! one of the volunteers signed to them. THERE'S ROGER IN YOUR NEW HOME! ROGER SHOWS YOU BED. HE GOES IN DOOR—BIG PLAYROOM. LOOK! DOOR THERE—YOU CAN GO OUT! SEE GRASS. YOU CAN RUN, CLIMB, PLAY. YOU WILL LOVE NEW HOUSE! WE ALL GO WITH YOU!

The chimps were entranced and they visibly relaxed at the sight of familiar faces in the new house. When the tape was finished, they asked to see it again.

Now, two mornings later, I was standing near Washoe in her new house as she gradually woke up from a drugged sleep. She stared at the sunlit field just outside her door. In an instant she seemed to realize that she was in the HOME on the videotape, and she started screaming with delight, just like on Christmas morning. She leapt up, turned to Loulis, and hugged him. Then she staggered toward the glass doors, looked right at me with a glint in her eye, and signed OUT OUT!

Our plan was to keep the chimps inside for two weeks, but they spent the first two days begging to go out. So on the third day, after breakfast, I told them, TODAY YOU GO OUT. Washoe leapt up and parked herself by the hydraulic door that leads to the outside upper deck. She waited there for more than an hour, with Loulis right behind her. He seemed a little nervous and needed his mother's reassurance.

Finally, the door slid up. Loulis swaggered, then seemed to think better of it and sat back down. Washoe waited for him patiently, but Dar squeezed by and exploded out the door and down the stairs to the ground. He raced across the grass field with such an ecstatic movement that he looked like he was skipping, quadrupedally. He headed directly for the far terrace, climbed to the top of the thirty-two-foot-high fence, and gazed out over Ellensburg. Then he turned toward us and let out a loud pant-hoot of happiness.

Washoe was the next one out. She stood upright and surveyed the terraces, the garden, and the familiar human faces at the observation window below. Stretching out her leg, she touched her toes to the first step and pulled them back quickly. Then she noticed Debbi standing at the fence near her. She walked over, reached through the fence, and kissed Debbi through the wire. This was clearly her way of saying "thank you."

Now Loulis edged out of the doorway. Washoe climbed down a few steps to encourage him, then looked back and signed HUG to Loulis. He was clinging to the fence. Tired of waiting, Washoe climbed the rest of the steps to the ground. Standing upright, she stamped her feet and thumped the back of her hand against the observation window to display her territoriality. Then she plastered her lips to the glass and delivered kisses to a few friends, including Dr. Fred Newschwander, her veterinarian.

Dar climbed down the fence, ran across the field, and embraced Washoe. She ambled leisurely across the grass, climbed a pole, and perched at the top, watching Moja, who was now coming out of the doorway. Moja's movements were stiff and slow, but she managed to make her way down the steps and up the terrace. Then, to my amazement, she grasped the fence and climbed to its highest point. From there she began strolling around the ledge of the garden, high above the ground.

Tatu crept onto the deck. She just gazed out at all the excitement for a while. Finally, she moved her crippled body slowly down the steps. At the bottom she settled into a nice spot in the garden where she spent the whole afternoon examining blades of grass and tasting various weeds.

Loulis was still stranded up on the deck, so Washoe climbed down from her perch and joined him. He crept back inside the door and nodded to Washoe, as if he was trying to convince her that it was time to go back inside. She reached in and pulled him out. Now Dar climbed up on the deck, apparently trying to get Loulis down. Dar sat next to his friend, staring up at the sky with the biggest chimpanzee smile I've ever seen. Loulis seemed buoyed by having Dar nearby, and, at long last, he walked cautiously down the steps. At the bottom his feet touched the earth for the first time since he was a baby.

Dar pant-hooted loudly, took a running leap off the deck, and jumped eight feet to the ground below. He landed in a tuck, rolled onto a tire, and jumped up to hug Washoe, who then hugged Loulis.

To our utter amazement, Loulis then went over to the fence and climbed to the very top. He had never in his life been more than seven feet off the ground. Now he looked down at the grass, thirty feet below, and screamed in absolute terror. Washoe raced up the fence to rescue him.

When Loulis made it down he went over to Dar, who was up on the deck basking in the sunlight. Dar put his arm around his little brother's shoulder and held him close. "This is the life, little buddy," he seemed to be saying. "This is the life."

FOR WEEKS MOJA AND TATU REFUSED to come inside, even for meals. We had to beg and cajole to get them to eat. They spent so much time in the sun that their pale skin turned bright red. But Moja and Tatu didn't seem to mind being sunburned. They lived for the sun. By August, only three months after the move, Moja and Tatu were not only tan, they were physically and psychologically transformed. Moja, who had been so hobbled in the old lab that she spent most days lying on a bench covered with dress-up clothes, was now moving skillfully across the walls and ceiling of the outdoor area. Tatu, who had been near death, was once again acting like a seventeen-year-old— running, climbing on the fire hoses, and initiating tickle and chase games. She even regained all the weight she lost. Every day she was signing BLACK BLACK about everything she saw in the garden. She and Moja were joining in territorial displays, an unthinkable event in the old lab.

We were overjoyed at their miraculous recovery, but it wasn't until weeks later, when I was reading yet another medical text, that we solved the mystery of their illness. Tatu and Moja had been suffering from rickets, a bone disease caused by a lack of vitamin D. Although the chimps had been getting plenty of vitamin D in their diet and supplements, they had not been getting exposure to direct sunlight, which is necessary for converting vitamin D into a usable form for bone growth. The sunshine that came through the windows in the old lab was useless because the sun's UV rays were filtered out by the glass panes. Tatu and Moja were starved for direct sunlight.

Today, I still stop whatever I am doing whenever I see Tatu and Moja chasing one another upside down and thirty-two feet above the ground. Their acrobatics are enough to give me vertigo. Who would believe that in 1993 they were lying crippled on the concrete floor?

Even on rainy or snowy days Tatu and Moja plead to go OUT. One freezing morning during her first winter in their new home, Tatu waited by the door and signed GO OUT.

SORRY VERY COLD, one of the volunteers responded. YOU MUST WAIT.

GIMME CLOTHES, Tatu demanded. So we gave Tatu sweaters and out she went.

Tatu and Moja were not the only ones reveling in their newfound freedom. Washoe, who was now twenty-eight years old and 160 pounds, had become something of a couch potato in the old lab, routinely lying on the bench and browsing food and fashion magazines. Now each morning she was waiting by the door, as she used to do as a child in Reno. Washoe never seemed to stop moving: one minute she was chasing Loulis, and the next she was climbing to the ceiling and gazing out at the snowcapped mountains to the north. Sometimes she seemed to run back and forth for no reason other than that she *could* run. She shed quite a few pounds in the process.

Dar, the seventeen-year-old, finally found a home where he could express his chimpanzee maleness. He looked even bigger outdoors and he showed off his strength in remarkable exhibitions of running, leaping from great heights, and slamming his feet into the observation windows. These territorial displays did not seem to indicate displeasure. As Debbi put it, "He was happily aroused"—confident, dominant, and having fun. Dar still greeted us with kisses each morning, but he was far too busy being "king of the hill" to spend too much time with humans.

FROM THE DAY WE BEGAN DESIGNING the Chimpanzee and Human Communication Institute in 1985, my overriding goal was that it should meet the psychological and biological needs of Washoe's family. Chimpanzees need a supportive and communicative family environment as much as humans do. But unlike us they are forest dwellers by nature. A wild chimpanzee sometimes climbs fifty or eighty feet up into a tree before building her sleeping nest. I knew we couldn't re-create an African forest, but I wanted to do the next best thing and create an environment that functions like a forest from the chimpanzee's point of view.

This functional approach is the opposite of what you might normally see in a zoo. Most zoos are designed to meet the needs of their human visitors, not their nonhuman residents. The chimpanzee enclosure is usually built so that people can see the chimps from any angle and the chimps have no place to hide. The terrain itself is designed to please human sensibilities. Many modern zoos place chimpanzees on islands surrounded by moats, rather than in the old cages. These big islands may look great to humans, but they're useless to chimps unless there is something to climb on. Chimpanzees do not need lots of square footage on the ground; they need lots of vertical climbing space.

Washoe's home is a three-dimensional version of a tropical forest. There are 5,400 square feet of ground area covered in grass and vegetation where the chimps can explore and play. Above that is three times as much area in vertical climbing space on the walls. And the entire space is covered by a three-story-high, open-air mesh roof that the chimps can swing across just as they would swing across a rain forest canopy in the wild. Washoe's family would strip real trees bare, so we erected telephone poles for climbing and platforms for branches, all connected by old fire hoses that serve as vines. There are multilevel terraces for them to explore, cargo nets where they lounge, a treat mound where they use tools to dig out goodies, and a cave where they can find privacy.

In addition, there are two large indoor exercise areas—each one is six hundred square feet and three stories high—that are naturally lit and arrayed with terraces, climbing structures, hanging fire hoses, and tractor tires. The fenced sleeping quarters face the glass-walled kitchen, so the chimps can watch their meals being prepared and can converse in sign with their cooks.

Except to clean, make repairs, and provide medical care, humans never enter Washoe's home, and her family is allowed to live naturally within their social community. When the chimps were younger they needed our physical affection, but now they have one another. Besides, once they grew to their full size, we got more out of the playing than they did. Dar has the strength of a 750-pound human in excellent physical condition. He had to be extremely careful when he played with me, or he could easily hurt me. It would be like me playing football with a child suffering from brittle bones. The chimps have always been careful with us, even through the wire mesh, but it is still not a good idea for me or anyone else to enter their home. If a family fight broke out while I was with them, they might forget themselves and treat me like any other family member—with disastrous consequences.

Other than eating their meals at set times and coming inside to sleep, the chimps are free to do what they want, when they want, and where they want. Human volunteers are still available if the chimps want toys, magazines, dress-up clothes, and so on. But in their new outdoor home, where they can run, climb, and play freely, they don't need the kind of constant activities we used to provide to keep their lives interesting.

The main scientific purpose of the Chimpanzee and Human Communication Institute is studying the use of sign language within Washoe's family. Our students still collect data on signing as they serve the family meals and observe their activities, but it's all done from positions outside the chimps' home. We also continue to use the remote video cameras. None of our studies involve drilling, testing, or training. If a study requires any human interaction with Washoe's family, the only way it can proceed is with the consent of the chimps. If the chimps want to converse with humans for a study, then they do. If the study does not interest them—if they wander off to look at their magazines or climb a pole—then they're excluded from it. Occasionally, a graduate student will complain that he can't get one of the chimps to take part in a study. "Too bad," I reply. "Think up a study that's more fun."

Much of the cost of Washoe's home was paid for by the state of Washington, through the university budget. In return, we promised to educate the public, especially schoolchildren and college students, about the biology, communication, family life, and culture of chimpanzees. I wanted Washoe's home to be a model for young people of noninvasive, compassionate scientific research. We placed the visitors' area behind glass panels that wind in and out of the chimps' outdoor area for about seventy feet. This gives people the visual sensation of standing among the chimps, but only when the chimps choose to be there. Washoe's family can greet their visitors at the observation window or they can stay in other areas where they can't be seen. The thick glass prevents the sound of human voices from disturbing the family.

This design follows a single principle: chimpanzee needs come first, human education comes second. The handful of zoo directors who are also following this principle—a complete reversal of the old zoo philosophy—are discovering that their visitor attendance is rising. People are attracted to humane environments where chimpanzees can live more energetic, interesting, and social lives, even if it means that sometimes they are out of the visitors' sight altogether. Over the past decade, Debbi and I have encouraged hundreds of zookeepers to enrich their chimpanzee environments. A new generation of zookeepers is working from this mind-set, and today you will see things in chimpanzee zoo enclosures—raisin boards, water hoses, crayons, Kool-Aid balloons—that would have been unthinkable ten years ago.

Over the years our budget for building Washoe's home had soared from $500,000 to $2.3 million, which was far more than the university had expected. We began lobbying state legislators with hundreds of letters and phone calls, and thanks to a university trustee named Ron Dotzauer, the governor and his wife, Booth and Jean Gardner, visited Washoe and became strong supporters of our proposed facility. And at a critical point, Jane Goodall traveled to Olympia, Washington, to address a joint session of the Washington state legislature. All other business came to a halt for twenty minutes while Jane made her emotional appeal on behalf of Washoe's family. The legislature voted to pay more than 90 percent of the cost of Washoe's home, and we raised the rest in individual donations to Friends of Washoe.

Moving into a new home seemed like the perfect time to transact one other piece of business that was long overdue. Although Loulis had lived as Washoe's adopted son for fourteen years, he was still owned by the Yerkes Regional Primate Center. They had given him to us in 1979 on "indefinite loan." If something happened to us, Loulis would most likely be taken from Washoe and returned to Yerkes and biomedical research.

That wasn't our only concern, though. With the boom in AIDS research, labs were scrambling for "clean" chimps who hadn't been infected by HIV, hepatitis, and other infectious diseases. If that wasn't enough reason for Yerkes to want Loulis back, I was not exactly endearing myself to them by my vocal opposition to NIH policies.

Debbi and I decided not to wait for Yerkes to come knocking. With Washoe and her family moving into their new home, it was the perfect time to make Yerkes an offer for Loulis. Buying Loulis would secure his place with Washoe and protect him forever. I called Fred King, the director of Yerkes, who acknowledged Loulis's contribution to signing research and agreed to sell him for ten thousand dollars, the going rate for a "clean" chimp. Over the next several months we raised the money, most of it in five- and ten-dollar contributions from people around the nation who had read about Washoe and Loulis. Just when we were about to send the money to Yerkes, our accountant told us we were $750 short; we had to pay a 7.5 percent sales tax on Loulis, as if we were buying a car. After some more fund-raising, we sent off the $10,750, and Loulis became the legal property of Friends of Washoe, an organization dedicated to his family's well-being.

IN JUNE 1993, I TURNED FIFTY YEARS OLD. Debbi and I celebrated with our children and the chimps. Each of the chimps downed a few doughnuts and a cup of coffee with lots of cream—their favorite drink. Then there was a silent chorus of HAPPY BIRTHDAY ROGER and the chimps helped me blow out the candles on my giant birthday doughnut. That night my friends threw a party. They gave me a gift-wrapped wheelchair and a bottle of Geritol.

I didn't feel *that* old, but I did feel like I was entering a new phase of my life. Erik Erikson, the renowned developmental psychologist, has written that the seventh stage of life ends around the age of fifty, when we have already created a career or family and begin asking, "What am I leaving for the next generation? How will my life benefit those who follow?"

I think that for scientists especially, who are expected to make breakthroughs in their twenties and thirties, there is a moment of reckoning that comes when you realize that the most productive years of research have come and gone. You start wondering how significant a contribution you've really made in the big scheme of things.

For me this moment of reckoning was a great opportunity. For starters, I looked at my own curriculum vitae and decided that, significant or not, it was long enough. I had already been scaling back my research now that I was spending more and more time advocating the well-being of chimpanzees outside Washoe's family. My students were still conducting important studies of the signed conversations in Washoe's family, and their work was published in professional journals. I supervised these studies and happily gave my students first authorship and a chance to build their own careers. There was nothing unusual about this. Aging professors naturally start doting on their academic offspring.

But the things I decided to do outside the lab were a lot less orthodox. Except for my teaching, I decided to move even farther out of the realm of theory and into the realm of action. I was speaking less often to linguists, psychologists, and anthropologists. Debbi and I were speaking more and more, as a team, to conferences of zookeepers, wildlife conservationists, and biomedical researchers. When she and I did publish, the topic was usually the humane treatment of captive chimpanzees or the ethical dilemmas of experimenting on them.

By the time I turned fifty, I knew I wanted to be judged not by what I wrote in scientific journals about chimpanzees but by what I *did* for them. I didn't think of this as particularly noble or righteous—just the necessary and natural culmination of my own rather unusual career in science.

I readily admit that I am emotionally attached to Washoe and her family, but that alone is not what has driven me into the political arena. I feel I have no choice but to take action because of what I've *learned* from Washoe and other chimpanzees. After thirty years of conversing with and observing them, I'm more convinced than ever that the chimpanzee mind and the human mind are fundamentally alike. And that only makes sense, because the chimpanzee brain and the human brain both evolved from the same brain—that of our common ape ancestor. The mental processes inside these two brains have become specialized as they adapted to different social needs over six million years, but they still share the same underlying ancestral intelligence.

I believe that the evolution of the human brain was driven in large part by the tongue. As I mentioned earlier, our ancestors probably began speaking about 200,000 years ago when their increasingly precise gestures and toolmaking led them to make similarly precise movements with their tongues. And, oddly enough, all this tongue wagging shaped the evolution of the human brain.

Before our hominid ancestors started mouthing words, their brains probably functioned like the brains of other ape species, with the right side of the brain controlling the left side of the body and vice versa. Specialized functions like communication were handled globally, not in one hemisphere or the other. One of the enduring mysteries of human evolution is why the human species, *alone among the apes,* developed a dominant hemisphere.

I believe this came about in the following way. Our hominid ancestors, like their ape cousins, communicated with gestures that involved bilateral brain activity. The two hemispheres controlled the two gesturing hands—the right hand was primarily controlled by the left hemisphere and the left hand was primarily controlled by the right hemisphere. But when the human tongue began moving precisely and producing words, the human brain faced a major neurological problem. With two hemispheres competing to control one tongue, the result would be a kind of vocal paralysis—like two drivers fighting over a steering wheel. (Indeed, this kind of hemispheric competition is one of the causes of stuttering in modern humans.) The human brain solved this by assigning control of the tongue's speech movements to one hemisphere. In most people, this is the left hemisphere.

As the left hemisphere developed mechanisms to control sequences of spoken words, naturally it also took control of other fine motor movements, such as toolmaking, which also require a person to follow complex sequences of chopping, cutting, and flaking. This link between speech and toolmaking helps explain why only our species developed a dominant hemisphere that is specialized for both language and the control of a favored hand. Individual chimpanzees can be right-handed or left-handed, but there are just as many lefties as righties. But 90 percent of humans are right-handed; their left hemisphere contains the neural mechanisms that control both speech *and* the favored right hand. (Among left-handers, 80 percent also handle language in the left hemisphere, even though their right hemisphere controls the favored left hand.)

The human brain's powerful bias toward one dominant hemisphere profoundly affected the way human intelligence developed. Our evolving ability to move the tongue and hands in long sequences led humans to begin *thinking* in long chains of thought, in which one idea logically follows another. Charles Darwin referred to this as "complex trains of thought," and he guessed correctly that our ability to think sequentially grew out of our sequential language skills. About five or six thousand years ago, our human ancestors figured out how to *write down* these sequences of thoughts, and that was soon followed by a veritable explosion of logical thinking, including mathematics, astronomy, and engineering.

Of course, as the left hemisphere became more dominant, the right hemisphere began taking on more of a supporting role. The right hemisphere is specialized for tasks that involve *simultaneous* processing of information. We think simultaneously when we read another person's body language instantaneously. For example, when I'm talking to a friend and I unintentionally offend him, I do not think to myself logically, "Okay, he's silent, his face is reddening, he's looking away, and he's moving his hands nervously. I must have hurt his feelings." No, I process *all* of this information simultaneously without any conscious train of thought at all.

We are all familiar with activities that we handle sequentially at first and only later simultaneously. For example, when you learn to drive a car you follow a checklist of things to do: turn on ignition, put foot on brake, let out emergency brake, check rearview mirror, put car in gear, step on gas, and so on.

You must focus consciously and totally on each action in the sequence if you are to link the actions together and drive. But pretty soon, you are driving without any attention to this sequence of actions. Your brain's simultaneous process takes over and you find yourself responding to every bend in the road, every approaching car, and every stoplight without thinking about it at all. In fact, driving becomes so automatic that you can engage in sequential thinking while handling the car. More than once on a Saturday, I have been driving to the store, and soon started daydreaming or talking to Debbi, and then suddenly realize I have driven to the university parking lot because that is where my brain usually takes me on weekdays.

Similarly, when you first learn to hit a baseball, golf ball, or tennis ball, you must focus on an exact sequence of movements, each dependent on the one that came before. But with plenty of practice, the whole process becomes more than the sum of its parts. The ability to hit a ball "dead solid" is a feat of simultaneous processing in which the brain lets go of sequential thinking altogether and reacts intuitively and instantaneously to a vast amount of incoming information. When this process is working effortlessly, without any conscious thought, athletes refer to it as being "in the zone." This mental power to synthesize reality in a single moment and respond without thinking has been prized for centuries and centuries as the secret to great art and spiritual insight.

Countless pop psychology books have described these two mental processes—sequential and simultaneous—as being synonymous with "left brain" and "right brain," but the human brain does not divide up this neatly. In 2 percent of right-handers and nearly 20 percent of left-handers, language and other sequential processing is handled by the right side of the brain and simultaneous processing by the left. A person can even have all of his thought processes controlled by one hemisphere, as I mentioned earlier in the case of the English boy named Alex.

Instead of seeing our mental processes as left brain–right

brain, it is more accurate to envision actual brain matter. There are two kinds of brain matter—gray and white—and they are both found on each side of the brain, though in different proportions. The gray matter is used for sequential processing, and the white matter is used for simultaneous processing. In most people the left hemisphere has proportionally more gray matter than the right. This led to the generalization about left-brain thinking and right-brain thinking. Two percent of right-handers and 20 percent of left-handers defy the left-right rule because these people have proportionally more gray matter in the right side of their brains.

Both types of brain matter and both types of thinking—sequential and simultaneous—are necessary for every human being to survive. We have to be able to plan sequentially so that we can make tools and grow food, among other things. And we have to respond to social cues simultaneously so that we can court and mate. More to the point, we use both sequential *and* simultaneous thinking in almost everything we do, thanks to the bundles of fiber that interconnect the two hemispheres of the brain and enable them to function in concert.

Unfortunately, ever since the time of Plato, philosophers have given simultaneous intelligence second-class status or just ignored it entirely. When René Descartes was faced with both reason and intuition intermingling in the same brain, he simply cut them in two. He defined the human mind by its analytical prowess and said that nonverbal processing belonged to the lower realm of brutish animals. And ever since, most psychologists have dismissed the natural interaction between sequential and simultaneous thinking and have focused instead on linear thinking that is easily measured. (A notable exception to this were the Gestalt psychologists, who studied simultaneous perception.)

The approaches that separate the mind from the body and the verbal from the nonverbal ignore the rich interaction between the simultaneous and sequential that occurs within every human being. Our modern obsession with measurable, sequential intelligence is what leads many scientists to believe that human intelligence is different from chimpanzee intelligence. In fact, both the human and the chimpanzee minds depend on these same two mental processes.

When two chimps court one another through hand gestures they are using the same simultaneous processing as two human lovers who communicate by gazing into one another's eyes across a candlelit dinner. When chimps pick out specific stones to use later as hammers in cracking open nuts, they are using the same sequential processing and planning as humans who take the trash cans out to the curb the night before the garbage truck arrives.

The human brain, thanks to the tongue, has come to rely much more heavily on sequential thought. The evolution of human thinking can be seen most easily in the development of the human infant. The human infant (like the chimpanzee infant) is born with neither brain hemisphere asserting dominance. Children rely heavily on simultaneous processing, especially when learning to interpret their mothers' facial expressions. But once the tongue begins moving precisely and speech kicks in, around age two to four, the infant's rapidly expanding gray matter becomes dominant on one side of the brain, usually on the left. However, preschool children's thinking is still strongly simultaneous, as evidenced by their free-form scribblings, expansive imagination, and intense physical connection to nature. The process of learning to read and write is what pushes the human child's brain even further into sequential and analytic thinking, in the same way that the invention of these activities did in our species about five thousand years ago.

The simultaneous and sequential thought processes are forever inseparable, but individuals have varying amounts of each. One of my colleagues was so strongly sequential that he couldn't think and drive at the same time; he kept crashing and the police finally took his license away. On the other hand, I've known poker players and trial lawyers who, like chimpanzees and children, were absolute geniuses at reading and exploiting body language.

There is no bold line distinguishing chimpanzee intelligence from human intelligence. Each of them blends simultaneous and sequential thinking in distinctive proportions. The fact that these proportions can be altered in each species is the most powerful evidence yet for how closely related the chimpanzee and human minds really are. Our brains are still so similar that their mental processes can be shaped to resemble one another.

That is exactly what happened with Washoe. When we raised infant Washoe among humans we biased her brain away from the simultaneous processing that is so crucial in the jungle and toward the sequential processing needed for human language. The circuitry of each brain—the pathways that connect billions of neurons—is distinctly shaped by the kind of experience and information it is processing in the first, formative years. Washoe's brain and thought processes did not become typically human, but they weren't that of a wild chimpanzee either. Washoe has the distinctive intelligence of a signing chimpanzee. And the reason she was able to develop in this way was because chimps and humans inherited their brains and intelligence from the same apelike ancestor.

Humans and chimpanzees differ in their intelligence by degree, not in the *kind* of mental process. Chimps are not as "intelligent" as humans when it comes to sequential thinking. It is unlikely that a chimpanzee could ever use American Sign Language with the full syntax—the complex sequential patterns—of a deaf, signing human adult. But a chimpanzee *can* learn enough signs to communicate sequentially to a surprising degree. Similarly, we are not as "intelligent" as chimpanzees when it comes to simultaneous thinking. It is unlikely that a human could ever read nonverbal signals to the same extent as a chimpanzee. But a human raised among wild chimpanzees would surely develop impressive simultaneous processing skills.

Saying that human intelligence is superior to chimpanzee intelligence is like saying that our way of walking on two feet is better than their quadrupedal method. Chimpanzees spend their entire lives in small intimate groups. Their simultaneous intelligence is perfectly suited to their social environment. They don't need a heavily sequential language, much less a global electronic network.

There is no question that human sequential thinking has produced many riches, such as sublime literature and breathtaking architecture, but we've given up plenty in return. All we have to do is enter a child's world for ten minutes to see that children have a grasp of reality so immediate, so physical, and so emotional that an entire novel might not do it justice. My own children are grown now, but I can enter this lost world any time I like by visiting Washoe, Loulis, Dar, Moja, and Tatu, who share our ancestors' admirable powers of simultaneous perception.

Heavily sequential intelligence is a new evolutionary experiment, and its adaptive value is still in doubt. Since the invention of agriculture twelve thousand years ago, and then written language five thousand years ago, sequential intelligence has produced a "cascade effect" of one technological innovation after another, at an ever increasing pace. In just the last three or four generations, our disproportionate gray matter has invented many things to benefit humankind while also giving us nuclear weapons, massive pollution, and the steady collapse of the earth's life-support systems. Any serious student of evolution will want to wait at least a half million more years before passing judgment on this experiment called the human mind.

IN THE FALL OF 1993 Debbi and I faced yet another financial crisis. Central Washington University had promised us $210,000 in state funds each year to operate the new Chimpanzee and Human Communication Institute. That money would pay for food, enrichment activities, animal care technicians, as well as our own salaries, as codirectors. But the first year we received only ninety thousand, which was barely enough to operate. And over the next couple of years the university changed its priorities, as universities will do, and our funding dropped to almost nothing.

We had the state-of-the-art chimpanzee facility we'd always wanted, but no money to run it. What would we do with the forty undergraduate students and ten graduate students who were signed up to collect observational data on the chimps? How would we educate the public, as we'd promised, if we didn't have enough money to open our doors?

In this new moment of crisis I came up with the "Chimposium," a new program that would open up Washoe's home to paying visitors and raise the income needed to keep the Institute self-supporting. I wanted these one-hour educational workshops to be run by members of the local community, so we began training teachers, ranchers, farmers, local businesspeople, homemakers, even a police chief. These enthusiastic volunteers became proficient in observing chimpanzee behavior and in communicating with basic American Sign Language.

Every Saturday morning, the one-hour Chimposium introduced members of the public to signing chimpanzees for the price of a ten-dollar ticket. The workshops began with a short film and lecture about the plight of chimpanzee cultures in Africa and the history of Project Washoe. Visitors then learned such chimpanzee etiquette as how to walk low to the ground, cover one's teeth, and sign FRIEND before they were taken to the soundproof observation area outside Washoe's home. There the visitors might see the chimps playing, talking to one another in sign, or signing to themselves while looking through magazines or books.

In the first two months, five hundred people attended the series of Chimposia. By 1995, we had fifty docents welcoming eight thousand visitors from around the world each year, providing the Institute enough funds for a bare-bones operating budget. The chimps have taken all of this in stride. If they're not in the mood for guests, they disappear to their favorite private spots. More often they will greet their visitors first by displaying—stomping, charging, and thumping the glass—as if to remind them, *this is our home*, and then by signing to them. (Chimpanzees in zoos rarely display, because they don't feel dominant in their "home.")

Loulis especially likes schoolchildren and will often pick one out of the crowd and ask him to play CHASE. All of the chimps have commented, at one time or another, on the visitors' body types, clothing, T-shirt pictures, bald heads, beards, Band-Aids, scars, and funny behavior. They are as fascinated by humans as humans are by chimps.

The Chimposium program has fulfilled my longtime dream of enabling others to learn as much from chimpanzees as I have. Thousands of people have now had the experience of looking into the eyes of our evolutionary siblings in an atmosphere of mutual respect and understanding. In Washoe's home human visitors and chimpanzees regularly converse and acknowledge one another as related beings.

Of all the people who visit Washoe's family, deaf children are the first to recognize the chimpanzee as our next of kin. To see a deaf child, who struggles daily to be understood by fellow humans, talking animatedly in sign with a chimpanzee is to recognize the absurdity of the age-old distinction between "thinking human" and "dumb animal." When deaf children look at Washoe, they don't see an animal. They see a person. It is my fondest hope that, one day, every scientist will see as clearly.

IN EARLY 1995 I GOT A CALL from Dean Irwin, a producer of *20/20*, the ABC newsmagazine. While planning a show about the morality of conducting biomedical experiments on chimpanzees, he had learned about Booee and my other former chimpanzee students at LEMSIP, the biomedical lab owned by New York University. He asked me if I would be willing to visit the lab and be reunited with Booee in front of television cameras.

I wanted to say no. I had deliberately avoided LEMSIP since Booee and the others were transferred there in 1982. Seeing them would be agonizing because I knew there wasn't a damn thing I could do to rescue them. As for helping them, I'd tried that already. In 1988, Jane Goodall and I had sent a student of mine, Mark Bodamer, to LEMSIP to start an enrichment and activity program for all 250 chimps. The program was a great success, but, unfortunately, it was scuttled after Mark left.

Mark never got to see Booee—Booee had been transferred to another lab temporarily—but he did visit Bruno for me. When Mark began signing to him, Bruno answered with two signs of his own: KEY OUT. I wasn't sure if Booee would remember me if we saw each other after more than a decade. But if he did remember, he might well think I was there to free him, something I could not do. It would break my heart and his.

But I also knew that visiting Booee would make for some very good television. We could take millions of people inside a biomedical laboratory, which I'd dreamed of doing since I toured Sema seven years earlier. If there was any chance that this wide exposure might improve conditions or help Booee, then I would do it.

A few months later, I found myself in the backseat of a long black limousine, sitting next to anchorman Hugh Downs as we drove to LEMSIP. A soundman and cameraman, seated across from us, taped our conversation. I couldn't help thinking that the back of the limo was bigger than Booee's cage. Hugh Downs wanted to know if Booee would remember me. I didn't know.

I couldn't begin to guess what thirteen years alone in a cage would do to someone's mind and personality. But the closer we got to LEMSIP the more I hoped that Booee would *not* remember me, that he would see me as just another lab-coated visitor passing through. I didn't want to sign GOOD-BYE to Booee. I was sure I would break down.

When we got to the lab, we were instructed to put on white gowns and caps. Then Dr. James Mahoney escorted Hugh Downs, the camera crew, and me to Booee's windowless barrack. Booee lived in a "hot unit," where all the inmates were infected with one virus or another. He was infected with hepatitis C, a

virus that can cause progressive liver disease. Through the door
I could see my friend sitting alone in his cage.

He looks the same, but bigger, I thought.

The last time I saw Booee he was a young teenager like
Loulis. Now he was twenty-seven.

This is really happening. It's too late to turn back.

I hesitated for another moment, then entered the room in a
low crouch. I approached Booee's cage uttering gentle chimpanzee greetings.

A big smile lit up Booee's face. He remembered me,
after all.

HI, BOOEE, I signed. YOU REMEMBER?

BOOEE, BOOEE, ME BOOEE, he signed back, overjoyed that
someone had actually acknowledged him. He kept drawing his
finger down the center of his head in his name sign—the one
I had given him in 1970, three years after NIH researchers had
split his infant brain in two.

YES, YOU BOOEE, YOU BOOEE, I signed back.

GIVE ME FOOD, ROGER, he pleaded.

Booee not only remembered that I always carried raisins for
him, but he used the nickname he had invented for me twenty-
five years earlier. Instead of tugging the ear lobe for ROGER, he
flicked his finger off the ear. This was like calling someone
"Rodg" instead of "Roger." Seeing him sign my old nickname
floored me. I had forgotten it, but Booee hadn't. He remembered
the good old days better than I did.

I gave Booee some raisins, and the years just melted away,
the way they do between old friends. He reached his hand
through the bars and groomed my arm. He was happy again. He
was the same sweet boy I met on that autumn day decades
earlier when Washoe and I first stepped onto the chimpanzee
island at Lemmon's Institute. That was before everything, before
the stun guns and Dobermans, before the adult colony and Sequoyah's death, before Yerkes and Sema. I was a young know-
it-all professor then, right out of graduate school. I yelled at
Booee one day, and he humbled me in front of my very first

college students by lifting me off my feet and letting me dangle there. For twenty-five years I'd been telling my students about how Booee embraced me and forgave my anger toward him.

Look at him now, I thought. *Thirteen years in a hellhole and he's still forgiving, still guileless.* Booee still loved me, in spite of everything that humans had done to him. How many people would be so generous of spirit?

As we signed back and forth and played CHASE and TICKLE through the iron bars, I forgot about the cameras and the millions of people who would be watching this. For one wonderful moment I even forgot where we were. But only for a moment.

I MUST GO NOW, BOOEE, I signed after a while. Booee's grin changed to a grimace, and his body sank. I MUST LEAVE, BOOEE. Booee moved to the back of his cage. GOOD-BYE, BOOEE.

As we left LEMSIP, I shook hands cordially with the director, Dr. Jan Moor-Jankowski, as if we were two colleagues who had just transacted some mundane piece of business. I was overwhelmed by shame. I was ashamed of Booee's hepatitis, ashamed of the professionalism of Moor-Jankowski and myself, ashamed of the respectability that hung over all this suffering.

After our limousine pulled out through the heavily barred security gate, no one spoke for the whole drive back to the hotel.

The *20/20* show was broadcast on May 5, 1995. Its portrayal of Booee, a nonhuman person trapped in biomedical research, affected a nationwide audience more powerfully than I ever could have imagined. For most people it was their very first glimpse into this secretive world, and they were outraged to see a thinking, loving, signing chimpanzee dangling in a cage without companionship or comfort. The fact that Booee was going to spend the rest of his life in that cage—perhaps thirty more years—was unimaginable.

Donations poured into ABC from sympathetic viewers hoping to fund Booee's retirement from research. LEMSIP was besieged once again by a public that demanded amnesty for chimpanzees. In October 1995, five months after the *20/20*

broadcast, LEMSIP gave Booee and eight other adult chimpanzees their freedom. They were shipped by truck to the nonprofit
Wildlife Waystation, in California, where they settled into a
new "retirement home" that has large, airy, and sunlit rooms
with sagebrush views. There are climbing ropes, and enrichment
activities, including music, books, television, magazines, and
toys. Booee still carries the incurable disease that LEMSIP
researchers gave him, but he has no symptoms yet that we
know of.

A few months later, Debbi and I went to see Booee in his
new home. He was so happy to see us. We spent the morning
grooming, playing, and signing. When it was time for us to
leave, Booee wasn't upset. He stood at his enclosure and calmly
signed GOOD-BYE.

FIFTEEN

BACK TO AFRICA

AFTER VISITING BOOEE, Debbi and I came home and spent the afternoon with Washoe's family. As we watched them playing and signing, I was struck by the randomness of their good fortune. Their generation of chimpanzees has been shuttled around America for forty years, with little rhyme or reason for where any one individual ended up.

Washoe was captured in the jungle but ultimately found a loving family. Thelma was also captured in the jungle but she ended up in a biomedical laboratory. Her daughter, Tatu, was born in Oklahoma and has never even seen the inside of a lab, while Bruno, who was also born in Oklahoma, landed in LEMSIP, where he died several years ago. Moja, Loulis, and Dar were all born as "laboratory animals" but today they find themselves living with Washoe in a wonderful home. Ally and Lucy were cross-fostered in loving families, then disappeared forever. And as for Booee, his life has been a microcosm of his whole generation: he was born in a lab, then adopted and cross-fostered, then sent to Oklahoma where he first met chimpanzees, then sold *back* into biomedical research, and finally was freed for the *second* time—all this before he turned thirty, with half his life still to live.

I tried to make sense of all these stories, but I couldn't. There was a clear pattern, however, when it came to happy endings and unhappy endings. Washoe, Loulis, Moja, Dar, Tatu, and

Booee got lucky because Debbi and I happened to have the power, at some point, to help each one of them.

But what about those who didn't get lucky? What obligation do we have to a chimpanzee who has spent his entire life in service to human health but has outlived his usefulness?

Many biomedical laboratories with chimpanzees now find themselves housing a growing number of surplus chimps who are too old, too diseased, or too psychotic for further research. Infection with HIV or hepatitis is usually "the end of the line" for a chimpanzee. Once a chimp like Booee completes vaccine testing protocols he will most likely not be used again. He may be only seven or eight years old. That chimp is facing another forty or fifty years, probably of solitary confinement, in a regulation-size cage as small as a broom closet. What will we do with these survivors?

Consider the fate of America's most famous chimpanzees—the "space chimps"—who were forcibly recruited from Africa by the United States Air Force in the 1950s and 1960s. Ham and Enos, the two chimponauts who paved the way for John Glenn and Alan Shepard, are long dead, and a few fortunate others like Washoe have moved on, but there are still 115 space chimps and their offspring living in cages at the Holloman Air Force Base, in New Mexico, as well as 29 more being held at other labs around the country.

It has been years since the Air Force stopped slinging them around on the end of a centrifuge arm to test the effects of superhigh gravity forces. These chimps have spent the last thirty years being transferred back and forth among various researchers conducting biomedical experiments. In one protocol I was told about, a group of these chimps had their teeth bashed in by a steel ball so that dental students could practice reconstructive surgery on them.

The space chimps are still owned by the Air Force, but since 1994 they have been controlled by the Coulston Foundation under a five-year lease. Dr. Frederick Coulston, the founder and owner of the Coulston Foundation, is a toxicologist who admits that his company's view of chimpanzees in research is rather unusual. While leading biomedical researchers use chimpanzees only for infectious disease or genetic research, Coulston has advertised his chimps for the development of insecticides and cosmetics. He calls chimps "the best possible model to test the fate and effects of foreign chemicals in man."

Coulston runs the White Sands Research Center, where Ally was reportedly sent in 1982. (White Sands was merged into the Coulston Foundation in 1994.) His labs have a record of multiple violations of the Animal Welfare Act and the unintended deaths of several chimpanzees. To cite just one recent and gruesome example: three of Coulston's chimpanzees were cooked to death when a space heater outside their cages became stuck on its highest setting and sent the room temperature soaring to 140 degrees. (This and other episodes led the USDA to file formal charges against the Coulston Foundation.)

In 1995 the Air Force quietly tried to make its arrangement with Dr. Coulston permanent. It asked Congress to transfer legal ownership of the space chimpanzees to the Coulston Foundation, which would have absolved the Air Force of any further ethical or financial responsibility toward the chimps. Our military's chimpanzee problem would have been "solved" once and for all by handing them over to a researcher who recently suggested in *The New York Times* that chimpanzees could be raised like cattle to be used as living blood and organ banks.

Thanks to the determined efforts of a few animal welfare groups and the crucial intervention of Jane Goodall, the proposed transfer of the space chimps to Coulston was brought to public attention and ultimately scrapped by Congress in 1995. Congress has instead ordered the Air Force to open a public bidding process on a contract to take permanent ownership of the chimps. This opens the door for an animal welfare group to bid on the chimps and, if successful, to retire them altogether from research. In the meantime, the space chimps will remain under Dr. Coulston's control at Holloman until his lease with the Air Force runs out in 1999.

The space chimps are not the only ones from Washoe's past who may find themselves sitting in one of Dr. Coulston's labs. After fifteen years inside LEMSIP, Thelma, Cindy, and the rest of my old students from Oklahoma are on the move again. In 1995 New York University, the owner of LEMSIP, announced that it was not willing to invest the funds necessary (seven to eight million dollars) to provide for the long-term care and retirement of its chimpanzees, and it struck a deal to transfer its chimpanzee population to Coulston. Although that deal is not yet finalized, about 100 of the 225 chimps have already been shipped to Coulston Foundation facilities. Except for Booee, who was freed, and Bruno, who died, it is likely that William Lemmon's chimps will wind up where America's chimpanzee saga began forty years ago—in a research facility in New Mexico.

If Coulston takes over LEMSIP, he would control 750 chimpanzees, which accounts for half of the total chimpanzee population in American biomedical research. (Even if he doesn't take over LEMSIP, he'll still control more than 600 chimps.) He is well on his way to becoming, as he puts it, "the sole source of chimpanzees for research." Coulston has mocked the idea of chimpanzee retirement and has called for expanding their use in research. His labs have injected chimpanzees with everything from trichloroethylene, an industrial solvent used in dry cleaning, to benzene, a known carcinogen, so we can only assume that the chimpanzees in Coulston's growing laboratory empire may be facing deprivation and experimentation for a long time to come. That empire is bound to grow as more and more labs look for ways to get out of the chimpanzee business entirely.

A decade ago the National Institutes of Health was complaining that there was such a critical shortage of chimps, they might even need to buy wild chimps. But today the chimp market is glutted, and NIH has scaled back its chimpanzee breeding program. Most laboratory chimps are on birth control. The director of one of the biggest primate labs recently told a reporter,

"If you said I could have one hundred chimps for free, I would say no thanks."

The reason the demand for chimpanzees has crashed is that they turned out to be a lousy model for AIDS research. After spending thirty-two million dollars in direct costs alone on chimpanzee breeding since 1986 and millions more on an untold number of studies, we have learned virtually nothing about AIDS from the chimpanzee. Every major advance in AIDS research—from understanding how the virus causes disease to the development of crucial new drugs (AZT, 3TC, and protease inhibitors) to identifying possible genetic factors that may provide resistance—has come from human studies.

Since 1984 only three or four chimpanzees out of more than one hundred infected have developed AIDS-like symptoms, and their illnesses may have been caused by a mutated virus or an entirely new virus. The fundamental differences between the human and chimpanzee immune systems continue to make any data from chimpanzees "virtually uninterpretable in human terms," according to a recent report from the Medical Research Modernization Committee.

For the most part, the biomedical community now agrees that chimpanzees are not needed in AIDS research. In 1986 I warned the government that breeding chimpanzees and infecting them with HIV might waste millions of AIDS research dollars and create a population of chimpanzee outcasts. To my great sorrow, I was right. Today, promising human studies are still starved for funding while labs are filled with HIV-infected chimpanzees, who will live for decades in expensive isolation, cared for by workers wearing biocontainment space suits. Some grant-hungry labs want to infect these surplus chimps with new viruses like "mad cow disease." Others are looking to Frederick Coulston to bail them out. It is a lot easier to discard chimpanzees than to take on the moral and financial responsibility of caring for them.

In the past, NIH has proposed euthanasia as an economical way of getting rid of surplus chimpanzees. The agency abandoned this option in the face of intense controversy, but some lab directors continue to raise euthanasia as a possibility. More recently, NIH set up a committee to make recommendations, but this committee doesn't include a single advocate for chimpanzees with any depth of experience. Obviously, neither NIH nor the biomedical community can be trusted to look out for the best interests of the chimps.

For this reason, a group of concerned scientists, including myself, are proposing our own solution to Congress: the National Chimpanzee Sanctuary System. We envision a network of refuges where chimpanzees who are no longer needed in research can live out their lives in social groups, with grass under their feet, room to climb, and freedom to play.

A scientific advisory committee is at work right now designing the sanctuaries. Its thirteen members include some of the world's foremost primate experts, among them Jane Goodall, zoologist Vernon Reynolds, and anthropologist Richard Wrangham. Debbi and I were selected to cochair this committee, and we intend to work as hard—if not harder—on these sanctuaries as we did on building Washoe's own home.

Our sanctuary design calls for large enclosures where each chimpanzee will have at least one acre of space, and groups of eleven to twenty chimps will have twenty acres. These outdoor areas will be enclosed by natural barriers and fencing, so chimps can socialize and play. Even those chimps with infectious viruses will be able to socialize; HIV chimps will live with HIV chimps, hepatitis chimps with hepatitis chimps, and so on. Special efforts will be made to rehabilitate those chimpanzees who are suffering the psychological effects of years or decades spent in solitary confinement. Individuals who are unable to form social bonds will receive plenty of enrichment from human caretakers. The sanctuaries will provide their chimpanzees with meals, sleeping quarters, and medical care. Educational programs will enable human visitors to observe the chimps, but only from overhead catwalks that don't compromise the chimps' space, activities, and privacy.

Our goal is to have several hundred chimpanzees in the National Chimpanzee Sanctuary System by the year 2000. Of course, we hope that the 144 Air Force "space chimps" will be among the very first residents of the sanctuary. But Congress must first pass an act that mandates the sanctuary system. Then we must raise money, buy land, and build facilities. The sanctuary system would function as a private, nonprofit organization and initial funding would come from a combination of federal funds (including money already set aside by the government for chimp retirement), private donations, and contributions from biomedical laboratories that will no longer have to care for surplus chimps. The sanctuaries will cost several million dollars to build and millions more in an endowment for operations, but they will still be far cheaper than warehousing chimps in single cages in laboratories.

Some biomedical researchers are already attacking our approach. Their main goal is to ensure a supply of chimpanzees for future biomedical experiments, and they would like to shuttle chimps back and forth between sanctuaries and biomedical labs, as well as breed chimps in the sanctuaries to supply their future needs. We think chimpanzee retirement ought to be final and absolute. The labs can decide which chimps to retire and when to do it, but once those chimps enter a sanctuary, they should be protected forever from experimentation.

Over the past forty years we have spun chimpanzees in centrifuges and shot them into space. We have smashed their skulls with steel pistons and used them as crash test dummies. We have deprived them of all maternal contact and driven them to psychosis. We have used them to test lethal pesticides and cancer-causing industrial solvents. We have injected them with massive doses of polio, hepatitis, yellow fever, malaria, and HIV.

Those chimpanzees who have survived deserve to be left alone in peace and quiet for the final years of their lives.

.   .   .   .

EVEN IF WE SUCCEED IN PROVIDING SANCTUARY for "used-up" chimpanzees we must still address the more fundamental question of whether it is morally acceptable to conduct harmful experiments on chimpanzees at all.

Just forty years ago, when the Air Force was plucking chimpanzees from the rain forests of Africa, scientists told us with certainty that these hairy beasts were devoid of any mental or emotional lives. African tribes who claimed that chimpanzees made tools were dismissed as superstitious. If an anthropologist had dared to utter the words "chimpanzee culture," he would have been laughed out of the academy.

Today we know that most of our scientific knowledge about the chimpanzee, prior to 1960, was nothing more than medieval superstition. Since Jane Goodall first observed the chimpanzees of Gombe fashioning tools, a cascade of further discoveries has shown that chimpanzee communities have their own unique hunter-gatherer cultures, just as communities of pretechnological humans do.

The Swiss ethologist Christophe Boesch has studied the chimpanzee stone tool culture that thrives across West Africa. Some of these chimps' hammers and anvils, used to crack nuts, are identical to the tools of our own hominid ancestors, and their style of toolmaking differs from community to community, also like that of early hominids.

The anthropologist Richard Wrangham has documented the chimpanzee's use of medicinal plants. The Mende people of West Africa have long supplemented their own knowledge of herbal medicine, which they call "leaf," by following and learning from chimpanzees. Now Western medical researchers are doing the same thing. Chimpanzees have led scientists to a variety of formerly unknown plant species that have pharmaceutical uses that range from antibiotics to antiviral agents. Richard Wrangham believes that there may be scores of local chimp medicine cultures scattered across Africa. The cultural gap between humans and chimps has turned out to be as illusory as the cognitive gap between our two species.

Four decades ago, when we began locking chimpanzees away, we didn't know any better, but today we do. We know that chimpanzees are not mindless beasts but highly intelligent and inventive beings who have been transmitting complex cultures for millions of years. They are our evolutionary brothers and sisters. What are the moral implications of this scientific revelation?

It is a recurring fact of human history that we draw moral universes to include those who are like us and to exclude those who are unlike us. We grant certain rights and liberties to those inside our moral sphere, and we feel free to exploit those who stand outside. How do we determine who is an "insider" and who is an "outsider"? Historically, these distinctions have been based on bigotry, superstition, religious doctrine, cultural habit, legal precedent, or scientific "evidence"—and sometimes all of the above.

We like to think of science as the noble pursuit of objective knowledge, always marching forward in service to truth. But scientists embody the prejudices of their time. And scientists are far more dangerous than the average bigots because they can pass off ignorance as knowledge, and their "facts" can be used to erect and support moral boundaries. Unfortunately, as history has shown, when ignorance is married to arrogance the results are usually deadly for those outside a culture's moral universe.

Science has been the handmaiden to morality ever since the time of Aristotle, the philosophic father of Western science. His *Scala Naturae* ranked Greek males as the most perfect beings, followed by elephants, dolphins, and women—in that order. It took another two thousand years to revoke a husband's right to beat his wife. In the meantime, generations of scientists had "proved" that women were witches, demoniacs, or hysterics. Women were joined outside the Western moral order by blacks, Asians, and aboriginal peoples, and the inferiority of all these groups was "proved" by the nineteenth-century European pseudoscience of neuroanatomy. Specious laboratory findings provided a rationale for enslaving Africans, exterminating

Aborigines, and denying legal rights to Asians. This shameful chapter of science was put on full display at the Saint Louis World's Fair of 1904, where Pygmies and other races of "inferior culture and intelligence" were housed with the chimpanzees and monkeys in a kind of cross-species zoo.

With the rise of modern biomedical experimentation, the commerce between science and morality became a two-way street. Science was used to justify excluding certain groups from the moral order, and these outcasts were then thrown back to science for exploitation in the laboratory. African Americans, European Jews, and mentally disabled children were treated as "laboratory animals."

Beginning in 1932, white doctors funded by the U.S. Public Health Service studied the course of syphilis in four hundred African-American men, without offering treatment for forty years. This was the longest involuntary experiment ever performed on humans in medical history. In the 1940s Nazi doctors subjected Jews in concentration camps to gruesome and often lethal medical experiments. In the 1950s, doctors at New York's Willowbrook School injected mentally disabled children with the hepatitis virus.

All of these researchers considered their experiments to be ethical. More to the point, opposing or impeding these experiments was considered immoral. If the life of a single Nazi pilot could be saved by studying the lingering death of Jews in freezing water, then it was unethical in the Nazi moral universe *not* to perform such research. If the life of a single white suffering from syphilis could be saved by experimenting on black cotton-field workers, then it was immoral in the white mind not to proceed.

We look back in horror at these experiments, and today we draw moral boundaries to include all people no matter what their cultural, racial, and cognitive differences are. In some cases these differences are vast. I have worked with children so profoundly mentally disabled that they resemble normal children only in appearance. In some cases they are, by their own parents' observation, less alert and responsive than the family pets. Yet we have finally acknowledged that our moral universe must include these children who so urgently need our love and our legal protection.

Now that medical researchers can no longer experiment on African Americans, Jews, and disabled children, they have turned to chimpanzees, our closest relation outside our moral universe. They didn't do this with any evil intent. They believed that there were mental and emotional barriers separating human from nonhuman.

Those barriers have proven illusory, but they still govern our actions. As a result, we maintain double standards that make a mockery of science: it is illegal to experiment on a brain-dead human being who can think and feel nothing, but it is perfectly legal—indeed, it is morally righteous—to perform that same experiment on a conscious, thinking, feeling chimpanzee. If an experiment will prolong the life of a single human being inside our moral community, then we are justified in inflicting suffering on countless chimpanzees who are outside our moral community.

In short, we are living by a moral code that is based on an arbitrary distinction between insiders and outsiders—in this case, two different species. This fact should trouble any person who believes that moral principles should be applied rationally and universally.

For example, we could expand our moral universe to encompass anyone capable of a certain kind of intelligence, self-awareness, family relations, and the capacity to suffer mental anguish. The fair application of these principles would immediately recognize all the great apes—chimpanzees, gorillas, and orangutans—as belonging in our moral community because they have demonstrated all of these traits.

This is the goal of the Great Ape Project, an international coalition of scientists and philosophers of which I am a member. We believe that apes should be entitled to certain basic rights, such as the right to life, liberty, and freedom from torture. To put it another way, humans should not be allowed to kill, imprison, or inflict pain on apes without pursuing due legal process.

These limited rights would not make Washoe and other apes full-fledged members of our society. We cannot expect apes to obey laws, serve on juries, and vote for president. But we don't expect children and mentally disabled adults to bear these responsibilities either, and yet we still protect them from imprisonment, torture, and wrongful death.

An obstacle to our goal is that many people still believe that human superiority over apes is self-evident, and they reject anything science says to the contrary.

We have souls—or at least "higher" souls. God made us superior—it says so in the Bible. We control the planet—that's proof of our higher nature.

These same arguments have all been used to assert the superiority of men over women, whites over blacks, Europeans over Aborigines. Most people who espouse human superiority don't realize that their view stems from the same ancient idea that produced nineteenth-century racism: that lower life-forms are here to serve higher life-forms. The top rung of this ladder of perfection, white male supremacy, has simply been replaced by human supremacy over all other species.

The Greek and Near Eastern philosophers who invented this ladder of perfection were ignorant of the biological relations between life-forms. They didn't know that the white race is related to the black race, and that all humans are cousins, descended from the same hominid ancestors. They didn't know that chimpanzees even existed, much less that chimpanzees are related to humans through shared ancestors. The belief in racial supremacy and in human supremacy spring from the same ancient illusion that nature is a collection of unrelated life-forms.

I am well acquainted, from my own family's history, with the terrible price paid for the illusion of racial supremacy. My great-grandfather, William Henry Harrison Jones, was a slaveholder. Although I never met my great-grandfather, I feel a special debt to him. He raised my mother from infancy after her own mother died in childbirth. He was, by her loving account, a decent, honorable, hardworking, and compassionate man. He just happened to believe that blacks were "subhuman." As a result he felt entitled to own them and to ignore their pain and suffering. My great-grandfather's economic interest in that suffering only reinforced his belief that blacks were inferior. And in defense of that self-interest he was willing to die, and he nearly did, fighting in the Civil War.

The fact of evolution that nineteenth-century white supremacists like my great-grandfather could never accept was that every human on earth is related to every other human; we are all family. And the evolutionary fact we still do not accept today is that every human on earth is related to every chimpanzee, gorilla, and orangutan on earth; we are all part of the same hominid family. You and I are connected to Washoe by an unbroken chain of mothers and daughters, fathers and sons, that stretches back six million years.

Why, then, do we think that human suffering outweighs chimpanzee suffering? Why is human life more valuable than chimpanzee life? At best, we experiment on chimpanzees not out of moral principle but out of naked self-interest. Interestingly, more and more biomedical researchers are acknowledging this. It is common now for researchers to justify their experiments by saying, "We may not have the right to experiment on chimpanzees, but we do have the need."

Now self-interest may not be as lofty as morality, but it's a survival impulse we all share. We look out for ourselves and our families first. When researchers pose the rhetorical question *Would you experiment on a chimpanzee if it would save the life of your child?* most parents immediately answer yes. We are encouraged to choose between them or us, and so we naturally choose us.

But as a society we routinely and rightfully set moral limits on this kind of blatant self-interest. For example, let's say my daughter has heart disease and can be saved only by a transplant of my neighbor's heart. I have a damned good need to take out my neighbor's heart, and if the choice is between my neighbor and my daughter, I will choose my daughter every time. After all, my daughter is genetically closer to me than my neighbor is, and she means more to me than my neighbor does.

But society won't let me take my neighbor's heart because beneath our differences my neighbor and I are very similar. He may not be part of my immediate family, but we are related through a common ancestor. We are cousins. Whatever genetic line I draw between us is arbitrary, and I must thwart my natural inclination to save my child by killing my neighbor. Of course this is not very difficult to accept because I *know* my neighbor and feel empathy for him. Therefore even if the law allowed me to kill him, my conscience would prevent me.

From an evolutionary point of view, taking a chimpanzee's heart is like walking next door and ripping out my neighbor's heart. The chimpanzee is not as close to me as my daughter is, but we are related through a common ancestor. Like my neighbor, the chimpanzee is my cousin. If morality stops me from killing my human cousin, then it must also stop me from killing my chimpanzee cousin. Why is it so hard for us to accept the logic of *this* moral prohibition? I believe it's because most people do not know chimpanzees in the way they know their neighbors. It is easier to objectify the chimp as different and unworthy of compassion. It is easier to see chimps as "them."

But why stop at chimpanzees, gorillas, and orangutans? We have common ancestors with dogs, too. Will we extend rights to them? And what about mice? Where will it all stop? I don't know, but we can't bolt the door to our moral universe out of fear that other "less desirable" groups may later gain entrance. Time marches on, and our moral sphere grows ever larger, and rarely smaller. This is a *good* thing. Otherwise, we would still live under legal systems where white people alone had rights and African Americans, Jews, and the mentally disabled were fodder for biomedical research, as was the case only fifty years ago.

372 / THE SEARCH FOR SANCTUARY

If the biological sciences have taught us one thing over the last one hundred years, it is that drawing all-or-nothing lines between species is completely futile. Nature is a great continuum. With every passing year we discover more evidence to support Darwin's revolutionary hypothesis that the cognitive and emotional lives of animals differ only by degree, from the fishes to the birds to monkeys to humans.

I personally believe it is senseless to draw moral boundaries where scientific ones do not exist. It makes no sense to elevate chimpanzees to the topmost, supernatural rung of the ladder of perfection, alongside humans, but to exclude baboons, dolphins, and elephants, who are all highly intelligent, social, and emotional creatures. In an ideal world I would get rid of this ancient ladder altogether. But our current system of law and morality is based on an imagined gap between humans and nonhumans. The great apes are the likeliest candidates to bridge that gap. And once they have done so, we humans may be more inclined to give up our godlike throne above nature and assume our rightful place as part of the natural world.

In the long run we should conduct our morality the same way we conduct our biological science, by assuming the continuity of all forms of life. For this reason I believe our ultimate goal should be the removal of all animals from research. The animal we experiment on today will surely be inside our moral universe tomorrow. Why not work toward that inevitable day instead of delaying it?

Compassion does not and should not stop at the imagined barriers between species. Something is wrong with a system that exempts people from anticruelty laws just because those people happen to be wearing white lab coats. Science that dissociates itself from the pain of others soon becomes monstrous. Good science must be conducted with the head *and* the heart. Biomedical doctors have strayed too far from the guiding principle of the Hippocratic oath, "First, do no harm." Hippocrates was not referring only to humans. "The soul is the same in all living creatures," he said, "although the body of each is different."

Charles Darwin understood this. He believed that our moral sense springs from our "social instinct" to care about others— an instinct we share with other animal species. At first we care only about those closest to us. But with time we show concern for more and more of our fellow creatures. Our moral progress will not be complete, Darwin said, until we extend our compassion to people of all races, then to "the imbecile, the maimed, and other useless members of society," and finally to the members of all species. It is not surprising that the two practices of his time that Darwin detested more than all others were the enslavement of African Americans and cruelty to animals. He worked avidly to abolish both.

Darwin was a man of science and of religion, but when it came to a reverence for all forms of life he saw no conflict in his affiliations. Neither do I. The reverence for life that follows from recognizing the evolutionary continuity of all species is no different from the reverence for the unity of God's Creation as taught by all of the major Eastern and Western religious traditions. They evoke the same sense of awe and the same feelings of compassion. Like my great-grandfather, I am a churchgoing Christian. But were he alive today, my great-grandfather and I would differ drastically on one essential point. He drew boundary lines through Creation, including those beings he identified with, excluding those he didn't. I have tried my best to stop drawing such lines. I did not learn this lesson of humility from my pastors or from my professors. I learned it from Washoe.

WHEN ALLEN AND TRIXIE GARDNER brought Washoe home from Holloman Air Force Base in 1966, they named her for the county in Nevada where she would live. At the time they had no idea what *Washoe* meant. Years later we discovered that in the language of the Washoe Indians, the first inhabitants of northern Nevada, *Washoe* means "people."

Thanks to Washoe I have gotten to know many chimpanzee

people. I have watched Washoe grow from a feisty, rabble-rousing two-year-old into a strong-willed and loving matriarch. And through her I've met other chimpanzee people. Proud Bruno. Sweet-natured Booee. Sensitive Lucy. Antic Ally. Pragmatic Tatu. Easygoing Dar. Insecure Moja. And of course Loulis, Washoe's little prince. All of these people share a full range of emotions, from joy to sadness, fear, anger, compassion, love, and remorse. But like humans, they differ dramatically in how they outwardly express their emotional and mental lives. For example, Moja and Washoe both paint, but I would never confuse their paintings.

I have spent a lifetime exploring chimpanzee individualism, but it turns out that I was only half right about Washoe's name. In 1996 I returned to Africa, where I discovered that chimpanzees are not only individual people, they are *a* people, which *The American Heritage Dictionary* defines as "a body of persons sharing a common religion, culture, or language." Of course, I knew intellectually that chimpanzees are a people, but actually seeing a chimpanzee culture firsthand was one of the peak experiences of my life.

For years Debbi and I had talked about visiting Africa with our kids, and whenever Jane Goodall came to Ellensburg she would urge us to visit her at Gombe Stream in Tanzania. But we kept putting if off. Our kids were too young, or the chimps needed us, or we just didn't have the money. Then, in 1996, we finally realized that if we kept waiting for an opportune time we'd never do it, so we took Jane up on her most recent offer.

Our son, Josh, who majored in film production, was thrilled at the chance to shoot some footage of wild chimpanzees. And our daughter Hillary, who is going to graduate school in anthropology, was excited about handling the still photography and getting a firsthand look at the places she'd only read about in books. Unfortunately, our daughter Rachel was unable to come with us because she was teaching.

Debbi and I had two main scientific goals in mind for our trip. We wanted to videotape chimpanzees communicating gesturally in the wild and begin comparing those gestures to the communication in Washoe's family. And we also wanted to visit some African chimpanzee sanctuaries to help us in designing similar sanctuaries in America.

The anticipation I felt on the way to Africa was almost overwhelming. After thirty years working with chimpanzees in captivity, I couldn't believe that I was finally going to see them where they really belong, living freely in the African jungle. On the morning of June 4, 1996, the four of us crawled out of bed at the Gombe Stream research station. Led by a tracker and a "behavior recorder," we searched for the sleeping nests of Freud, the dominant male of the Gombe chimpanzee community, and his friend Gimbel. We followed the two male chimps to a clearing where they were soon joined by Fifi, Freud's mother and the community's highest-ranking female. Fifi was accompanied by her two youngest sons, Faustino, five years old, and Ferdinand, two years old. As the boys played in the trees, the three adults sat down below for a leisurely grooming session.

This tranquil family scene looked like something out of a *National Geographic* video. *What's so hard about observing chimpanzees in the wild?* I thought to myself as I snapped another picture on my camera. Then, as if on cue, all five chimps suddenly got up and headed off into the jungle.

Over the next several hours, as we tracked Fifi's family, I experienced the most grueling physical challenge of my entire life. We climbed straight up rocky hills, holding on to vines for dear life, crawled on our stomachs under impassable brush, and hacked through a dense sea of thorns with our bare hands. We slid, stumbled, fell, and cursed. All of us were bleeding from the arms, legs, and head. Debbi was bleeding from the sternum, as well, where she had gouged herself on a sharp rock that broke her fall as she slid down a ravine.

There were moments when a precarious toehold or loose vine seemed like a matter of life and death, and my intense desire to survive blotted out any interest at all in the chimpanzees. I was going on adrenaline now, and all of my sensory neurons were firing in unison for perhaps the first time in my life. I began to smell warthogs before I saw any trace of them. I could hear the call of chimpanzees greeting one another in the jungle a mile ahead. I could taste the sweet-salty mixture of blood and sweat as it streamed into my mouth from a cut on my face. My mind was processing all of this simultaneously, just as our own ancestors must have. Some long-buried awareness emerged from the most primitive part of my brain, controlling my consciousness. *Do not stop and focus on one sight, smell, or sound,* this voice said, *or you will lose the group and lose your way.*

The chimps, who moved swiftly and surefootedly through the jungle on all fours, could have easily lost us in a second. But they didn't seem to be in any hurry. When we managed to catch up to them, they paid us no more attention than they did the other jungle pests, such as baboons and insects. We were on their turf now, and all our sequential intelligence was absolutely useless. Some scientists love to measure an animal's mind by comparing it to the human I.Q. In these tests chimpanzees come off like mentally disabled children or adults. But when *we* are dropped in the jungle, we suddenly test like mentally disabled chimpanzees, and the chimpanzees look like certified geniuses.

Moving through the jungle we have no time to string together two thoughts in linear fashion. Survival comes down to simultaneous processing, in which we must maintain moment-to-moment awareness of the whereabouts of our family and social group, any potential predators, and the direction to fruiting trees. I could easily see that even little Ferdinand and Faustino were beautifully adapted to thinking and surviving perfectly in the jungle. I was not.

After three hours of trudging we reached the top of a hill. As we staggered into a clearing, there were no fewer than twenty-two chimpanzees, virtually the entire Gombe community, all assembled in this one sunny spot. While most of the adults groomed and socialized, a few females in estrus courted males and copulated. Nearby, the infants and juveniles romped in the trees, swinging from vines and play-fighting with each other. One adult male chimp was acting as a kind of playground supervisor, disciplining any of the children who got too rowdy or dropped on top of grooming adults. This was a wonderful opportunity to observe gestural communication, and we did see quite a bit of it, especially solicitations for grooming. (We almost certainly missed a lot of subtle gestures, so this coming winter Debbi and I will be returning to Africa for several months to study wild chimpanzee communication in greater depth.)

This socializing went on for a few hours until the group broke up, and we were again following Fifi and her children. At one point Fifi disappeared into the jungle but came back a few minutes later with a long branch. I wondered what she might use it for. She walked to an open grassy area, stripped her branch of leaves, and began dipping it into an anthill. When she pulled her stick out of the hill it was covered with a writhing mass of red driver ants. In one smooth motion she then sucked the ants off the stick, swiftly chewing and swallowing them before they could bite her mouth. Her son Ferdinand sat and carefully watched what his mother was doing.

Suddenly we heard monkeys screaming, and we raced to another clearing where a group of red colobus monkeys were shrieking from the treetops at a chimpanzee hunting party. It was too late. Using our binoculars we could see that Beethoven, an adult male chimp, had captured a monkey. Now all the other chimps were racing up the tree to get a share of the kill. They gathered in the branches about fifty feet up the tree and tore the monkey apart. Two-year-old Ferdinand threw a tantrum to get his share. He shrieked and moved to hurl himself out of the tree, catching a branch at the very last moment before falling. His performance was so effective that Fifi gave him a piece of meat, and he calmed down. Our cameraman, Josh, caught this rare scene on videotape.

We continued to follow Fifi's family through the jungle, stopping only when they paused to eat some small orange berries along the trail. When they came to another clearing, Fifi met and socialized with another female, Patti, while Patti's children,

Tanga and Titan, played with Ferdinand and Faustino. At one point the play grew too rough and Faustino began screaming. Immediately, Fifi rushed over to comfort her son.

For two decades I'd been reading Jane Goodall's descriptions of Fifi's mothering, and there was something surreal about watching this chimpanzee legend in the flesh, parenting her two boys. It was Fifi's close relationship, thirty years ago, with her own mother, Flo, that first showed Jane the true nature of chimpanzee mothering and family ties. There is no mistaking the impact Fifi has had on her own people. She is the most influential and respected female of the Gombe community.

As I watched her comfort her youngest son it was hard not to think about Washoe. Like Fifi, Washoe was a natural-born matriarch. What kind of family might Washoe have raised here in the jungle? What mark would she have left on her people? We will never know. When the animal poachers kidnapped infant Washoe for the United States space program in 1965, no doubt killing her mother to do so, they took away more than Washoe's future. They robbed her people of a loving mother, a nurturing sister, and an indomitable spirit.

I suddenly felt a terrible sadness as I thought about Washoe's life back in Ellensburg. She finally had an outdoor space to run and climb in, but how puny that space seemed after a day in the jungle. This rain forest was where she truly belonged. And as we packed our bags to leave Gombe, I couldn't help wondering again if there was a way to bring Washoe and her family back to Africa, after all.

The next day we flew to Kenya and visited Sweetwaters, a game preserve with a large sanctuary for wild-born chimpanzees. These chimps—about twenty adults and twenty young ones—had been rescued from poachers before they could be shipped overseas to be used as pets, circus animals, or biomedical subjects.

Debbi and I wanted to see Sweetwaters because it is a good model for the National Chimpanzee Sanctuary System we want to build in America. The chimps at Sweetwaters live on more than one hundred acres of savanna, all of it surrounded by an electric fence that keeps the residents in and predators and poachers out. During the day, the chimps play, socialize, and travel in their territory. They are given fresh fruit by human caretakers, and at night they move inside and sleep in hammocks.

If there is any hope of reintroducing chimpanzees to the wild then it rests with places like Sweetwaters and a handful of African sanctuaries run by the Jane Goodall Institute. These young wild-born chimpanzees may not be living in their natural community, but many of them still carry the survival skills of foraging, hunting, and toolmaking that remain from their own culture. They, in turn, can pass on this culture to even younger chimps rescued from poachers. Ideally, these African sanctuaries might function as halfway houses until the chimps could be moved to a piece of rain forest that is protected and patrolled by rangers. These wild reserves would have to be managed by local people, so that they will have an economic stake in the habitat's protection.

Unfortunately, even wild-born chimps at places like Sweetwaters may never be able to return to the rain forest if the human population in Africa keeps growing at its current explosive pace, creating an ever increasing demand for farmland and timber. Humans and chimps are now competing for the same virgin forest, and the humans are winning. But in the long run, the interests of our two species overlap; Africans will not escape their cycle of crushing poverty and chimpanzees will not survive at all unless local governments can control population growth and foster sustainable development.

The obstacles facing the wild-born chimps at Sweetwaters made it clear to me just how futile the chimpanzee breeding programs in American and European zoos are. In the name of species conservation, more and more zoos are breeding endangered animals, creating what they like to call an "ark" of surviving animals that can be reintroduced to the wild if and when a species goes extinct. But wild chimpanzees will never be replenished from a zoo. Chimps that are bred in zoos, or elsewhere in captivity, have lost all of their cultural traditions and survival skills. They and their offspring will be marooned on the zoo ark forever. Zoos should instead focus their efforts on preserving African habitat and supporting the wild-born chimps who have a prayer of surviving in the rain forest.

By the time we finished traveling through East Africa, I was pretty much over my romantic notion of returning Washoe's family to the wild. As alluring as the jungle looked, I had to consider Washoe's individual needs and history. Though she might flourish in a sanctuary like Sweetwaters (as long as she had her toys and magazines along), I knew that I could never return Washoe to the kind of rain forest where she was born. I could never give her back the freedom that was taken from her thirty years ago.

WHEN DEBBI AND I RETURNED FROM AFRICA, Washoe's family greeted us with heartwarming screams and pant-hooting—a kind of chimpanzee cheering squad. But it was a bittersweet reunion. I couldn't stop thinking about Gombe and the way Fifi and her children carried themselves with such nobility and grace. For thirty years I'd known that Washoe had been born in the wild. Now I'd seen exactly how much she'd lost.

Our reunion was significant in another way. In the past, whenever Debbi and I had left the chimps for days or weeks, they would let us know they were mad at us as soon as we returned. Washoe, especially, used to mope around during my absence, and when I got back she would give me the cold shoulder, except to sign comments like DIRTY ROGER! But this time Washoe and the others didn't seem to have missed us much at all.

In fact, they seemed downright happy and content. They had a big outdoor home that they loved. They had the freedom to get up every morning and go outside, even in winter, and they did. Our decision to stay out of their new house—to withdraw

from them physically—had paid off. They were emotionally dependent upon each other, not on us. They still needed us for
food and many other things, but they had many familiar and
loving human caregivers. Best of all, a former student of mine,
Dr. Mary Lee Jensvold, has been around the chimps since 1986
and she runs operations flawlessly in our absence.

It took a long trip to Africa for Debbi and me to realize this,
but in some very real way the chimps, like our children, had
finally grown up. It is very easy to fall into the trap of treating
captive apes as if they are perpetual children. So for years we
had gone out of our way to respect Washoe, Loulis, Moja, Dar,
and Tatu as the full-grown, capable adults they really are. And
now it was with a mixture of sadness and relief that we finally
accepted their emotional independence from us.

At the same time, we felt great admiration for our three
grown children. They have always been our number one priority,
but when we started bringing chimpanzees into the family thirty
years ago, it meant that our kids had to share parental love with
all the funny-looking newcomers. It says a lot about Josh, Rachel, and Hillary that they have always viewed the chimps as
brothers and sisters with special needs, not as sibling rivals. It
strikes them as odd when people ask what it's been like to grow
up with Washoe; they've never known anything else. Coming
home to celebrate holidays with the chimps is just a natural part
of their lives. Their feelings and respect for nonhuman animals
were shaped early and profoundly by their experience on the
job with us. We've had our share of ups and downs just like any
family, but we've taken this journey together, and from what
our children tell us, it has greatly enriched their lives.

Now that the chimps are grown and more independent, my
fondest goal is to make our Institute eventually obsolete. I believe that any research, including my own, that depends on captivity ought to be phased out. To realize that goal we will not
allow breeding within Washoe's family. So far the chimps have
shown occasional sexual curiosity about one another but they've
never come close to mating. Having grown up together they are

probably observing an incest taboo, as wild chimps do. But if two members of Washoe's family do begin mating, we will immediately administer birth control. There is no happy scenario for their children in our society.

A sentimental part of me wishes Washoe, Moja, and Tatu could have the opportunity to bear and mother children. It is unspeakably tragic that we must limit births among the captive members of an endangered species when its native population is crashing in Africa. But chimpanzees like Washoe are not prepared to live self-reliantly in any North American or African ecosystem. Without the necessary cultural traditions and survival skills they will always live in a kind of halfway house at best, dependent upon humans for food, protection, and medical care. That is not a humane solution to the plight of Washoe's people.

Debbi and I dream of moving Washoe's family to an even better, semi-independent chimpanzee environment, which might be a protected forest habitat in Hawaii or some other tropical climate where they could roam free, pick fruit, and socialize with friends. We would go with Washoe's family to their new home, temporarily or permanently—whatever is best for them. I would then happily make my living teaching and working with children who have problems communicating, the people who first inspired me to become a psychologist thirty years ago.

DURING THE SAME WEEK Debbi and I came back from Africa, on June 21, 1996, we celebrated Washoe's thirtieth birthday and the thirtieth anniversary of Project Washoe, the longest ongoing study of ape language. Washoe was really born in 1965, but we don't know her actual birthday so we celebrate it on the day she arrived at the Gardners home in 1966. We had been planning to hold a large scientific conference on Washoe's thirtieth birthday, but we decided that a small party for family and friends was more appropriate.

On the morning of her birthday Washoe went outside and found twelve long-stemmed roses—her favorite flowers. She grew very excited. SMELL GOOD, she signed to no one in particular as she held up the roses to her nose. Then, cradling them carefully, she carried them up to a high cargo net, where she ate them with great delicacy, petal by petal, signing GOOD FOOD after every few bites.

Twenty-four bananas, wrapped in yellow cellophane, were also delivered by the local florist, sent by some of Washoe's human friends. These treats sent the chimps into gales of delighted shrieks and pant-hoots. Then we had a rousing, signed chorus of HAPPY BIRTHDAY TO WASHOE. The birthday girl hugged her family members, and then they all rushed around looking for the gifts and treat bags we had hidden in the tall grass. There was a big box from Allen Gardner and his students containing kites, hats, stuffed toys, tasty treats, and shoes of every description. (Trixie Gardner, the gentle and compassionate heart of Project Washoe, had died suddenly one year earlier while on a European lecture tour.) There were green coconuts, crayons, dress-up clothes, and a football from other friends around the country.

Washoe was in heaven. She lugged around a plastic bucket for hours, picking up her treasures. Moja took some dress-up clothes, a few books, and a treat bag high up to a platform where she spent the morning trying on clothes and leafing through the pictures. After chewing all her gum and eating all her candy, Tatu brushed her teeth with a brand-new toothbrush and some toothpaste she found in her treat bag. Loulis rifled through both his own treat bag as well as Dar's. Ever the patient brother, Dar waited until Loulis was finished and then took the bag up to his favorite spot, thirty feet above the ground. An hour later he put on a swimming mask and chased Loulis around in circles.

For me, Washoe's thirtieth birthday was a time for reflection, a rare moment in my life with chimpanzees. The last thirty years have been full of great joy, intense anguish, and amazing discovery. But never boredom. Like a canoeist running the whitest rapids, I never get to look backward. But on Washoe's birthday I did.

Of the hundreds of moments that came flooding back to me, from Washoe's first signs to her trials with motherhood, there was one that stood out. It was that terrible morning in 1970 when five-year-old Washoe woke up inside the chimpanzee colony at the Institute for Primate Studies in Oklahoma. For the first time since her infancy she was face to face with other chimpanzees, and she referred to them, disdainfully, as BLACK BUGS. From that moment forward Washoe could have held on to her "human superiority" and ignored or mistreated the other chimps. After all, they were strange looking, ill-mannered, and didn't even know sign language. But Washoe let go of her cultural arrogance, and she seemed to care deeply for her long-lost kin. She mothered the young, defended the weak, and saved the life of a newcomer.

I have often wondered what it would be like to wake up one day, as Washoe did, and discover that you are not the superior being you thought you were. How would my great-grandfather have reacted, for example, if he had learned that he was part black? Would he have acknowledged his true self and embraced his newfound kin—his own slaves? Or would he have oppressed them all the more out of self-hatred and fear he might be found out? What would you or I do if confronted by such a dilemma? *That will never happen to me*, you are probably saying.

But it already has. When Charles Darwin told us that we are related to apes we all woke up to a terrible nightmare: *They are us*. And in the hundred years since, we have wiped out millions of chimpanzees in a fury born of denial, arrogance, and self-interest. This fratricide is almost complete. And if we do not halt it now, then we will wake up one day soon only to discover that we've destroyed the living link to our own evolutionary past. Washoe has been a constant reminder to me that humans do not travel upon this earth alone. For the past six million years we have been accompanied on our journey by the biological and spiritual kin we call chimpanzees.

If Washoe and I can help in some small way to save the chimpanzee from extinction, then all that we have been through together will have served a purpose. The ancestral chain of mothers and daughters, fathers and sons that links every human being to every chimpanzee being will remain intact. My own grandchildren will come face-to-face with their own chimpanzee cousins, the grandchildren of Washoe's sisters in the jungle. They will reach across that six-million-year-old divide, not to enslave or destroy their sibling rivals, but to embrace their next of kin.

AUTHOR'S NOTE

WASHOE'S FAMILY IS SUPPORTED through private contributions from people around the world. If you would like to join the nonprofit Friends of Washoe organization and receive regular updates about the family, please send a donation to:

Friends of Washoe
Chimpanzee and Human Communication Institute
Central Washington University
400 East Eighth Avenue
Ellensburg, WA 98926-7573

You can visit us on the Internet at:
www.cwu.edu/~cwuchci/

Nearly two thousand chimpanzees in captivity continue to suffer, without any legal protection, because our judicial system views them as things—"inanimate property"—instead of as the thinking, feeling individuals they truly are. The most effective, enforceable, and lasting way to protect our next of kin is to change their legal status. Toward that end, the Great Ape Project, on whose board Debbi and I serve, has joined forces with the Animal Legal Defense Fund to create the Great Ape Legal Project. Our goal is to win legal rights for all nonhuman great apes, including the right to life, liberty, and freedom from cruel treatment. At this writing we are in the process of bringing a series of court cases that, if successful, will compel the judicial system to recognize our fellow apes as "beings," entitling them to the legal protections they deserve.

If you would like to be a part of this humane and historic effort, please send a donation to:

Great Ape Legal Project
40 Fourth Street, Suite 256
Petaluma, CA 94952

Hundreds of "surplus" chimpanzees, who have outlived their usefulness to biomedical research, are now facing a lifetime of confinement in small laboratory cages with little, if any, companionship or comfort. Recently, the Institute for Captive Chimpanzee Care was founded in order to realize our dream of building a sanctuary that will offer humane lifetime care for these chimpanzee survivors, as well as for the 144 Air Force space chimps. The sanctuary will provide spacious outdoor enclosures, an enriching environment, social housing, and special rehabilitation for those chimps suffering the ill effects of years spent in solitary confinement.

If you would like to play a role in building this national chimpanzee sanctuary, please send a donation to:

Institute for Captive Chimpanzee Care
Sanctuary Program
P.O. Box 3746
Boynton Beach, Florida 33424

ACKNOWLEDGMENTS

IT IS THE CONVENTION TO THANK FAMILIES LAST, but *Next of Kin*, like Project Washoe, was a family affair, so we begin with our life partners. Debbi Fouts was truly the third author of this book. She has lived the story with devotion and determination, and her memories, wisdom, and warmth are present on every page of the retelling. The life and deeds written about in these pages would never have transpired without her boundless love and encouragement. Susan Emmet Reid helped plant the seed for this book, recognized its potential from the start, and helped it to grow day by day, page by page, through her nurturing spirit, limitless curiosity, and keen insight. Both of us thank both of you from the bottom of our hearts.

Next, our children. Joshua, Rachel, and Hillary Fouts cannot be thanked enough for being such uncomplaining, supportive, and wonderful companions throughout all the years of this adventure. They have been indispensable to this book as well by generously sharing some of their earliest childhood memories. Sky Reid-Mills was a never-ending inspiration and a veritable walking encyclopedia of primate behavior, ages three to four. His imprint is here in story after story.

We are uniquely indebted to Joshua Horwitz of Living Planet Press for envisioning this book, bringing the authors together, then playing midwife through a grueling eighteen-month labor. We couldn't have done it without you. We thank Steve Ann

Chambers of the Animal Legal Defense Fund, who also helped pave the way for our collaboration, and our agent, Gail Ross, who found the best possible home for our book.

Our editor at Morrow, Henry Ferris, deserves enormous credit for keeping Next of Kin on course, both dramatically and scientifically. His hundreds of penciled notes and thoughtful E-mail messages proved indispensable to the book's ultimate clarity and readability.

In the course of researching Next of Kin we were assisted by many people who shared their stories, materials, and expertise. We would particularly like to acknowledge the help of Bob Fouts, Don Fouts, Mark Bodamer, Valerie Stanley, Raymond Corbey, Michael Aisner, George Kimball, Anne Flynn, Shawna Grant, Christiane Bonin, and Eric Kleiman of In Defense of Animals.

We want to thank several people who read and commented so helpfully on the manuscript in its various stages: Richard Johnson, Geraldine Brooks, Helen Saxenian, Tania Rose, Barbara Newell, Jeffrey Moussaieff Masson, Jeffrey Norman, and Brian Carrico. Special thanks to Linda Lopez, Sydell Tukel, and Kenneth Tukel, not only for their feedback but for their unwavering moral support.

We are very indebted to three scientists whose critiques made this a much better book: Drs. Agustin Fuentes, Lisa Weyandt, and Mary Lee Jensvold. Any errors that remain are, of course, our own.

The work of Project Washoe and the Chimpanzee and Human Communication Institute has been made possible through the dedication of thousands of caring people. We wish to acknowledge the hundreds of student volunteers who, over the past thirty years, have given freely of their time to better understand and care for Washoe's family. Many of them, too numerous to name here, have contributed personal anecdotes and observations to the Friends of Washoe Newsletter that ultimately found their way into this book. We are equally grateful to

CHCI's volunteer docents who help to educate the public on a daily basis, raising the consciousness of our species as well as funds to support the chimps.

We extend heartfelt thanks to the citizens of Norman, Oklahoma, and Ellensburg, Washington—from the members of La Leche League to the produce department employees at Albertsons—for donating the resources that kept Washoe's family well fed and safe during the hardest of times. To the Friends of Washoe who have so generously supported her family with contributions over the years, we salute you. We also want to acknowledge Washoe's academic supporters around the world who have defended this research even when it was not the popular thing to do. And to all the children who have visited Washoe and accepted her personhood without hesitation, our prayers are with you and your efforts for a future that includes chimpanzees.

A special debt of gratitude is owed to Allen and Beatrix Gardner for demonstrating that sound science can be conducted in a compassionate and caring fashion. And to Jane Goodall, who imagined this book a long time ago, we extend our heartfelt thanks—not only for your gentle prodding, but for helping Washoe's family at so many crucial junctures over the years and for all that you have done for chimpanzees everywhere.

We cherish the memory of two of our kin, Ed Fouts and Milton Mills, who died during the writing of this book. They may never read it, but they are surely a part of it.

Finally, most happily, we pay tribute—with thank-yous, hugs, and pant-hoots—to the five people who inspired this book: Washoe, Loulis, Moja, Tatu, and Dar. We have tried our very best to tell your story in the beautiful spirit with which you have lived it.

NOTES

Page   **PART ONE**

1. *"I am inclined to conclude from the various evidences that the great apes have plenty to talk about"*: See Robert M. Yerkes, Almost Human, The Century Company, 1925.

TWO: BABY IN THE FAMILY

19. *Washoe's body was designed for this dual-purpose lifestyle*: See J. R. and P. H. Napier, The Natural History of the Primates, MIT Press, 1994; John G. Fleagle, "Primate Locomotion and Posture," and Matt Cartmill, "Nonhuman Primates," in The Cambridge Encyclopedia of Human Evolution, Cambridge University Press, 1994.

21. *This approach was called cross-fostering*: See R. Allen Gardner and Beatrix T. Gardner, "A Cross-Fostering Laboratory," in Teaching Sign Language to Chimpanzees, eds. R. Allen Gardner, Beatrix T. Gardner, and Thomas E. Van Cantfort, State University of New York Press, 1989.

23. *"It is not unreasonable to suppose, if an organism of this kind is kept in a cage"*: See W. N. and L. A. Kellogg, The Ape and The Child, Hafner Publishing Co., 1933.

24. *Then in the late 1940s psychologist Keith Hayes and his wife, Cathy, home raised a newborn chimpanzee named Viki*: See Cathy Hayes, The Ape in Our House, Victor Gollancz LTD, 1952.

26. *Yerkes concluded that an animal that doesn't imitate sounds "cannot reasonably be expected to talk"*: See Robert M. Yerkes, Almost Human, The Century Company, 1925.

29. *Washoe was now making steady and dramatic progress*: See Beatrix T. Gardner and R. Allen Gardner, "Two-Way Communication with an Infant Chimpanzee," in Behavior of Nonhuman Primates, Vol. 4, eds. Allan M. Schrier and Fred Stollnitz, Academic Press, 1971.

THREE: OUT OF AFRICA

40. *The Air Force trained its sixty-five chimponauts on a simulated flight panel by means of operant conditioning:* See Michael Aisner, "The Astro Chimps on Their 30th Anniversary," in *Gombe 30 Commemorative Magazine*, The Jane Goodall Institute, 1991. Most of the chimponaut material is based on Aisner's research.

48. *The Oubi people of the present-day Ivory Coast refer to chimpanzees as "ugly human beings":* See Eugene Linden, *Apes, Men & Language*, Penguin Books, 1981.

48. *The Mende people of the Upper Guinean forests refer to the chimpanzee as "different persons":* See Paul Richards, "Local Understandings of Primates & Evolution: Some Mende Beliefs Concerning Chimpanzees," in *Ape, Man, Apeman: Changing Views since 1600*, eds. Raymond Corbey and Bert Theunissen, Department of Prehistory, Leiden University, 1995. (To obtain this book contact: Dr. R. Corbey, Dept. of Prehistory, Leiden University, P.O. Box 9515, NL 2300 RA Leiden, Netherlands; fax (011-31)71-5272928.

48. *One group of the Gouro people believe that they are the descendants of chimpanzees:* See Frédéric Joulian, "Représentations Traditionnelles du Chimpanzé en Côte d'Ivoire," in *Ape, Man, Apeman: Changing Views since 1600*, eds. Raymond Corbey and Bert Theunissen, Department of Prehistory, Leiden University, 1995. The material on the Baoulé, Bakwé, and Bété peoples also comes from Joulian's study.

49. *He elevated humans above other animals, and then designed a Great Chain of Being, at the top of which stood the free man:* See Steven M. Wise, "How Nonhuman Animals Were Trapped in a Nonexistent Universe," *Animal Law*, 1, no. 1, 1995.

49. *The first reports of great apes began arriving in Europe in 1607:* See Emily Hahn, *On the Side of the Apes*, Thomas Y. Crowell Company, 1971. The Battell account was given to Samuel Purchas in 1613 and published in the book "Purchas His Pilgrimes."

50. *In 1699, England's best-known anatomist, Edward Tyson, performed the first dissection of a chimpanzee:* See J.M.M.H. Thijssen, "Reforging the Great Chain of Being," in *Ape, Man, Apeman: Changing Views since 1600*, eds. Raymond Corbey and Bert Theunissen, Department of Prehistory, Leiden University, 1995.

50. *But Tyson was a good Cartesian and he assumed that a thinking, talking animal was simply not possible:* See Robert Wokler, "Enlightening Apes," in *Ape, Man, Apeman: Changing Views since 1600*, eds. Raymond Corbey and Bert Theunissen, Department of Prehistory, Leiden University, 1995.

51. *Thomas Huxley, a naturalist who was known as "Darwin's bulldog," was the first to argue, in 1863, that our anatomical resemblance to apes was no coincidence:* See Thomas Henry Huxley, *Evidence as to Man's Place in Nature*, Williams and Norgate, 1863.

51. *"Man is descended from a hairy, tailed quadruped, probably arboreal in its habits":*

Quoted in *The Essential Darwin*, ed. Robert Jastrow, Little, Brown and Company, 1984.

54. *The mystery began to unravel that very same year, in 1967, when two biologists named Vincent Sarich and Allan Wilson:* See V. Sarich and A. Wilson, "Immunological Timescale for Human Evolution," *Science*, 158, 1967. For a popular summary of the evidence and the Sarich quotation about gophers, see V. Sarich, "Immunological Evidence on Primates," in *The Cambridge Encyclopedia of Human Evolution*, Cambridge University Press, 1994.

55. *But in the early 1980s two scientists named Charles Sibley and Jon Ahlquist confirmed the genetic similarity between humans and chimps by studying DNA itself:* See C. G. Sibley and J. E. Ahlquist, "The Phylogeny of the Hominoid Primates, as Indicated by DNA-DNA Hybridization," *Journal of Molecular Evolution*, 20, 1984. For the best popular summary of the Sibley-Ahlquist evidence, see Jared Diamond, *The Third Chimpanzee*, HarperCollins, 1993.

56. *Their DNA evidence, which is now widely accepted, favors the following branching pattern:* See C. G. Sibley, "DNA-DNA Hybridisation in the Study of Primate Evolution," in *The Cambridge Encyclopedia of Human Evolution*, Cambridge University Press, 1994. The dates on the family tree are taken from this article and the Sibley evidence as presented by Jared Diamond. The hominid classification follows the Smithsonian (see next note). For a more skeptical view of the DNA clock and its application to the ape family tree, see J. Marks, "Chromosomal Evolution in Primates," in *The Cambridge Encyclopedia of Human Evolution*. For a survey of the scientific papers and positions on this topic, see the "Further Readings" section in Jared Diamond, *The Third Chimpanzee*.

57. *As a result, if you look in the most recent edition of the Smithsonian's definitive classification:* See *Mammal Species of the World*, Second Edition, eds. Don. E. Wilson and DeeAnn M. Reeder, Smithsonian Institution Press, 1993.

57. *Vincent Sarich, the pioneer of this molecular anthropology, says that if we could go back in time five million years and observe them, we would consider our ancestors to be small chimpanzees:* See V. Sarich, "Immunological Evidence on Primates," in *The Cambridge Encyclopedia of Human Evolution*, Cambridge University Press, 1994.

FOUR: SIGNS OF INTELLIGENT LIFE

69. *Descartes summed up this argument when he said that even "depraved and stupid" men could tell other people what they were thinking:* Descartes is quoted in Jeffrey Moussaieff Masson and Susan McCarthy, *When Elephants Weep*, Delacorte Press, 1995. Original passage is from *Discours de la Méthode*.

69. *"No one is more strongly convinced than I am of the vastness of the gulf between . . . man and the brutes":* Huxley is quoted in Sue Savage-Rumbaugh and Roger Lewin, *Kanzi*, John Wiley & Sons, Inc. 1994. Original passage is from *Evidence as to Man's Place in Nature and Other Anthropological Essays*, D. Appleton and Company, 1900.

70. *Darwin suggested that what makes human language distinctive are its abstract cognitive features:* See Merlin Donald, *Origins of the Modern Mind*, Harvard University Press, 1991. This is an excellent exploration of Darwin's thesis of language origins.

70. *In the early 1950s Keith and Cathy Hayes reported that whenever their chimpanzee foster daughter, Viki, wanted a ride in the car:* See K. Hayes and C. H. Nissen, "Higher Mental Functions of a Home-raised Chimpanzee," in *Behavior of Nonhuman Primates*, Vol. 4, eds. Allan M. Schrier and Fred Stollnitz, Academic Press, 1971.

74. *To understand what happened next, it helps to know a bit about ASL:* See B. T. Gardner, R. A. Gardner, and S. G. Nichols, "The Shapes and Uses of Signs in a Cross-Fostering Laboratory," in *Teaching Sign Language to Chimpanzees*, eds. R. Allen Gardner, Beatrix T. Gardner, and Thomas E. Van Cantfort, State University of New York Press, 1989.

75. *The building blocks of ASL are cheremes, which are meaningless hand configurations, placements, and movements:* See Beatrix T. Gardner and R. Allen Gardner, "Two-Way Communication with an Infant Chimpanzee," in *Behavior of Nonhuman Primates*, Vol. 4, eds. Allan M. Schrier and Fred Stollnitz, Academic Press, 1971.

76. *The Gardners decided to teach Washoe ASL signs as if they were teaching a rat to press a lever by using an operant conditioning technique called shaping:* See R. Allen Gardner and Beatrix T. Gardner, "A Cross-Fostering Laboratory," in *Teaching Sign Language to Chimpanzees*, eds. R. Allen Gardner, Beatrix T. Gardner, and Thomas E. Van Cantfort, State University of New York Press, 1989. The Gardners also relate how Washoe learned the signs for MORE, TOOTHBRUSH, and SMOKE.

78. *During a full year of positive reinforcement for babbling, this method produced a grand total of one sign:* See Beatrix T. Gardner and R. Allen Gardner, "Two-Way Communication with an Infant Chimpanzee," in *Behavior of Nonhuman Primates*, Vol. 4, eds. Allan M. Schrier and Fred Stollnitz, Academic Press, 1971.

79. *After observing many discussions of flowers in the fall of 1967, Washoe began using the flower sign but in her own childlike form:* See Beatrix T. Gardner and R. Allen Gardner, "Two-Way Communication with an Infant Chimpanzee," in *Behavior of Nonhuman Primates*, Vol. 4, eds. Allan M. Schrier and Fred Stollnitz, Academic Press, 1971.

80. *The Gardners were somewhat appalled by my unorthodox technique and cautioned me against using guidance:* See Beatrix T. Gardner and R. Allen Gardner, "Two-Way Communication with an Infant Chimpanzee," in *Behavior of Nonhuman Primates*, Vol. 4, eds. Allan M. Schrier and Fred Stollnitz, Academic Press, 1971.

80. *Christophe Boesch, a primatologist who studies the stone tool culture of chimpanzees in the Tai Forest of the Ivory Coast:* See C. Boesch, "Aspects of Transmission of Tool-Use in Wild Chimpanzees," in *Tools, Language and Cognition in Human Evolution*, eds. K. R. Gibson and T. Ingold, Cambridge University

Press, 1993. Boesch also relates the story of chimpanzee maternal guidance in *The New Chimpanzees*, a National Geographic videotape.

82. *She referred to her toilet as* DIRTY GOOD *and the refrigerator as* OPEN FOOD DRINK: See B. T. Gardner, R. A. Gardner, and S. G. Nichols, "The Shape and Uses of Signs in a Cross-Fostering Laboratory," in *Teaching Sign Language to Chimpanzees,* eds. R. Allen Gardner, Beatrix T. Gardner, and Thomas E. Van Cantfort, State University of New York Press, 1989.

82. *On one occasion she signed* THAT FOOD *while looking at a picture of a drink in a magazine:* See R. Allen Gardner and Beatrix T. Gardner, "Feedforward Versus Feedbackward: An Ethological Alternative to the Law of Effect," *Behavioral and Brain Sciences,* 11, 1988.

82. *Washoe was also capable of inventing entirely new signs:* See Beatrix T. Gardner and R. Allen Gardner, "Two-Way Communication with an Infant Chimpanzee," in *Behavior of Nonhuman Primates,* Vol. 4, eds. Allan M. Schrier and Fred Stollnitz, Academic Press, 1971.

83. *As the Gardners finally conceded, "Young chimpanzees and young children have a limited tolerance for school":* See Beatrix T. Gardner and R. Allen Gardner, "Two-Way Communication with an Infant Chimpanzee," in *Behavior of Nonhuman Primates,* Vol. 4, eds. Allan M. Schrier and Fred Stollnitz, Academic Press, 1971.

84. *Freehand drawing, an example the Gardners often point to, is one of the favorite activities of nursery school children:* See R. Allen Gardner and Beatrix T. Gardner, "Feedforward Versus Feedbackward: An Ethological Alternative to the Law of Effect," *Behavioral and Brain Sciences,* 11, 1988. The Desmond Morris quote about chimpanzee art appears in this article. The original passage is from Desmond Morris, *The Biology of Art,* Knopf, 1962.

85. *In 1967, the same year I met Washoe, the Dutch ethologist Adriaan Kortlandt published a breakthrough study of wild chimpanzee communication:* All Kortlandt quotations appear in Emily Hahn, "Chimpanzees and Language," *The New Yorker,* December 11, 1971. Original passages are from Adriaan Kortlandt, "The Use of the Hands in Chimpanzees in the Wild," in *The Use of the Hands and Communication in Monkeys, Apes and Early Hominids,* eds. B. Rensch, Verlag Hans Huber, 1968. (The article and book are in German; an English summary of Kortlandt's article is available.)

86. *In 1978 William McGrew and Carolyn Tutin observed that two chimpanzee communities, which were only eighty kilometers apart in Tanzania, used slightly different gestures in order to ask for grooming:* See W. C. McGrew and C.E.G. Tutin, "Evidence for a Social Custom in Wild Chimpanzees?" *Man,* 13, 1978.

86. *In 1987 Toshisada Nishida observed the use of a "leaf-clipping" gesture that, again, was found only in the Mahale chimpanzee culture:* See T. Nishida, "Local Traditions and Cultural Transmission," in *Primate Societies,* eds. B. B. Smuts, D. L. Cheney, R. M. Seyfarth, R. W. Wrangham, and T. T. Struhsaker, University of Chicago Press, 1987.

87. *Like a human, she comes into the world with certain postures, gestures, and calls*

but learns how to use those signals properly only after years of experience in her community: See Jane Goodall, *The Chimpanzees of Gombe,* Harvard University Press, 1986.

FIVE: BUT IS IT LANGUAGE?

92. *In fact, said Chomsky, language is acquired independently from all other learning processes and cognitive abilities:* See Noam Chomsky, *Knowledge of Language: Its Nature, Origin, and Use,* Praeger, 1986.

93. *One attempt to describe French in this logical fashion required twelve thousand items just to classify its simple predicates:* See Philip Lieberman, *The Biology and Evolution of Language,* Harvard University Press, 1984.

93. *Chomsky claimed that the universal grammar was part of a child's genetic makeup, making language as unique to humans as dam building is to beavers or waggle dancing to bees:* See Noam Chomsky, *Cartesian Linguistics,* Harper and Row, 1966.

96. *Linguists who have studied such interaction say that as much as 75 percent of the meaning in a face-to-face conversation is communicated through body language and intonation:* See R. L. Birdwhistell, "Background to Kinesics," ETC, 13, 1955; R. L Birdwhistell, *Kinesics and Context: Essays on Body Motion Communication,* University of Pennsylvania Press, 1970.

96. *In fact, during the first half of this century, educators tried mightily to eradicate American Sign Language because they thought its gestures were too "monkeylike":* See Douglas C. Baynton, *Forbidden Signs,* The University of Chicago Press, 1997.

98. *The Gardners introduced a procedure to test Washoe's knowledge and prevent cueing:* See B. T. Gardner and R. A. Gardner, "A Test of Communication," in *Teaching Sign Language to Chimpanzees,* eds. R. Allen Gardner, Beatrix T. Gardner, and Thomas E. Van Cantfort, State University of New York Press, 1989. This article discusses the Gardners' test procedures, the use of fluent deaf signers as independent observers, and Washoe's test results and errors.

102. *For example, in the "doll test," Susan would "accidentally" step on Washoe's doll:* These observations are reported in Beatrix T. Gardner and R. Allen Gardner, "Two-Way Communication with an Infant Chimpanzee," in *Behavior of Nonhuman Primates,* Vol. 4, eds. Allan M. Schrier and Fred Stollnitz, Academic Press, 1971.

103. *The best evidence that humans and chimpanzees learn language in the same way was the fact that Washoe developed language in the same exact sequence as a human child:* See Beatrix T. Gardner and R. Allen Gardner, "Development of Phrases in the Utterances of Children and Cross-Fostered Chimpanzees," in *The Ethological Roots of Culture,* eds. R. A. Gardner, B. T. Gardner, B. Chiarelli, and F. X. Plooij, Kluwer Academic Publishers, 1994. This article traces the parallel development of language in children and cross-fostered chimpanzees.

104. *In 1969 the Gardners published the first report of Washoe's linguistic progress in*

the respected journal Science: See R. A. Gardner and B. T. Gardner, "Teaching Sign Language to a Chimpanzee," *Science*, 165, 1969.

105. *Motherese is marked by dozens of features that make it appropriate only for children who are learning language:* See S. E. Snow, "Mother's Speech to Children Learning Language," *Child Development*, 43, 1972. For a study of English-speaking and Spanish-speaking families see B. Blount and W. Kempton, "Child Language Socialization: Parental Speech and Interactional Strategies," *Sign Language Studies*, 12, 1976. For a study of ASL-using families see J. Maestas y Moores, "Early Linguistic Development: Interactions of Deaf Parents with Their Infants," *Sign Language Studies*, 26, 1980.

106. *They suggested a new "checklist" definition of language:* See J. S. Bronowski and Ursula Bellugi, "Language, Name, and Concept, *Science*, 168, 1970.

106. *By 1969 Washoe certainly had what is classified in children as Stage I language:* See R. A. Gardner and B. T. Gardner, "Comparative Psychology and Language Acquisition," *Annals of the New York Academy of Sciences*, 309, 1978.

PART TWO

115. *"Speak and I shall baptize thee":* Polignac is quoted in Robert Wokler, "Enlightening Apes," in *Ape, Man, Apeman: Changing Views since 1600*, eds. Raymond Corbey and Bert Theunissen, Department of Prehistory, Leiden University, 1995. Original passage is reported in Diderot, *Suite du Rêve de d'Alembert*. Diderot writes that Polignac was speaking to an "orangutan," a name that referred to the two species we now call orangutans and chimpanzees. Polignac must have seen a chimp, because the first living orangutan did not arrive in Europe until 1776.

115. *"It's no wonder that these animals, when confronted with the prospect of salvation":* Rousseau is quoted in Robert Wokler, "Enlightening Apes," in *Ape, Man, Apeman: Changing Views since 1600*, eds. Raymond Corbey and Bert Theunissen, Department of Prehistory, Leiden University, 1995. Original passage is in a letter from Rousseau to David Hume, March 29, 1766.

SIX: THE ISLAND OF DR. LEMMON

126. *"In our experience," the Doctor advised, "an adopted chimpanzee infant is clearly not the solution of choice for a precarious marriage":* Quoted in Emily Hahn, "Chimpanzees and Language," *The New Yorker*, December 11, 1971.

132. *Lemmon took Bruno away from his mother soon after the birth and turned him over to Dr. Herbert Terrace:* See Herbert S. Terrace, *Nim*, Knopf, 1979.

145. *Here are the results as they appeared in the journal Science in June 1973:* See Roger Fouts, "Acquisition and Testing of Gestural Signs in Four Young Chimpanzees," *Science*, 180, 1973.

147. *This resiliency was most dramatically demonstrated in the recent case of Alex:* See F. Vargha-Khadem, L. Carr, E. Isaacs, E. Brett, C. Adams., and M. Mishkin,

"Onset of Speech After Left Hemispherectomy in a Nine-year-old Boy," *Brain*, 120, 1997.

148. *Then, in 1987, the psychologist Melissa Bowerman made the most radical proposal:* Bowerman is discussed in Philip Lieberman, *Uniquely Human*, Harvard University Press, 1991. For the original study, see Melissa Bowerman, "What Shapes Children's Grammars," in *The Cross-linguistic Study of Language Acquisition*, ed. D. I. Slobin, Lawrence Erlbaum Associates, 1987.

SEVEN: HOUSE CALLS

150. *She was born into a colony of carnival chimpanzees in 1964 and was sold, at two days old, to Lemmon, who gave her to the Temerlins to raise:* See Maurice K. Temerlin, *Lucy: Growing Up Human*, Science and Behavior Books, 1975. Maury Temerlin describes how Jane retrieved Lucy from the carnival in Florida and brought her back to Oklahoma. See also Dale Peterson and Jane Goodall, *Visions of Caliban*, Houghton Mifflin Company, 1993. Mae Noell, the co-owner of the carnival, told Dale Peterson that Lemmon promised in writing to return Lucy to the carnival once the behavioral experiment was finished.

151. *"Shortly after we adopted Lucy I began to love her without reservation":* See Maurice K. Temerlin, *Lucy: Growing Up Human*, Science and Behavior Books, 1975. The anecdotes about Lucy's home life can be found in Temerlin's book.

155. *"Lucy came and sat close beside me on the sofa, and simply stared into my eyes for a long, long time. It gave me a strange feeling. . . . What, I kept wondering, is she thinking about me?":* See Dale Peterson and Jane Goodall, *Visions of Caliban*, Houghton Mifflin Company, 1993.

156. *But even more interesting was what happened when Lucy used her limited vocabulary to describe some of the new foods:* See R. S. Fouts, "Communication with Chimpanzees," in *Hominisation and Behaviour*, eds. G. Kurth and I. Eibl-Eibesfeldt, Gustav Fischer Verlag, 1975.

158. *For example, Ursula Bellugi and Jacob Bronowski, two critics of Washoe, said in 1970:* See J. S. Bronowski and Ursula Bellugi, "Language, Name, and Concept," *Science*, 168, 1970.

163. *In our first study I wanted to see if Ally could understand sentences he had never seen signed before:* See R. S. Fouts, G. Shapiro, and C. O'Neil, "Studies of Linguistic Behavior in Apes and Children," in *Understanding Language Through Sign Language Research*, eds. P. Siple, Academic Press, 1978; R. S. Fouts and R. L. Mellgren, "Language, Signs and Cognition in the Chimpanzee," *Sign Language Studies*, 13, 1976.

168. *The way that Maybelle and Salomé died in the absence of their mothers was reminiscent of a death that Jane Goodall observed among the wild chimpanzees of Gombe around this very same time:* See Jane Goodall, *The Chimpanzees of Gombe*, Harvard University Press, 1986.

170. *But once they are reunited, Goodall says, "there is none of the frenzied greeting, none of the hugging and kissing, that one would expect":* See Dale Peterson and Jane Goodall, *Visions of Caliban,* Houghton Mifflin Company, 1993.

175. *Lemmon described their "anaclitic depressions and atypical neurological states which in a human infant would have been indicative of severe central nervous pathology":* Quoted in Emily Hahn, "Chimpanzees and Language," *The New Yorker,* December 11, 1971.

177. *Later studies showed that children altered their own speech to suit different listeners:* See M. Shatz and R. Gelman, "The Development of Communication Skills: Modifications in the Speech of Young Children as a Function of Listener," *Monographs of the Society for Research in Child Development,* 38, 1973; M. Tomasello, M. J. Farrar, and J. Dines, "Children's Speech Revisions for a Familiar and an Unfamiliar Adult," *Journal of Speech and Hearing Research,* 27, 1984.

177. *My observations on the island led me to undertake, with the help of two speech pathology students, the first comparative study of how chimpanzees and deaf children communicate with American Sign Language:* See D. Gorcyca, P. H. Garner, and R. S. Fouts, "Deaf Children and Chimpanzees: A Comparative Sociolinguistic Investigation," in *Nonverbal Communication Today,* ed. M. R. Key, Mouton Publishers, 1982.

EIGHT: AUTISM AND THE ORIGINS OF LANGUAGE

184. *Accompanied by an old friend and clinical psychologist named George Prigatano, I passed through the double doors of the hospital, proceeded to a small room on the second floor, and met a nine-year-old boy named David:* I have changed the names of both of the autistic children discussed in this chapter.

185. *Meanwhile, I had turned up several studies from the late 1960s that showed that many autistic children responded to facial expression, gestures, and being touched:* See B. A. Ruttenberg and E. G. Gordon, "Evaluating the Communication of the Autistic Child," *Journal of Speech and Hearing Disorders,* 32, 1967; W. Pronovost, P. Wakstein, and P. Wakstein, "A Longitudinal Study of the Speech Behavior of Fourteen Children Diagnosed as Atypical or Autistic, *Exceptional Children,* 33, 1966.

189. *By the time I reported the results of this study in the* Journal of Autism and Childhood Schizophrenia *in 1976:* See R. Fulwiler and R. S. Fouts, "Acquisition of American Sign Language by a Noncommunicating Autistic Child," *Journal of Autism and Childhood Schizophrenia,* 6, no. 1, 1976.

189. *I had discovered that at least two other teams of researchers were trying sign language with autistic children and had reported results similar to mine:* See A. Miller and E. E. Miller, "Cognitive Developmental Training with Elevated Boards and Sign Language," *Journal of Autism and Childhood Schizophrenia,* 3, 1973; C. D. Webster, H. McPherson, L. Sloman, M. A. Evans, and E. Kuchar, "Communicating with an Autistic Boy by Gestures," *Journal of Autism and Childhood Schizophrenia,* 3, 1973.

190. *One of my hosts was a neurologist, Dr. Doreen Kimura, who had just done some interesting research on aphasics:* See D. Kimura, "The Neural Basis of Language Qua Gesture," in *Studies in Linguistics*, Vol. 2, eds. H. Whitaker and H. A. Whitaker, Academic Press, 1976.

191. *Hewes said that early hominids communicated with their hands, which naturally led them to develop other skills, like toolmaking, that also required precise hand movements:* See Gordon Hewes, "Primate Communication, and the Gestural Origin of Language," *Current Anthropology*, 14, nos. 1–2, 1973.

194. *The linguist Derek Bickerton has said that "syntax must have emerged in one piece, at one time—the most likely cause being some kind of mutation that affected the organization of the brain":* See Derek Bickerton, *Language and Species*, University of Chicago Press, 1990.

194. *"If you will, swing your right hand across in front of your body and catch the upraised forefinger of your left hand":* See David F. Armstrong, William C. Stokoe, Sherman E. Wilcox, *Gesture and the Nature of Language*, Cambridge University Press, 1995.

196. *"If all adults were stuck with the kinds of speech deficiencies normal enough in early childhood, we would probably still be using a well-developed sign language":* See Gordon Hewes, "The Current Status of the Gestural Origin Theory," in *Origins and Evolution of Language and Speech*, eds. S. R. Harnad, H. D. Steklis, and J. Lancaster, *Annals of the New York Academy of Sciences*, 280, 1976.

208. *Although the female chimp in the wild begins menstruating at age ten, nature gives her a kind of grace period by keeping her sterile for the first one to three years of her puberty:* See Jane Goodall, *The Chimpanzees of Gombe*, Harvard University Press, 1986. The material in this chapter on wild chimpanzee sexuality, courtship, and incest is drawn from Goodall.

212. *"I saw him as infallible," wrote Temerlin, "and I literally believed his most outlandish statements":* See Maurice K. Temerlin, *Lucy: Growing Up Human*, Science and Behavior Books, 1975. The material quoted in the following two paragraphs is from Temerlin's book as well.

TEN: LIKE MOTHER, LIKE SON

236. *"One chimpanzee is no chimpanzee":* See R. M. Yerkes, *Chimpanzees: A Laboratory Colony*, Yale University Press, 1943.

241. *Imagine what would happen if every member of our species over age one suddenly disappeared from the earth and the remaining infants somehow managed to survive:* See Gordon Hewes, "The Current Status of the Gestural Origin Theory," in *Origins and Evolution of Language and Speech*, eds. S. R. Harnad, H. D. Steklis, and J. Lancaster, *Annals of the New York Academy of Sciences*, 280, 1976.

244. *And beginning in 1982, Debbi and I published a series of scientific articles documenting Loulis's accomplishments:* See R. S. Fouts, A. D. Hirsch, and D. H. Fouts, "Cultural Transmission of a Human Language in a Chimpanzee

Mother-Infant Relationship," in *Psychobiological Perspectives: Child Nurturance Series*, 3, eds. H. E. Fitzgerald, J. A. Mullins, and P. Gage, Plenum Press, 1982; R. S. Fouts, D. H. Fouts, and T. Van Canfort, "The Infant Loulis Learns Signs from Cross-Fostered Chimpanzees," in *Teaching Sign Language to Chimpanzees*, eds. R. Allen Gardner, Beatrix T. Gardner, and Thomas E. Van Cantfort, State University of New York Press, 1989.

256. *Poor Lucy was still emaciated and pleading in sign language for Janis to find food for her—MORE FOOD, JAN GO*: See Janis Carter, "A Journey to Freedom," *Smithsonian*, April 1981.

PART THREE

259. *"Though the difference between man and the other animals"*: See Galileo, *Dialog Concerning the Two Chief World Systems*, translated by Stillman Drake, University of California Press, 1967.

259. *"How smart does a chimpanzee have to be"*: See Carl Sagan, *The Dragons of Eden*, Random House, 1977.

ELEVEN: AND TWO MORE MAKES FIVE

272. *Dar and Tatu had both been signing since infancy and each had a reliable vocabulary of more than 120 signs*: See B. T. Gardner, R. A. Gardner, and S. G. Nichols, "The Shapes and Uses of Signs in a Cross-Fostering Laboratory," in *Teaching Sign Language to Chimpanzees*, eds. R. Allen Gardner, Beatrix T. Gardner, and Thomas E. Van Cantfort, State University of New York Press, 1989.

273. *Herbert Terrace, a student of B. F. Skinner, had a very different approach*: See Herbert S. Terrace, *Nim*, Knopf, 1979. All the Terrace quotations, unless otherwise noted, are from this book.

274. *Nim's learning environment was so devoid of natural human interaction that the linguist Philip Lieberman described Nim as "the Wolf-Ape"*: See Philip Lieberman, *The Biology and Evolution of Language*, Harvard University Press, 1984.

275. *"Nim had fooled me," Terrace later wrote*: See "Why Koko Can't Talk," *The Sciences*, 8–10, December 1982.

276. *Signers, unlike speakers, overlap in their conversation about 30 percent of the time*: See C. Baker, "Regulators and Turn-Taking in ASL Discourse," in *On the Other Hand*, ed. L. Friedman, Academic Press, 1977.

277. *For example, the linguist Philip Lieberman concluded that Terrace was guilty of "the systematic misrepresentation of other investigators' work, particularly that of the Gardners"*: See Philip Lieberman, *The Biology and Evolution of Language*, Harvard University Press, 1984.

277. *Two comparative psychologists, Thomas Van Cantfort and James Rimpau, published a fifty-page article in the journal* Sign Language Studies *that detailed Terrace's distortions of the scientific record*: See T. Van Cantfort and J. B. Rimpau,

"Sign Language Studies with Children and Chimpanzees," *Sign Language Studies*, 34, Spring 1982.

277. *After he returned to Oklahoma in 1977, a new study showed that his spontaneous signing increased dramatically when he was allowed to socialize naturally under relaxed conditions:* See C. O'Sullivan and C. P. Yeager, "Communicative Context and Linguistic Competence: The Effects of a Social Setting on a Chimpanzee's Conversational Skill," in *Teaching Sign Language to Chimpanzees*, eds. R. Allen Gardner, Beatrix T. Gardner; and Thomas E. Van Cantfort, State University of New York Press, 1989.

277. *These ongoing, specious attacks on Washoe and other signing chimps were effectively refuted back in 1983 by the pioneering linguist and ASL authority William Stokoe:* See W. Stokoe, "Apes Who Sign and Critics Who Don't," in *Language in Primates*, eds. H. T. Wilder and J. de Luce, Springer-Verlag, 1983.

278. *"There can be little doubt," Stokoe wrote in his most recent book, "that chimpanzees have well-developed abilities to communicate using signs":* See David F. Armstrong, William C. Stokoe, and Sherman E. Wilcox, *Gesture and the Nature of Language*, Cambridge University Press, 1995.

278. *They learned to sign, Stokoe says, in the same way that the deaf children of deaf parents do, through spontaneous interaction with signing human adults:* See W. C. Stokoe, "Comparative and Developmental Sign Language Studies: A Review of Recent Advances," in *Teaching Sign Language to Chimpanzees*, eds. R. Allen Gardner, Beatrix T. Gardner, and Thomas E. Van Cantfort, State University of New York Press, 1989.

280. *Although painting and drawing are not a part of natural chimpanzee culture (as far as we know), cross-fostered chimps love to make art:* See K. Beach, R. S. Fouts, and D. H. Fouts, "Representational Art in Chimpanzees," *Friends of Washoe Newsletter*, Summer 1984 (Part 1 of the study), and Fall 1984 (Part 2 of the study).

TWELVE: SOMETHING TO TALK ABOUT

285. *Officials at White Sands have never acknowledged receiving a chimpanzee named Ally:* See Eugene Linden, *Silent Partners*, Times Books, 1986.

297. *One computer-using chimp at Yerkes named Lana typed out sentences like "Please machine tickle Lana" when her human companions went home at night:* See Boyce Rensberger, "Computer Helps Chimpanzees Learn to Read, Write and 'Talk' to Humans," *The New York Times*, May, 29, 1974.

299. *They string together what the linguist Virginia Volterra calls "sequences of communicative signals":* See V. Volterra, "Gestures, Signs, and Words at Two Years," *Sign Language Studies*, 33, 1981.

300. *The most significant finding of this study of chimpanzee conversations was that chimpanzees do not use language to get rewards as Terrace and others had charged:* See R. S. Fouts, D. H. Fouts, and D. Schoenfeld, "Sign Language Conversational Interactions Between Chimpanzees," *Sign Language Studies*, 42, 1984.

THIRTEEN: MONKEY BUSINESS

315. *Harlow began by separating infant monkeys from their mothers and ended years later by raising monkeys totally alone in the bottom of a V-shaped metal chamber that he called the "pit of despair" or the "hell of loneliness"*: See Deborah Blum, *The Monkey Wars*, Oxford University Press, 1994.

316. *The president of Sema, Dr. John Landon, denied that there were any problems at his lab and chided Jane for lending credence to a videotape from an animal rights group without ever visiting the lab herself*: See Dale Peterson and Jane Goodall, *Visions of Caliban*, Houghton Mifflin Company, 1993.

320. *Not only did Jane not write NIH their hoped-for letter of endorsement, she published a scathing account of our visit to Sema in* The New York Times Magazine: See Jane Goodall, "Prisoners of Science," *The New York Times*, May 17, 1987.

326. *We submitted our recommendations to the USDA, which was just then writing its regulations to promote the psychological well-being of primates, as Congress had ordered*: The complete set of our recommendations is reprinted in Appendix B of Dale Peterson and Jane Goodall, *Visions of Caliban*, Houghton Mifflin Company, 1993.

327. *He said that the government's refusal to enlarge cages and set standards for psychological well-being was "arbitrary and capricious and contrary to law"*: See 813 F. Supp. 888–890 (D.D.C. 1993). For the appeals court decision see 29 F. 3d 720 (D.C. Cir. 1994).

328. *Our lawsuit against the government was just the beginning, not the end, of the legal fight to protect chimpanzees*: In March 1996 the Animal Legal Defense Fund again sued USDA for not issuing adequate regulations for the psychological well-being of primates under the Animal Welfare Act. One of the primates the suit sought to protect was a chimpanzee named Barney, who was languishing in solitary confinement in a caged cell in a government-licensed game farm. I visited Barney and documented that he had been deprived of all companionship and as a result was suffering from severe psychological and physical distress. On October 30, 1996, Judge Charles Richey again ruled that the USDA had violated the law and must rewrite its rules to prevent animal suffering. This landmark victory, if it stands, could vastly improve conditions for captive primates in all settings, including research laboratories. As a result, the National Association of Biomedical Research recently announced that it will join USDA in appealing Judge Richey's ruling. For his latest ruling, see *ALDF* v. *Madigan*, 943. F. Supp. 44 (D.D.C. 1996).

329. *At the same time we were battling over laboratory conditions in America, the Jane Goodall Institute produced a landmark scientific report on wild chimpanzee populations in Africa*: See Geza Teleki, "Population Status of Wild Chimpanzees and Threats to Survival," in *Understanding Chimpanzees*, eds. Paul G. Heltne and Linda A. Marquardt, Harvard University Press, 1989. Jane Goodall now says there may be 250,000 chimpanzees in Africa, most of them living in

very small groups and spread out across twenty-one nations. The World Wildlife Fund estimates the number at 100,000 to 200,000.

329. *NIH officials began claiming that a critical shortage of research animals was creating an urgent need to acquire more wild chimpanzees from Africa, something that hadn't been done since the early 1970s:* See Geza Teleki, "Testimony Submitted to The Subcommittee on Oversight and Investigations of the House Committee on Merchant Marine and Fisheries Concerning Implementation of CITES," July 13, 1988, on behalf of the Committee for Conservation and Care of Chimpanzees and the Jane Goodall Institute. For an excellent discussion of NIH's secretive attempts to circumvent the ban on importing chimpanzees, see Chapter 11 in Dale Peterson and Jane Goodall, *Visions of Caliban,* Houghton Mifflin Company, 1993.

329. *The U.S. Fish and Wildlife Service received 54,212 letters in support of the endangered classification from private citizens, wildlife organizations, scientific experts, and African governments:* See Geza Teleki, "They Are Us," in *The Great Ape Project,* eds. Paola Cavalieri and Peter Singer, St. Martin's Press, 1993.

330. *Janis Carter found Lucy's skeleton by their old campsite:* See Janis Carter, "Freed from Keepers and Cages, Chimps Come of Age on Baboon Island," *Smithsonian,* June 1988.

FOURTEEN: HOME AT LAST

345. *As the left hemisphere developed mechanisms to control sequences of spoken words, naturally it also took control of other fine motor movements, such as toolmaking:* Doreen Kimura theorizes that it happened the other way around; speech came under the control of the left hemisphere because that hemisphere was already specialized for precise motor control in toolmaking. See Doreen Kimura, "Neuromotor Mechanisms in the Evolution of Human Communication," in *Neurobiology of Social Communication in Primates,* eds. H. D. Steklis and M. J. Raleigh, Academic Press, 1979.

346. *Charles Darwin referred to this as "complex trains of thought," and he guessed correctly that our ability to think sequentially grew out of our sequential language skills:* See Merlin Donald, *Origins of the Modern Mind,* Harvard University Press, 1991.

348. *In most people the left hemisphere has proportionally more gray matter than the right:* R. C. Gur, I. K. Packer, J. P. Hungerbuhler, M. Reivich, W. D. Obrist, W. S. Amarnek, and H. Sackeim, "Differences in the Distribution of Gray and White Matter in Human Cerebral Hemispheres," *Science,* 207, 1980.

FIFTEEN: BACK TO AFRICA

360. *He calls chimps "the best possible model to test the fate and effects of foreign chemicals in man":* See *Regulatory Toxicology and Pharmacology,* 5, 1985.

360. *To cite just one recent and gruesome example: three of Coulston's chimpanzees were cooked to death when a space heater outside their cages became stuck on its*

highest setting and sent the room temperature soaring to 140 degrees: See "King of the Apes," *U.S. News and World Report,* August 14, 1995. For the USDA complaint against the Coulston Foundation see Animal Welfare Act Docket No. 95-65. The complaint was settled by the Coulston Foundation agreeing to pay a $40,000 fine and to cease and desist from violating the Animal Welfare Act.

360. *Our military's chimpanzee problem would have been "solved" once and for all by handing them over to a researcher who recently suggested in* The New York Times *that chimpanzees could be raised like cattle to be used as living blood and organ banks:* See "Chimp Surplus Spurs Debate About Animals' Future," *The New York Times,* February 4, 1997.

361. *He is well on his way to becoming, as he puts it, "the sole source of chimpanzees for research":* See *Almagordo* (New Mexico) *Daily News,* October 2, 1994.

361. *Coulston has mocked the idea of chimpanzee retirement:* See "Apes on Edge," *The Boston Globe,* November 7, 1994.

361. *The director of one of the biggest primate labs recently told a reporter, "If you said I could have one hundred chimps for free, I would say no thanks":* See "King of the Apes," *U.S. News and World Report,* August 14, 1995.

362. *Every major advance in AIDS research—from understanding how the virus causes disease to the development of crucial new drugs (AZT, 3TC, and protease inhibitors) to identifying possible genetic factors that may provide resistance—has come from human studies:* See Neal D. Barnard and Stephen R. Kaufman, "Animal Research Is Wasteful and Misleading," *Scientific American,* February 1997.

362. *The fundamental differences between the human and chimpanzee immune systems continue to make any data from chimpanzees "virtually uninterpretable in human terms," according to a recent report from the Medical Research Modernization Committee:* See S. Kaufman, M. Cohen, and S. Simmons, "Shortcomings of AIDS-Related Animal Experimentation," *Medical Research Modernization Committee Report,* 9, no. 3, September 1996.

365. *The Swiss ethologist Christophe Boesch has studied the chimpanzee stone tool culture that thrives across West Africa:* See C. Boesch and H. Boesch, "Tool Use and Tool Making in Wild Chimpanzees," *Folia Primatologica,* 54, 1990; see also Frédéric Joulian, "Comparing Chimpanzee and Early Hominid Techniques: Some Contributions to Cultural and Cognitive Questions," in *Modelling the Early Human Mind,* eds. P. A. Mellars and K. A. Gibson, McDonald Institute for Archaeological Research, 1996.

365. *The anthropologist Richard Wrangham has documented the chimpanzee's use of medicinal plants:* See Michael A. Huffman and Richard W. Wrangham, "Diversity of Medicinal Plant Use by Chimpanzees in the Wild," in *Chimpanzee Cultures,* eds. R. W. Wrangham, W. C. McGrew, F. de Waal, and P. Heltne, Harvard University Press, 1994.

365. *The Mende people of West Africa have long supplemented their own knowledge of herbal medicine, which they call "leaf," by following and learning from chimpanzees:* See Paul Richards, "Local Understandings of Primates & Evolution: Some Mende Beliefs Concerning Chimpanzees," in *Ape, Man, Apeman:*

Changing Views since 1600, eds. Raymond Corbey and Bert Theunissen, Department of Prehistory, Leiden University, 1995.

367. *This was the longest involuntary experiment ever performed on humans in medical history:* See Marjorie Spiegel, *The Dreaded Comparison,* Mirror Books, 1996; also "Tuskegee's Long Arm Still Touches a Nerve," *The New York Times,* April 13, 1997.

373. *Our moral progress will not be complete, Darwin said, until we extend our compassion to people of all races, then to "the imbecile, the maimed, and other useless members of society," and finally to the members of all species:* Darwin quoted in James Rachels, "Why Darwinians Should Support Equal Treatment for Other Great Apes," in *The Great Ape Project,* eds. Paola Cavalieri and Peter Singer, St. Martin's Press, 1993. Original passage is in Charles Darwin, *The Descent of Man, and Selection in Relation to Sex,* John Murray, 1871.

INDEX

aboriginal peoples, and Western moral
   order, 366–367, 369
Africa, 204–206, 359
   chimpanzee populations of, 42–43,
      311–312, 329–331, 359, 374–
      380, 382
   chimpanzee rehabilitation project
      in, 212–214, 216, 256–257, 288,
      330–331, 359
   chimpanzee sanctuaries in, 378–
      379, 380
   collection methods in, 42–43, 204,
      359
   human population growth in, 379
   peoples of, 47–48, 58, 365
African Americans, and Western
   moral order, 366–367, 368, 369–
   370, 371, 373
aggressive behavior, 120, 121, 122,
   127–128, 129, 139–140, 182–
   183, 212
Agriculture Department, U.S.
   (USDA), 315, 321–323, 326–
   328, 360
Ahlquist, Jon, 55–56
AIDS research, 310, 311–315, 317,
   318–319, 326, 342, 359, 362,
   363, 364
Air Force, U.S., 39–44, 206, 252,
   263, 310, 359–360, 364
Ally, 161–167, 200, 224, 253, 256,
   257, 266, 273, 358
   action painting by, 161–162
   ASL taught to, 161, 162–167, 285
   baptism of, 161, 286

depression of, 175–176
grammar rules applied by, 162–167,
   191
human self-image of, 174–175, 210
hysterical paralysis of, 175, 203
identity crisis of, 175
Lemmon's ownership of, 215, 245,
   247–248
in *People*, 199, 284
signing style of, 161
sold for biomedical research, 247–
   248, 283–286, 289, 322
transfer of, 174–179
as Washoe's mate, 211, 215, 221,
   233, 239, 242, 244, 245, 247,
   286
at White Sands Research Center,
   285–286, 322, 340
Almost Human (Yerkes), 28
American Sign Language (ASL), ix–
   x, 4, 13, 14, 27, 29, 30, 34–35,
   71, 74–76, 87, 95–97, 105, 112,
   148, 211, 272–278, 289, 303, 352
   cheremes of, 75
   grammar of, 76
   history of, 74–75
   inflection of, 76, 275
   proper names in, 76
   shades of meaning in, 95–96
   syntax of, 350
   teaching of, at IPS, 133, 134, 142–
      146, 150, 153–167, 168, 176–
      178, 185, 285, 355, 356, 357
   teaching of, to autistic children,
      185–189

transmitted to Loulis, x, 217–218, 240–245, 272, 297–298, 299, 302, 303–304, 336
turn-taking device in, 275–276
visual references in, 75–76
see also Washoe
Amory, Cleveland, 285
Anderson, Linn, 117–118
Animal Legal Defense Fund, 315, 327
Animal Welfare Act (1985), 321–328, 360
Animal Welfare Institute, 327
anthropomorphism, 58
Ape and the Child, The (Kellogg and Kellogg), 23
"Apes Who Sign and Critics Who Don't" (Stokoe), 277–278
aphasia, 190
Aristotle, 48, 49, 50, 51, 69, 366
Armstrong, David, 194
Asians, and Western moral order, 366–367
Australopithecus, 53, 56
autistic children, 149, 184–192, 199, 204, 206
  ASL taught to, 185–189
  characteristics of, 184, 185
  cross-modal transfer in, 186–187
  hyperactivity of, 188–189
  speech and, 184, 185, 187, 188, 189–192, 195
  stereotypies of, 186
  theories about, 184–185

Babe, 22
Bakwé, 48
Baoulé, 48
Battell, Andrew, 50
Beach, Kat, 291
behaviorism, 73–84, 126
  see also operant conditioning; Skinner, B. F.
Bellugi, Ursula, 158
Bété, 48
Bettelheim, Bruno, 184
Bickerton, Derek, 194
biomedical research, xi, 42, 43, 44, 50, 133, 202, 236–237, 281, 283–286, 308, 309–331, 358–373
  AIDS in, 310, 311–315, 317, 318–319, 326, 342, 359, 362, 363, 364

Ally sold for, 247–248, 283–286, 289, 322
behavioral research vs., 278–279, 324–325
castration in, 281
code of silence about, 316, 320–323, 326, 356
dental reconstruction in, 359
Goodall and, 316–321, 323–324, 326, 328
government regulation of, 312–314, 321–330, 362, 363–364
in *Greystoke*, 289
hepatitis in, 220–221, 265, 283–284, 310, 315, 318–319, 342, 354–355, 357, 359, 363, 364, 367
implantation in, 237
insecticides in, 285, 286, 360, 361, 364
isolation in, 309–311, 315–316, 317, 318–320, 322, 328, 354, 359, 362, 363
legal action against, 315–316, 326–328, 360
Lemmon and, 137, 220–221, 247–248, 251, 256, 283–286
lobby for, 326, 329
media attention to, 283–285, 320–321, 353–357
morality of, 312–313, 320, 324–326, 344, 353, 365–373
pragmatic approach to, 325–326
self-interest in, 370–371
surplus chimps in, 359–364
suspect data produced by, 312, 313–314, 319–320, 362
20/20 show on, 353–357
wild chimpanzee populations and, 42–43, 311–312, 329–331, 359
  see also National Institutes of Health; *specific laboratories*
bipedalism, 56, 57, 193, 287
Blakey, Susie and Church, 168
Bodamer, Mark, 354
body language, 68, 90, 95, 96, 97, 275, 300, 346, 350
Boesch, Christophe, 80–81, 365
bonobo (pygmy chimpanzee), 55, 273
Booee, 131–146, 149, 174, 179, 200, 201, 256, 266
  ASL taught to, 142, 143–144, 145–146, 185, 285, 355, 356, 357
  dominance of, 176–178

brain of, 26, 50, 52, 57, 94, 193, 344–345, 346, 349–351
as endangered species, 312, 329–330
estrus of, 86, 153, 208–211, 268–269, 299, 307
facial expressions of, 68, 135
handedness of, 346
imprisonment of, 199–200, 204, 206, 216, 236–237, 255, 275, 289, 291, 325, 344, 359
individual personalities of, 33, 131–134, 143–146, 161–162
infant, 18, 42–43, 62, 80–81, 86–87, 138–139, 140, 141, 152, 167–169, 170, 242
laughter of, 271
lifespan of, 42
movie portrayals of, 286–288
pregnancy of, 208–209
prolonged childhood of, 58–59, 81–82, 244
as property, 251–253, 327–328
quietness of, 26
separation behavior of, 170
as siblings, 62–65
toolmaking of, 44, 48, 54, 70, 71, 80–81, 82, 105, 191, 242
vocalizations of, 26–27, 68–69, 85, 128, 135, 136, 159, 193
vocal tract of, 26, 50, 57, 197
water feared by, 131, 179–180
chimponauts, 39–44, 45, 54, 364
in biomedical research, 42, 359–361
operant conditioning of, 40, 41
performance of, 40, 41–42
"Chimpressionistic Works by Washoe and Friends" art exhibition, 280
Chomsky, Noam, 92–97, 101–102, 104–105, 106, 147, 164, 165, 167, 177, 273, 276, 277
Chown, Bill, 175
Cindy, 131–146, 149, 156, 162, 185, 200, 256, 283–285, 361
Clever Hans effect, 98–99, 275, 276, 284
"Clever Hans Phenomenon, The: Communication with Horses, Whales, Apes, and People," 276
Columbia University, 132, 273–275
communication, 66–90, 95, 153
body language in, 68, 90, 95, 96, 97, 275, 300, 346, 350
channels of, 67

closed vs. open systems of, 158–159, 164
dialects in, 70, 85–87
facial expressions in, 67–68, 76, 86, 96, 97, 135, 349
gestural, 27, 68, 85–87, 90, 96–97, 135, 146, 190–198, 241, 299, 345, 349, 377
at IPS chimp island, 135–136, 176–178
modes of, 66–67
need for, 84–85
social hierarchy and, 176–179
vocalizations in, 26–27, 68–69, 85, 128, 135, 136, 159, 193
see also language
computerized keyboards, 178, 273, 297
Conference on the Behavior of the Great Apes, 198–199
Congress, U.S., 312–314, 321–328, 360
National Chimpanzee Sanctuary System proposed to, 363–364, 378
Convention on International Trade in Endangered Species, 312
copulation, disruption of, 152
cosmetics testing, 285, 360
Coulston, Frederick, 359–361, 362
Coulston Foundation, 359–360, 361
courtship, 86, 208–211, 349
human males favored for, 209–210, 268, 272, 299, 306
cross-fostering, 21–30, 125, 126, 132, 133, 150–169, 236, 248–250, 273, 280, 358
of humans, 23, 350
human self-image produced by, 121–122, 160, 174–175, 206, 210, 212, 384
ill effects of separation from, 167–169, 175–176, 203, 238, 250–251, 265
by Lemmon's psychotherapy patients, 125, 126, 150–169, 175, 203, 204, 211–214, 220
obsolete subjects of, 174–179, 201–204, 211–214
see also Ally; Lucy; Viki; Washoe
cueing, 98–99, 163, 218, 245, 275, 302
Curious George (Rey), 3–5, 39
Curious George Gets a Medal (Rey), 39, 42, 203